LABORATORY STUDIES IN
Animal Diversity

SIXTH EDITION

Cleveland P. Hickman, Jr.

Professor Emeritus

Washington and Lee University

Lee B. Kats

Professor

Pepperdine University

Susan L. Keen

Senior Lecturer

University of California, Davis

Original Artwork by

William C. Ober, M.D., and Claire W. Garrison, R.N.

Connect
Learn
Succeed™

D0209633

LABORATORY STUDIES IN ANIMAL DIVERSITY, SIXTH EDITION

1 2 3 4 5 6 7 8 9 0 QDB/QDB 1 0 9 8 7 6 5 4 3 2 1

ISBN 978-0-07-734597-6
MHID 0-07-734597-5

Vice President & Editor-in-Chief: *Marty Lange*
Vice President of Specialized Publishing: *Janice M. Roerig-Blong*
Publisher: *Michael S. Hackett*
Sponsoring Editor: *Rebecca Olsen*
Executive Marketing Manager: *Patrick E. Reidy*
Lead Project Manager: *Jane Mohr*
Design Coordinator: *Brenda A. Rolwes*
Cover Designer: *Studio Montage, St. Louis, Missouri*
Cover Image: *© Comstock Images/PictureQuest RF*
Buyer: *Laura Fuller*
Compositor: *S4Carlisle Publishing Services*
Typeface: *10/12 Garamond*
Printer: *Quad/Graphics*

Some of the laboratory experiments included in this text may be hazardous if materials are handled improperly or if procedures are conducted incorrectly. Safety precautions are necessary when you are working with chemicals, glass test tubes, hot water baths, sharp instruments, and the like, or for any procedures that generally require caution. Your school may have set regulations regarding safety procedures that your instructor will explain to you. Should you have any problems with materials or procedures, please ask your instructor for help.

www.mhhe.com

BRIEF CONTENTS

CONTENTS

PREFACE

Laboratory Studies in Animal Diversity offers students hands-on experience in learning about the diversity of life. It gives students the opportunity to become acquainted with the principal groups of animals and to recognize the unique anatomical features that characterize each group as well as the patterns that link animal groups to each other. Although this manual was written to accompany a particular textbook, *Animal Diversity,* it can easily be adapted to use with any other introductory zoology text and with a variety of course plans. Every effort has been made to provide clear instructions and enough background material to create interest and an understanding of the subject matter. Many illustrations complement the written word.

Features

- Special projects entitled "Experimenting in Zoology" appear at the ends of certain exercises. Some of these activities can be completed within a single laboratory period; others extend for a longer time. In all these project exercises, the student follows experimental procedures, records and analyzes quantitative data, and draws conclusions from the results. Many instructors will want their students to gain additional experience by writing a laboratory report in which the student states the objectives, methods followed, results obtained, and conclusions that can be drawn from the results. The Experimenting in Zoology exercises are: Effect of Temperature on the Locomotor Activity of *Stentor* (Exercise 4); Planaria Regeneration Experiment (Exercise 7); Behavior of the Medicinal Leech, *Hirudo medicinalis* (Exercise 10); The Phototactic Behavior of *Daphnia* (Exercise 12); Aggression in Paradise Fish, *Macropodus opercularis* (Exercise 16); and Analysis of the Multiple Hemoglobin System in *Carassius auratus,* the Common Goldfish (Exercise 16).

- All exercises are designed to be interactive, with questions placed throughout the text, and spaces for students to write down their responses and observations. This "active learning" approach involves students in the exercise and encourages them to think about the information as they read. Some questions may require students to consult their textbook for the answers. A few exercises have questions in the figure legends, to be answered in the spaces provided when the student consults the figure. Examples of this interactive approach are found in Exercises 15 through 20.

- This laboratory manual provides many aids for the student. Throughout the exercises, working instructions are clearly set off from the descriptive material. Function is explained along with anatomy. Topic headings help the student mentally organize the material. Metric tables and definitions of special terms appear on the inside front and back covers for convenient use. Much of the artwork was designed to assist the student with difficult dissections.

New to This Edition

- With this edition, we divided exercises into "Core Study" and "Further Study" sections. This was done to make exercises more manageable for students and instructors having labs of varying lengths. The "Core Study" makes the critical concepts of each topic accessible to students in short lab periods or when only a single lab is provided for each topic. The "Further Study" section of each exercise permits in-depth exploration of materials when time permits.

- In many exercises, labels for illustrations have been expanded to provide functional information for the identified structures. Where appropriate, transfer of text explanation to an illustration allows more effective understanding of structural and functional relationships.

- Introductory cladograms for exercises have been revised

and enlarged to include shared derived characters and have been changed from "ladder" to "tree" depiction, the format of choice in scientific literature.

- The classification of the protozoan groups was extensively revised following new molecular phylogenies of some taxa, in particular the amebas. Phyla Foraminifera and Radiolaria are still in use, although some group members have been moved to other phyla. New phyla described here are Cercozoa, Centrohelida, and Amoebozoa, but readers familiar with protozoans will recognize many genera within these phyla. Common protozoans described in previous editions remain in the manual, albeit within new phylogenetic groups. Motile green algae, previously in phylum Chlorophyta, are now within Viridiplantae. The new name reflects the shared ancestry of green algae and land plants. Both the introductory cladogram and the pie diagram showing species numbers in each protozoan group were updated to reflect these revisions. As with all exercises for this edition, the revised classification with pronunciations, derivations, and descriptions was moved to the end of the exercise.

Acknowledgments

We are indebted to the following reviewers whose many suggestions were essential in guiding our revision for this edition:

Catherine Dumas
College of the Holy Cross
Michael Meyer
Christopher Newport University
Jan M. Reber
Taylor University

The authors also express their appreciation to the editors and support staff at McGraw-Hill Higher Education who guided this revision, especially Patrick Reidy, Executive Editor; Wendy Langerud, Developmental Editor; and Jane Mohr, Project Manager. Although we make every effort to bring to you an error-free manual, errors of many kinds inevitably find their way into a book of this scope and complexity. We will be grateful to readers who send comments or suggestions concerning content. Please address these remarks to Patrick Reidy, Executive Editor, The McGraw-Hill Companies, 501 Bell Street, Dubuque, Iowa 52001. Patrick may also be contacted by email at patrick_reidy@mcgraw-hill.com.

LABORATORY SAFETY PROCEDURES

1. Keep your work area uncluttered. Place unnecessary books, backpacks, purses, etc., somewhere other than on your desktop.

2. Avoid contact with embalming fluids. Wear rubber or disposable plastic gloves when working with preserved specimens.

3. Wear eyeglasses or safety glasses to protect your eyes from splattered embalming fluid.

4. Keep your hands away from your mouth and face while in the laboratory. Moisten labels with tap water, not your tongue.

5. Sponge down your work area and wash all laboratory instruments at the end of the period.

6. Wash your hands with soap and water at the end of the laboratory period.

GENERAL INSTRUCTIONS

Equipment

Each student will need to supply this equipment:

Laboratory manual and textbook
Dissecting kit containing scissors, forceps, scalpel, dissecting needles, pipette (medicine dropper), probe, and ruler, graduated in millimeters
Drawing pencils, 3H or 4H
Eraser, preferably kneaded rubber
Colored pencils—red, yellow, blue, and green
Box of cleansing tissues
Loose-leaf notebook for notes and corrected drawings

The department will furnish each student with all other supplies and equipment needed during the course.

Aim and Purpose of Laboratory Work

The zoology laboratory will provide you with "hands-on" experience in zoology. It is the place where you will see, touch, hear, smell—but perhaps not taste—living organisms. You will become acquainted with the major animal groups, make dissections of preserved or anesthetized specimens to study how animals are constructed, ask questions about how animals and their parts function, and gain an appreciation of some of the architectural themes and adaptations that emphasize the unity of life.

General Instructions for Laboratory Work

Prepare for the Laboratory. Before coming to the laboratory, read the entire exercise to familiarize yourself with the subject matter and procedures. Also read the appropriate sections in your textbook. Good preparation can make the difference between a frustrating afternoon of confusion and mistakes and an experience that is pleasant, meaningful, and interesting.

Follow the Manual Instructions Carefully. It is your guide to exploring and understanding the organisms or functions you are investigating. Its instructions have been written with care to help you work (1) in logical sequence, (2) with economy of time, and (3) with a questioning attitude that will stimulate interest and curiosity.

Use Particular Care in Making Animal Dissections. A glossary of directional terms used in dissections will be found inside the back cover. The object in dissections is to separate or expose parts or organs in order to see their relationships. Working blindly without the manual instructions may lead you to destroy parts before you have had an opportunity to identify them. *Learn the functions* of all the organs you dissect.

Record Your Observations. In a notebook, keep a personal record of everything that is pertinent, including the laboratory instructor's preliminary instructions and all your experimental observations. Do not record data on scraps of paper with the intention of recopying later; record directly into a notebook. The notes are for your own use in preparing the laboratory report later.

Take Care of Equipment. Glassware and other apparatus should be washed and dried after use. Metal instruments in particular should be thoroughly dried to prevent rust or corrosion. Put away all materials and equipment in their proper places at the end of the period.

Tips on Making Drawings

You need not be an artist to make laboratory drawings. You do, however, need to be *observant*. Study your specimen carefully. Your simple line drawing is a record of your observations.

Before you draw, locate on the specimen all the structures or parts indicated in the manual instructions. Study their relationships to each other. Measure the specimen. Decide where the drawing should be placed and how much it must be enlarged or reduced to fit the page (read further for estimation of magnification). Leave ample space for labels.

When ready to draw, you may want first to rule in faint lines to represent the main axes, and then sketch the general outlines lightly. When you have the outlines you want, draw them in with firm dark lines, erasing unnecessary sketch lines. Then fill in details. Do not make overlapping, fuzzy, indistinct, or unnecessary lines. Indicate differences in texture and color by stippling.

Stipple deliberately, holding the pencil vertically and making a neat round dot each time you touch the paper. Placing the dots close together or farther apart will give a variety of shading. Avoid line shading unless you are very skilled. Use color only when directed to do so.

Label the drawing completely. Print labels neatly in lowercase letters and align them vertically and horizontally. Plan the labels so that label lines do not cross each other. If there are to be many labels, center the drawing and label it on both sides.

Indicate the magnification in size beneath the drawing—for instance, "×3" if the drawing is three times the length and width of the specimen. For objects viewed through a microscope, indicate also the magnification at which you viewed the subject—for example, 430× (43× objective used with a 10× ocular).

Estimating the Magnification of a Drawing

A simple method for determining the magnification of a drawing is to find the ratio between the size of the drawing and the actual size of the object you have drawn. The magnification of the drawing can be determined by this formula:

$$\times = \frac{\text{Size of drawing}}{\text{Size of object}}$$

If your drawing of a specimen is 12 cm (120 mm) long, and you have estimated the specimen to be 0.8 mm long, then × = 120 ÷ 0.8, or 150. The drawing, then, is ×150, or 150 times the length of the object drawn.

This same formula holds true whether the drawing is an enlargement or a reduction. If, for example, the specimen is 480 mm long, and the drawing is 120 mm, then × = 120/480, or 1/4.

STATEMENT ON THE USE OF LIVING AND PRESERVED ANIMALS IN THE ZOOLOGY LABORATORY

The United States Congress has probably received more mail on the topic of animal research in universities and business firms than on any other subject. Do humans have the right to experiment on other living creatures to support their own medical, pharmaceutical, and commercial needs? A few years ago, Congress passed a series of amendments to the Federal Animal Welfare Act, a body of laws covering animal care in laboratories and other facilities. These amendments have become known as the three R's: **r**eduction in the number of animals needed for research; **r**efinement of techniques that might cause stress or suffering; and **r**eplacement of live animals with simulations or cell cultures whenever possible. As a result, the total number of animals used each year in research and in commercial product testing has declined steadily as scientists and businesses have become more concerned and more accountable. The animal rights movement, largely comprising vocal antivivisectionists, has helped create an awareness of the needs of laboratory research animals and has stretched the resources and creativity of the researchers to discover cheaper and more humane alternatives to animal experimentation.

However, these alternatives—computers and cell cultures—can only simulate the effects on organismal systems of, for instance, drugs, when the principles are well acknowledged. When the principles are themselves being scrutinized and tested, computer modeling is insufficient. Nor can a movie or computer simulate the visual and tactile experience provided by direct dissection of preserved or anesthetized animals. Such hands-on experience is the best way to achieve complete comprehension of anatomical relationships.

Medical progress depends on animal research. Every drug and every vaccine that you and your family have ever taken has first been tested on an animal.

Animal research has wiped out smallpox and polio; provided immunization against diseases that were previously common and often deadly, such as diphtheria, mumps, and rubella; helped create treatments for cancer, diabetes, heart disease, and manic depression; and helped develop procedures used in heart surgery, blood transfusions, and cataract removal, to name a few.

Animal research has also benefited other animals. The vaccine for feline leukemia that could threaten the life of your cat and the parvo vaccine given to your puppy were first introduced to other cats and dogs. Many other vaccinations for serious animal diseases have been developed through animal research—for example, rabies, distemper, anthrax, hepatitis, and tetanus.

The animal models used by the artist for the illustrations in this laboratory manual, and the animals you will dissect in this laboratory course, were prepared for educational use following strictly humane procedures. No endangered species have been used. No living vertebrate organisms will be harmed in this laboratory setting. Invertebrate animals that are to be dissected while alive are anesthetized before the procedure. The experiments selected are unoffensive, respect the integrity of the animal's evolutionary contributions, and often require only close observation. The experiments closely follow the tenets of the scientific method, which cannot dictate ethical decisions but can provide the structure for common sense. Do not be wasteful. Share the animals with your fellow students as often as possible. At the same time, observe the live animal in its natural setting, including its relationships to other species, for only in this manner will you gain a full appreciation of the unique evolutionary position and special structure and systems of each animal.

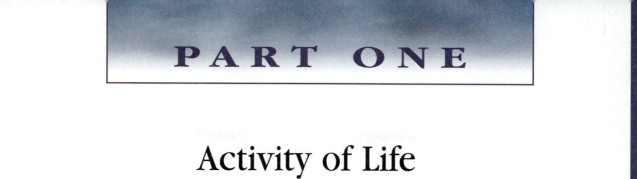

PART ONE

Activity of Life

1

Ecological Relationships of Animals

EXERCISE 1[1]

A Study of Population Growth, with Application of the Scientific Method

One goal of this course is to introduce students to the methods scientists use to gather knowledge. In this project you will apply the "scientific method" to the problem of determining what regulates animal populations.

The Scientific Method

People, scientists included, often acquire knowledge by applying a two-stage process—**conjecture** followed by **confirmation**—although few of us think of it this way. Conjecture consists of generating a general explanation of how the world is constructed, and is often based on general observations. Confirmation tests the validity of this conjecture. For example, in our everyday lives, we speculate on the quality of a future music concert based on our observations of recorded music, and then subsequently confirm, reject, or modify that speculation based on our experiences while attending the concert.

Scientists employ a similar method, although with a lot more rigor, in attempting to discover new facts. An idealized form of the method they use is known as the **scientific method,** and is broken down into four steps:

1. **Observation.** Observations may be based on direct examination of a system or on something we read, or may even be the result of discussions with others about a process or concept. Such observations frequently stimulate questions about why species exhibit certain traits, why internal organs interact the way they do, what the advantages of a particular body shape might be, or the role of certain genes in a particular process.

2. **Hypothesis formulation.** Formulating a hypothesis is like saying, "Let's suppose . . . ". Its objective is to explain, by induction, the observation. Typically, several alternative hypotheses are formulated, each a possible and reasonable explanation for the observation. Weeding out these hypotheses is the role of the third and fourth steps.

3. **Prediction.** Predictions are deduced from hypotheses, and are often based on some knowledge of the organisms or concepts being studied. They follow the form, "If hypothesis A is true, then I predict the following pattern." Predictions must be generated such that one set of hypotheses predicts one result, but alternative hypotheses predict another result. A prediction is worthless if it can be made for all of the hypotheses under consideration. Testing whether a prediction holds true allows one or more hypotheses to be rejected, thus reducing the number of hypotheses still under consideration.

4. **Testing of predictions.** The final step is to design a test so that a prediction, if incorrect, can confidently be rejected. Tests use observations or experimental manipulations. The confidence with which we can make a rejection is quantified by the use of inferential statistics, which we shall not discuss. However, note that our confidence in a result increases with the use of (1) treatments known as controls, in which all variables except for the one manipulated are held the same, and (2) several replicates of each treatment, to ensure that the observed result is due to conditions of the treatment and not simply due to variation among

---

[1] Exercise written by James C. Munger, Department of Biology, Boise State University, Boise, Idaho; and Richard S. Inouye, Department of Biological Sciences, Idaho State University, Pocatello, Idaho.

individuals. With skill (and perhaps some luck), all but one hypothesis will have been rejected. The unrejected hypothesis, however, is not proven true. Hypotheses can never be fully accepted; they can only be rejected (what lonely lives they must lead).

The next step is to repeat this process. With the results in hand from the tests of previous predictions, it is possible to fine-tune the hypothesis, and then set about testing the new one. Here is an example of this repetitive process:

Initial observation:	Roommate breaks dish
Generalization (hypothesis):	Roommate breaks everything
Prediction:	Will wreck borrowed car
Test/observation:	Didn't wreck car
New generalization/ hypothesis:	Only breaks dishes
Prediction:	Won't break borrowed camera
Test:	Does break camera
New generalization/ hypothesis:	Roommate breaks small objects

And so on . . .

The unrejected hypothesis at each stage is our best guess as to how the world works. If a hypothesis withstands repeated tests and has great explanatory value, it may be elevated to the level of a scientific theory. Note that a scientific theory is not an untested hypothesis, but is instead as close as scientists will come to calling a hypothesis proven. The theory of evolution is an example.

Whether a hypothesis is accepted or rejected, the observations made while testing the hypothesis frequently lead to more hypotheses, more predictions, more tests, more observations, and so on. It is often said that scientific investigation raises more questions than it answers; it is this aspect of science that many people find most exciting.

Application of the Scientific Method to the Study of Populations

If you were a scientist, you would make your own observations, formulate your own hypotheses, derive your own predictions, and perform your own tests of those predictions. A classroom situation, however, involves certain constraints, as you will see. Here we will apply the scientific method to the study of population growth.

Step 1—Observation

In 1798 a British economist, Thomas Malthus, published an essay in which he observed that populations do not grow indefinitely, but often tend to stay at relatively stable numbers. We can make similar observations: if we look around us, we do not see populations of organisms forever growing—instead, they are relatively stable.

Step 2—Hypothesis Formulation

Why do populations not grow indefinitely? Again we can look to Malthus, this time for one possible explanation (a hypothesis). Malthus reasoned that if a population had unlimited resources, it would grow **exponentially** to infinite size. However, since no population grows to infinite size, resources must be limiting. He therefore hypothesized that a limitation of resources is the cause of limited population growth. We can depict these two possible conditions graphically. If resources are unlimited, the population should grow exponentially:

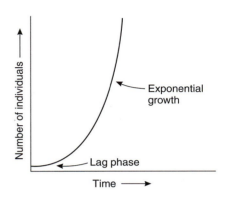

If resources are limited but in relatively constant supply, the population will experience **logistic growth,** growing rapidly at first and then eventually reaching an equilibrium, known as the carrying capacity. Logistic growth occurs in situations where resources are renewed (such as when plankton reach a barnacle) or where resources are not consumed (such as a nest site that can be used again and again).

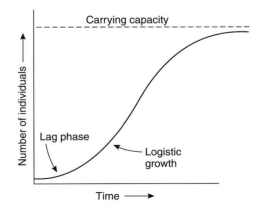

A third type of population growth occurs when resources are consumed but not renewed. For example, this sort of population growth occurs in a test tube bacterial culture in which the population increases until all the nutrients are consumed, and then crashes:

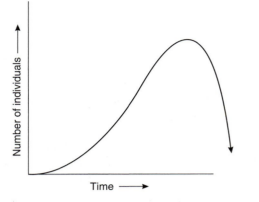

Now we can erect our first hypothesis regarding what limits population growth:

H_1: Limited food limits population size.

However, there are a number of factors that could limit population size, such as predation, climate, disease, limited nest sites, and intraspecific strife, including cannibalism. Can you think of others? We can cast these factors as this list of hypotheses:

H_1: Limited food regulates population size.
H_2: Predation regulates population size.
H_3: Disease regulates population size.
H_4: Limited nest sites regulate population size.
H_5: Intraspecific strife regulates population size.

Note that this list is not exhaustive. Also note that in this case the hypotheses are not mutually exclusive—that is, more than one may be true in a particular population; this is especially true if we consider a wide range of species.

Step 3—Prediction Derivation

What predictions logically follow from these hypotheses? The best predictions are those that (1) allow the investigator to decide between two or more competing hypotheses, and (2) are straightforward to test.

For the purposes of discussion, we will focus on our hypotheses as they apply to the setup to be used in this exercise: a population of flour beetles eating flour, living in flour, and laying eggs in flour, all contained in a small jar. In this system, we can discount the possibility of one hypothesis, predation, because we will not allow predators into the system.

A prediction that follows from H_1 is that if we limit food availability, we expect a smaller population to result (less flour, fewer beetles; more flour, more beetles). But do the other hypotheses make different predictions? If nest sites are limited, then adding more food will increase the availability of nest sites, giving the same prediction: more flour, more beetles. If intraspecific strife (e.g., cannibalism) is limiting, what will adding more

flour do? It will give the beetles more room to hide, meaning fewer encounters and more to eat, leading to less hunger; both mean less cannibalism. Again, more flour, more beetles. And if disease is limiting, more flour means fewer encounters among beetles and less disease transmission. Again, more flour, more beetles.

However, what if we were to vary the amount of food available while holding constant the total volume available for the beetles to roam? If H_1 were true, more food would lead to more beetles. But if H_3, H_4, or H_5 (but not H_1) were true, then more food would have no effect on beetle numbers, as long as the total volume was constant. So this prediction allows us to distinguish among competing hypotheses.

Step 4—Test of Predictions

Next, we need to create an experiment that allows us to vary food without varying volume. One way to accomplish this is to put various amounts of food into jars, and then add an inert filler (such as vermiculite) to maintain constant volume.

Step 5—Repeat the Process

When you look at the results from your experiment, you can consider what modifications to make to your hypotheses and what new predictions to use to test your new hypothesis.

Experimental Procedures

We will start cultures of *Tribolium confusum* (a flour beetle) with the same initial population size, but varying amounts of food and varying amounts of space. Near the end of the term, we will count the number of larvae, pupae, and adults in each container and compare age distribution and resulting densities.

T. confusum develops from egg to adult in about 28 days as follows: egg stage, 5 days; larval stages, 17 days; pupal stage, 6 days (Figure 1.1). The average life span of adult beetles is roughly 200 days.

☞ Work in groups of four students each. Each group should prepare:

A. One low-density, high-food jar, containing 50 g of resource (95% whole wheat flour; 5% brewer's yeast).
B. One medium-density, medium-food jar, containing 10 g of resource.
C. One high-density, low-food jar, containing 3 g of resource.
D. One low-density, medium-food jar, containing 10 g of resource with filler (such as vermiculite screened to standardize the size) added to bring the total volume to that of jar A.

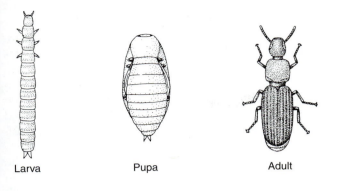

Larva   Pupa   Adult

Figure 1.1
Appearance of *Tribolium confusum* at larval, pupal, and adult stages.

E. One low-density, low-food jar, containing 3 g of resource with filler added to bring the total volume to that of jar A.

Next, sort through the culture that has been provided to you. Be careful not to damage the animals. Use a fine brush or small spatula to push them around. Each group should sort and count 250 healthy-looking adults, and put 50 into each of the half-pint Mason canning jars. We will assume that within each group of 50 there are plenty of both males and females. Cover the jar with the precut wire mesh (window screen) and then with a piece of paper towel; then screw on the top ring. Place the jars in a cabinet with a light (for warmth; about 30°C) and an open container of water (for humidity).

You and your lab instructor may decide to try other variations on this experimental setup. For example, you might try to see if temperature is important. How would you do this?

Why don't we use just one jar at each density for the whole class? If there were only one jar, it would be difficult to say (because of biological variation and experimental error) that the results from that jar were representative of any jars the class might start with the same density. For example, what if the only jar your class used for its medium-density culture had previously contained a toxic chemical? That could invalidate your results. However, if each group of four students prepares one jar at each density, a class of 24 students would have a total of six jars at the same density (known as **replicates**). If all six give approximately the same result, we can have substantial confidence that those jars are representative of all jars at that density.

During the term, make occasional observations of beetle behavior, and record those observations in your notebook. For example, do the beetles live on top of the flour or within it? Do the beetles congregate or space themselves out?

☞ At the end of the experiment, sort through each jar and count live adults, dead adults, pupae, and large (final instar) larvae. For the purposes of this experiment, we will count neither the eggs nor the early instar larvae. Compile data for the whole class and calculate averages for each age class for each jar. Construct an age distribution for each treatment.

Now that you have the totals for each jar, we can compare all five groups to each other using the totals from the entire class. When we compare two or more groups statistically we use an analysis of variance (ANOVA). An excellent online program (http://faculty.vassar.edu/lowry/anova1u.html) will allow you to enter the totals for the entire class and do a statistical comparison of the groups. This program will calculate the mean totals for each grouping (A–E) and tell you whether any group is statistically different from any other group. To use the online program indicate in the setup window that you are comparing five samples and click on "independent samples" in the setup as well. Move to the data entry window and enter the totals for all A groups under "Sample 1" and all B groups under "Sample 2" and so on. After you have entered all the data click "Calculate." The data summary will show you the mean for each group. The ANOVA summary will indicate the test statistics (F and P). If $P < 0.05$, at least one of your groups is significantly different from one of the others. Look at the summary under the Tukey test to find out which group means are different from other group means. If $P > 0.05$, then none of your groups are significantly different from one another and there will be no Tukey summary.

Questions

1. Which jars will test what hypotheses? For example, what hypothesis can be tested by comparing jars A, D, and E? What would you conclude if they had the same densities of beetles?

2. What other hypotheses can you formulate to explain the observation that populations do not grow indefinitely? What predictions can you derive from these hypotheses, and what tests could you perform?

3. Two life stages (eggs and pupae) are not mobile, and so are particularly vulnerable to cannibalism. Do you see evidence of this when comparing, for example, jar A with jar C?

4. An attitude commonly encountered in undergraduate science labs is that if you did not get the result the instructor expected, the experiment "did not work." What do you think of this view?

5. Given the results thus far, what will be your next step in this study, if your goal is to understand what determines the abundance of flour beetles?

Written Report

For your report, prepare appropriate graphs and write a summary statement of the experimental approach and an explanation of the results. Answer any of the preceding questions that your instructor may assign.

Alternatively, your instructor may want you to follow the format of a scientific paper for your report: introduction, materials and methods, results, discussion, and literature cited. The questions listed above are designed to bring up possible topics that might be included in the report. Be sure to think about your results and look at your data in original ways before writing the report.

References

Edmunds, J., J. M. Cushing, R. F. Costantino, S. M. Henson, B. Dennis, and R. A. Desharnais. 2003. Park's *Tribolium* competition experiments: a non-equilibrium species coexistence hypothesis. Jour. Anim. Ecology **72:**703–712.

Hasting, A., and R. Constantino. 1987. Cannibalistic egg-larva interactions in *Tribolium:* an explanation for the oscillations in population numbers. American Naturalist **130:**37–52.

Ho, F., and P. Dawson. 1966. Egg cannibalism by *Tribolium* larvae. Ecology **47:**318–322.

Lloyd, M. 1968. Self regulation of adult numbers by cannibalism in two laboratory strains of flour beetles *(Tribolium castaneum)*. Ecology **49:**245–259.

Lutherman, C., E. Miller, and T. Park. 1939. Studies in population physiology, IX. The effect of imago population density on the duration of larval and pupal stages of *Tribolium confusum* Duval. Ecology **20:**365–373.

Park, T. 1932. Studies in population physiology: the relation of numbers to initial population growth in the flour beetle *Tribolium confusum* Duval. Ecology **13:**172–181.

Park, T. 1933. Studies in population physiology, II. Factors regulating initial growth of *Tribolium confusum* populations. Jour. Exper. Zoology **65:**17–42.

Peters, M., and P. Barbosa. 1977. Influence of population density on size, fecundity, and developmental rate of insects in culture. Ann. Rev. Entomol. **22:**431–450.

Rich, E. R. 1956. Egg cannibalism and fecundity in *Tribolium*. Ecology **37:**109–120.

Stevens, L. 1989. The genetics and evolution of cannibalism in flour beetles. Evolution **43:**169–179.

Young, A. 1970. Predation and abundance in populations of flour beetles. Ecology **51:**602–619.

2

Introduction to Animal Classification

Taxonomy involves assigning scientific names to organisms and grouping or classifying them with reference to their exact position in the kingdoms of life. Scientific names are always latinized and are recognized internationally. This tends to prevent confusion—whereas one animal might be called by several different common names in different geographical areas, its scientific name is the same the world over.

Hundreds of thousands of different species of animals have been classified by comparing shared derived anatomical characters and, more recently, by using biochemical procedures such as DNA hybridization. Most of these species tend to fall into certain large groups because of similarities in structural organization. These primary groups are known as **phyla** (sing., **phylum**). Members of a phylum share certain distinctive characteristics that set members of that group apart from all other members of the animal kingdom.

In traditional Linnean classification, a phylum is subdivided into smaller groups called **classes;** classes are further subdivided into **orders;** orders into **families;** families into **genera** (sing., **genus**); and genera into **species** (sing. and pl.). In large groups, other categories, such as superclass, suborder, infraorder, and subfamily, also exist.

A **species** is a distinctive kind of living thing. The species name, or scientific name, is a **binomial**—that is, it consists of two parts, the genus name and the species epithet. We call this two-name system the Linnean system of **binomial nomenclature.** For example, the human species is *Homo sapiens; Homo* ("a man") is the genus, and *sapiens* ("mighty" or "wise") is the species epithet, actually an adjective that modifies the genus name. The genus name can be used alone when referring to a group of species included in that genus, such as *Rana* (a large genus of frogs) or *Felis* (a genus of cats including wild and domestic species). The specific epithet, however, would be meaningless if used alone because the same epithet may be used in combination with different genera. For example, the domestic cat is designated *Felis domestica; domestica* used alone is without significance, since it is a commonly used epithet that identifies no particular organism. Therefore, the species epithet must always be preceded by the genus name. However, you can abbreviate the genus name when it is

used in a context in which it is understood. *Felis domestica* might then be designated *F. domestica.*

In some cases, where geographical varieties or races of a species exist, three names may be used, in which case the last name indicates the **subspecies.** When three names are used, the method is called **trinomial nomenclature.** For example, one race (subspecies) of the long-tailed salamander, *Eurycea longicauda longicauda,* is classified as follows:

Phylum Chordata
    Subphylum Vertebrata
        Class Amphibia
            Order Urodela
                Family Plethodontidae
                    Genus *Eurycea*
                        Species *Eurycea longicauda*
                            Subspecies *Eurycea longicauda longicauda*

Note that all except the species and subspecies names are capitalized; species and subspecies names begin with lowercase letters. Genus, species, and subspecies names are printed in italics or are underlined when written or typed.

New species arise by speciation, also called **cladogenesis.** In cladogenesis, members of some populations of a pre-existing species cease exchanging genes with individuals of other populations. For example, this may occur when two subspecies diverge from each other to the extent that they no longer select individuals of the other subspecies as mates. This process of lineage splitting could be diagrammed in an evolutionary tree where each fork in the tree represents a speciation event. Such a diagram of lineage splitting is called a **phylogeny** or a **cladogram.** Producing a cladogram that would describe the pattern of lineage splitting that led to all species, both extant and extinct, is one of the goals of evolutionary biologists. The process used to infer the structure of an

---

> **EXERCISE 2A**
> Phylogeny Reconstruction—How to Make a Cladogram
> **EXERCISE 2B**
> **Use of a Taxonomic Key for Animal Identification**
> How to Use a Taxonomic Key
> Key to the Chief Phyla and Classes of the Animal Kingdom

evolutionary tree is called **phylogeny reconstruction.** Following is a short exercise that will help you understand the methods used in phylogeny reconstruction.

EXERCISE 2A
Phylogeny Reconstruction—How to Make a Cladogram

Core Study

A cladogram is a branching, or tree-like, diagram that represents a hypothesis about evolutionary relationships among the taxa shown. Taxa placed on a cladogram can come from any level of a Linnean hierarchy—species, genera, or phyla. Each cladogram illustrates a pattern of lineage splitting that produced the taxa we see now.

Figure 2.1 indicates that taxa A, B, and C shared a common ancestor at some point in the past prior to node 1. At node 1, the lineage of this ancestor split into two groups. Descendants of one group survive to the present as taxon A, whereas descendants of the other group form a lineage represented as the line that runs between node 1 and node 2. At node 2, this lineage split to form two extant taxa: B and C. Any pair of taxa resulting from a lineage split are called **sister taxa,** so B and C are sister taxa. What is the sister taxon to group A? It is the other half of the lineage-splitting event that occurred at node 1, so the sister taxon to group A is the combined group B plus C.

The pattern of branching (lineage splitting) that led to particular taxa is the critical element of a cladogram, but the relative left or right positions of taxa on a cladogram are of no importance. Thus, Figure 2.2 is identical in meaning to Figure 2.1. The pattern of branching has not changed (B and C are still sister taxa); instead, each taxon pair simply rotates at each node.

If a cladogram depicts a particular pathway that evolution followed to produce the three descendant taxa we now see, we should be able to map the evolution of characteristics present in the taxa onto the tree as well (Figure 2.3). For example, if the ancestor of all three taxa had a certain feature, such as three toes, it may have

passed this feature to all of its descendants. Features of organisms are called **characters,** and characters may come in different versions, called **character states.** In the example just given, the character is "number of toes" and the character states might be "one, two, or three toes." Another character might be "skull crest" and the two character states might be "skull crest present" and "skull crest absent." We map characters onto a tree with a horizontal line that indicates evolution of the character. Once a character evolves, it is passed to all descendants unchanged unless another horizontal line is marked to indicate a change of character state. The tree in Figure 2.3 says that the common ancestor of the three extant taxa had three toes and this state was passed to descendants A and C unchanged, but there was a loss of one toe in the lineage leading to taxon B after node 2. According to this diagram, "three toes" is the **ancestral state** of the character "number of toes," and "two toes" is the **derived state** for this character. Notice where the character "skull crest" evolved on the tree. Which taxa will have a skull crest? According to this tree, skull crests are present in taxa B and C because they both inherited the crest from a common ancestor. Given that B and C both have skull crests, it is perhaps possible that each lineage independently evolved skull crests, but this explanation would require two evolutionary changes (two separate events of skull crest evolution), instead of the one event of evolution of this trait in the common ancestor. When reconstructing phylogeny, biologists use the **Principle of Parsimony,** which states that the simplest explanation

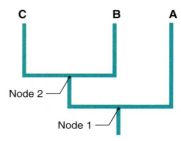

Figure 2.2
Cladogram depicting a branching pattern identical to that shown in Figure 2.1.

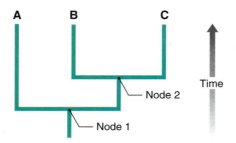

Figure 2.1
Cladogram depicting an evolutionary branching pattern for a lineage with three descendant taxa.

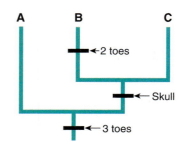

Figure 2.3
Cladogram depicting the evolution of two characters within the lineage leading to taxa A, B, and C.

that covers all the available evidence is the one to use. Thus, it is more parsimonious to assume that skull crests evolved in the common ancestor of taxa B and C.

For any group of taxa under study, we can identify characters that *are* likely to be passed from ancestors to descendants. Suitable characters are genetically based and show variation in character states within the taxa of interest. Character states are commonly recorded in a character matrix like the one shown here:

Taxon	Number of Toes	Skull Crests
A	3	Absent
B	2	Present
C	3	Present

Many potential characters exist for any group of taxa of interest. Noting each character state is called **scoring** the characters. Shortly, you will see how the distribution of character states among taxa can be used to predict the most likely path of evolution, but first we must outline all possible paths of evolution.

For three taxa, there are three possible paths that evolution could have taken. These three paths, all different, are depicted in the three cladograms shown in Figure 2.4. Examine the branching pattern in each to satisfy yourself that they differ in the arrangement of sister taxa.

How will we decide which of the three hypotheses about the path of evolution is most likely? We will use the observed distribution of character states to help us, but we also need to add an **outgroup** to our cladogram. An outgroup is a taxon that is distantly related to the taxa under study. For example, if the three taxa under study were members of the dog family, Canidae, we could use another four-legged vertebrate family as the outgroup. A member of the cat family or the bear family might be appropriate. If taxa A, B, and C were all canids, where would you place a line leading to the outgroup on each of the cladograms? The line leading to the outgroup (X) belongs at the base of each tree because the ancestor of all canids diverged from the ancestor it shared with bears *before* there was further subdivision of the canids into the taxa we see now (A, B, and C). A tree with the outgroup added has three nodes. The outgroup joins the tree at the lowest node, node 1 (Figure 2.5).

The need for an outgroup arises because we want to know which state for each character is ancestral and which states are derived. If a character has two or three states, how can we know which appeared first in evolutionary time? If a particular character state is present in one or several outgroups—for example, in members of the bear and cat families—it is likely to have evolved in the common ancestor of all four-legged vertebrates. Thus, if the state evolved in the common ancestor of all vertebrates, it must be the ancestral state for evolution within the dog family. We add the outgroup (X) to our character matrix along with another character represented by C3.

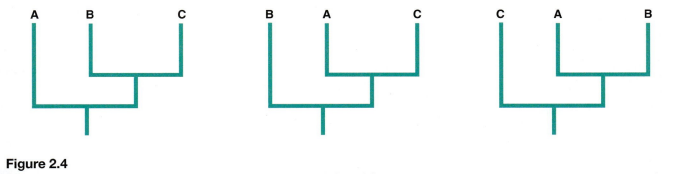

Figure 2.4
Three hypothetical branching patterns for the lineage leading to taxa A, B, and C.

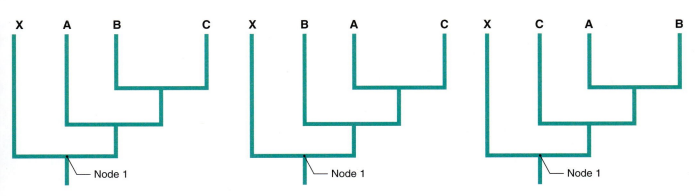

Figure 2.5
Three hypothetical branching patterns in relation to an outgroup for the lineage leading to taxa A, B, and C.

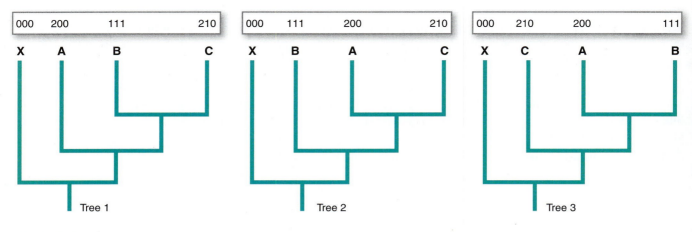

Figure 2.6

Three hypothetical branching patterns as in Figure 2.5, called trees 1 to 3, with added summary codes for each taxon.

Taxon	Character 1	Character 2	C3
A	3	Absent	Present
B	2	Present	Absent
C	3	Present	Present
X	1	Absent	Present

Notice that the character matrix can have both quantitative characters, represented here by states 1, 2, and 3 in character 1, and qualitative characters, represented here by the presence/absence states for characters 2 and 3. We are going to code these data to make it easier to see patterns. They can be coded in different ways as long as we are consistent, but for the sake of simplicity, we will code the outgroup state as zero, and the other state as 1. If the character has more than two states, other states are coded as 2, then 3, etc. Codes are shown to the right of the states in the next character matrix. The last column is a summary of all the character state information for each taxon. We have condensed a variety of kinds of data on each taxon to a simple three-digit code in the summary.

We place each summary above the appropriate taxon in each of the possible evolutionary pathways (Figure 2.6).

Now we map the characters onto each tree, *one character at a time,* noting where the characters must have changed to produce taxa with the character states that we observe now (Figure 2.7). For example, character 1 has state 0 in the outgroup, but it has state 1 or 2 in

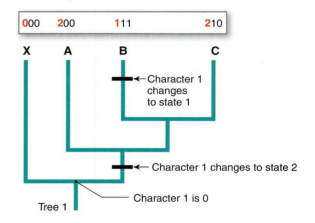

Figure 2.7

Tree 1, as in Figure 2.6, showing the evolutionary changes in character 1 that are required if this tree represents the true path of evolution.

the taxa of interest. Assume that character 1 has state 0 at node 1 as shown on the tree in Figure 2.7. If this tree represents the true path of evolution, then character 1 changed to state 2 between nodes 1 and 2—this change allows taxa A and C to have state 2. However, taxon B has state 1 for character 1, so we must add another evolutionary step to the tree: a change to state 1 in the lineage leading to taxon B. Thus, the most parsimonious mapping procedure requires two evolutionary changes for character 1 on this tree.

Taxon	Character 1	Character 2	C3	Summary
A	3 = 2	Absent = 0	Present = 0	200
B	2 = 1	Present = 1	Absent = 1	111
C	3 = 2	Present = 1	Present = 0	210
X	1 = 0	Absent = 0	Present = 0	000

Note that "absent" is coded as 0 for character 2 while "present" is coded as 0 for character 3.

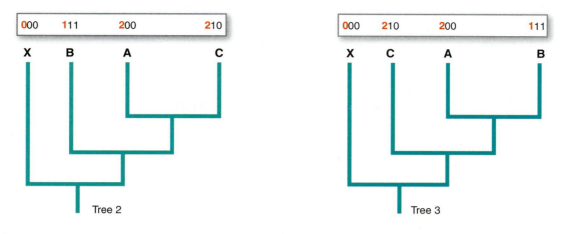

Tree 2

Tree 3

Figure 2.8
Trees 2 and 3, as in Figure 2.6. Use these trees to map the evolutionary changes in character 1 that are required if each tree represents the true path of evolution.

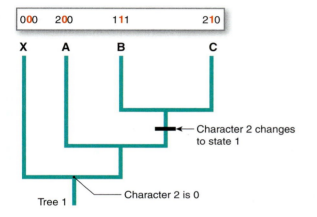

← Character 2 changes to state 1

Character 2 is 0

Tree 1

Figure 2.9
Tree 1, as in Figure 2.6, showing the evolutionary changes in character 2 that are required if this tree represents the true path of evolution.

To test your understanding, map character 1 onto the other two trees shown in Figure 2.8. It will take two evolutionary steps to map character 1 onto each tree, although the pattern of evolutionary change differs with each tree.

Continuing the procedure of mapping each character one at a time, we will turn to character 2 on tree 1 (Figure 2.9). From the outgroup, you know that character 2 has state 0 at node 1. Character 2 is still in state 0 at node 2 because taxon A has state 0 for this character. Character 2 requires only one change, between nodes 2 and 3, to correctly describe the pathway of evolution depicted on tree 1. Following this change, character 2 is in state 1 for taxa B and C.

Again, test your knowledge by mapping the changes needed to correctly depict the pathway of evolution for character 2 onto trees 2 and 3 (Figure 2.10). Did you discover that each of these trees required two changes to correctly map the evolution of character 2 onto them?

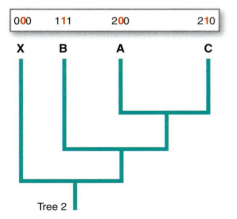

Tree 2

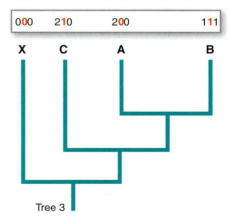

Tree 3

Figure 2.10
Trees 2 and 3, as in Figure 2.6. On these trees, map the evolutionary changes in character 2 that are required if each tree represents the true path of evolution.

Map the changes needed for character 3 onto each tree (Figure 2.11). You have now mapped all the character changes needed for each character on each tree. Count up the total number of evolutionary changes required to map *all three characters* onto all three trees. List the total number of evolutionary changes required for:

Tree 1: _____

Tree 2: _____

Tree 3: _____

Which tree represents the most likely path of evolution? _____ We again use the Principle of Parsimony to decide which tree is the simplest possible explanation that covers all the available facts. The tree with the least number of changes is the simplest explanation for the character states that we observe in the extant taxa. This tree is our best attempt to reconstruct the phylogeny for these taxa given the data we have. Tree 1 should have the fewest changes, so check your answer.

What if two trees were to have an equivalent number of changes? Is there a way to distinguish between them? With the given data, either tree represents a likely pathway, but more characters could be studied to distinguish between two equally parsimonious trees.

Now it is your turn to collect data on character states for a new set of organisms. We have selected some taxa in the following table, but if your instructor permits, you could choose other organisms and work your own example. We will use the sponge as the outgroup. We have filled in the first character for body symmetry. The next two characters can be described as present or absent, but the third character is quantitative. Score the remaining three characters. Code all characters using the system described previously and fill in the summary. Add more characters and make a bigger character matrix if your instructor permits.

Diagram the three possible relationships for these taxa (Figure 2.12). Map the characters onto each tree and discover which tree is most parsimonious.

Which two taxa are most closely related to each other (sister taxa) according to your analysis? _____

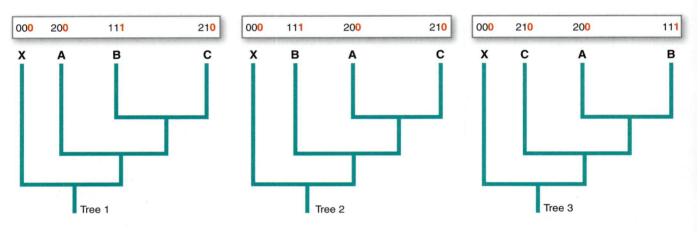

Figure 2.11

Trees 1, 2, and 3, as in Figure 2.6. On these trees, map the evolutionary changes in character 3 that are required if each tree represents the true path of evolution.

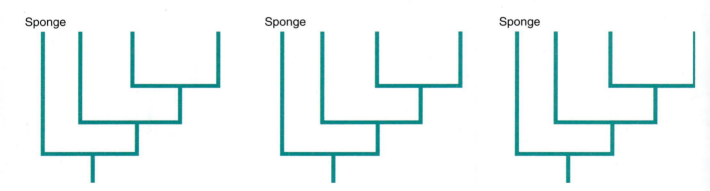

Figure 2.12

Three trees depicting a sponge as an outgroup. On these trees, diagram the three possible branching patterns for a lineage leading to the three taxa shown in the table below. The required evolutionary changes for the four characters listed in the table are to be mapped onto each tree.

Taxon	Symmetry	Head	Wings	Number of Legs	Summary
Frog	Bilateral				
Butterfly	Bilateral				
Jellyfish	Radial				
Sponge	Asymmetrical				0000

EXERCISE 2B
Use of a Taxonomic Key for Animal Identification

Further Study

How does a scientist identify an unknown specimen? One way is by direct comparison to specimens in a museum reference collection. However, few biologists have ready access to such collections. Even with such access, most nonspecialists would find this a tedious approach, involving working through thousands of museum specimens.

A practical alternative is to use a taxonomic key. A key is a convenient tabular device that enables us to identify a specimen by comparing it feature by feature with alternatives given in key couplets. Keys may be designed to identify species, genera, families, orders, or any other taxon.

In this exercise, you will identify specimens of animals (and animal-like protists) representing several different phyla and classes. Your instructor may assign this key or may substitute a different key based on forms common to your area. An alternative is the key to insect orders found in Exercise 13.

Once you have made an identification, it is important to verify its accuracy by consulting one or more references containing a text description of the distinctive characters of the species or group in question and a drawing or photograph of the species or of representatives of the group to which the species belongs.

Select a specimen and then, using the appropriate key and the following instructions, identify the specimen and record the phylum or class in the lab report on pp. 17 and 18. Verify the identification in one or more of the reference books provided by the instructor. Identify as many of the specimens as requested by the instructor. Return all specimens to the proper trays on the service desk.

How to Use a Taxonomic Key

A two-choice system serves as the basis of a **dichotomous key.** In the dichotomous key, two contrasting alternatives are offered at once, so you can choose the one that fits

your specimen. At the end of the choice, you will find a reference number to the next set of alternatives to be considered. Again make a decision and proceed in the same manner until you arrive at the scientific name of the animal or the taxon to which it belongs.

This key also has the capacity for reverse use, so that you can retrace your steps if you make a mistake. In each couplet, the number in parentheses refers to the number of the couplet from which that couplet was reached.

Keep in mind that individual variations exist; keys are based on the average, or "typical," adult specimen, whereas your specimen may be immature or somewhat abnormal. It is often very helpful to examine more than one specimen of a species or group, if available, when a particular descriptive character proves troublesome.

Written Report

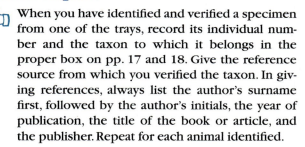

When you have identified and verified a specimen from one of the trays, record its individual number and the taxon to which it belongs in the proper box on pp. 17 and 18. Give the reference source from which you verified the taxon. In giving references, always list the author's surname first, followed by the author's initials, the year of publication, the title of the book or article, and the publisher. Repeat for each animal identified.

Key to the Chief Phyla and Classes of the Animal Kingdom

Following is a simple key to the more common phyla and classes of animals (and animal-like protists). The key, for the most part, uses external characters that can be visualized without dissection. It is designed for use with adult specimens. Like most keys, this key is utilitarian in the sense that the animal groups are not arranged in perfect phylogenetic sequence, and the characters used in the key may have no particular phylogenetic significance for the taxon. They are simply the characters that provide the best assurance of correct identification.

Key to the Major Animal Taxa

Numbers in parentheses refer back to the couplets from which these couplets were reached, making it possible to work backward if a selection is wrong, following the path of choices made.

1a   Free-living, single-celled organisms, some in colonies; chiefly microscopic, no tissues . . . . **Unicellular organisms** . . . . Go to 2

1b   Many-celled organisms, mostly macroscopic . . . **Metazoan organisms** . . . . Go to 4

2a (1a) Cilia or ciliary organelles in some stage; cilia used for locomotion; usually two types of nuclei; contractile vacuole and cytostome usually present . . . . . **Phylum Ciliophora** (ciliates)

2b   No cilia; locomotion by pseudopodia, flagella, or both . . . . Go to 3

3a   Locomotion by pseudopodia, body shape variable; body naked or with internal or external skeleton . . . . . **Amebas,** belonging to taxa not easily distinguished by observation.

3b   Flagellated, single-celled or colonial, cells containing colored bodies (chloroplasts) . . . . **Phyla Plantae or Euglenozoa**

4a (1b) Body with numerous pores; body radially symmetrical or irregular, with one or more large openings (oscula); no mouth or digestive tract . . . . . **Phylum Porifera** (sponges)

4b   Body without numerous pores or oscula; usually symmetrical in outline; mouth and digestive tract usually present . . . . . . Go to 5

5a (4b) Body with radial, pentaradial, or biradial symmetry . . . . Go to 6

5b   Body with bilateral symmetry . . . . . Go to 15

6a (5a) Body mostly soft and gelatinous; cylindrical, umbrella-shaped, or somewhat spherical; body parts usually in divisions of four, six, or eight . . . . Go to 7

6b   Body usually hard and spiny or with leathery skin; body parts in divisions of five (pentaradial); tentacles branched when present; usually with tube feet . . . . . **Phylum Echinodermata** . . . . Go to 11

7a (6a) Body cylindrical or umbrella-shaped; mouth or rim of umbrella usually encircled with unbranched tentacles; cnidocytes (stinging cells) present; mostly marine . . . . . **Phylum Cnidaria** (jellyfish, hydroids, corals, sea anemones, and relatives) . . . . . Go to 8

7b   Symmetry biradial; one pair of tentacles (not encircling mouth) or none; eight radial rows of ciliated comb plates . . . . **Phylum Ctenophora** (comb jellies, sea walnuts)

8a (7a) Body a gelatinous medusa in bell or umbrella shape; free-swimming . . . . Go to 9

8b   Body a cylindrical polyp, usually sessile or attached; single or colonial; tentacles surrounding the mouth . . . . Go to 10

9a (8a) Medusa small and possessing a velum; usually four to eight radial canals . . **Class Hydrozoa** (hydromedusae)

9b   Medusa usually large (2 to 20 cm or more) and lacking a velum; fringed oral lobes; highly branched radial canals; scalloped margins . . . . **Class Scyphozoa** (true jellyfish)

10a (8b) Polyps typically small, often in branching colonies with more than one type of polyp; gastrovascular cavity not divided by septa . . . **Class Hydrozoa** (hydroids)

10b   Polyps typically large (>0.5 cm diameter); gastrovascular cavity divided by septa . . . . . **Class Anthozoa** (sea anemones and corals)

11a (6b) Body without arms; body with hard endoskeleton of calcareous plates, or with soft, leathery skin . . . . Go to 12

11b   Body with branched or unbranched arms; body with hard endoskeleton of calcareous plates . . . . Go to 13

12a (11a) Body with rigid endoskeleton; body globular or flattened, with movable spines . . . . . **Class Echinoidea** (sea urchins, sea biscuits, sand dollars)

12b   Saclike body with leathery skin; body elongated in mouth-to-anus axis; mouth surrounded by branching tentacles . . . . . **Class Holothuroidea** (sea cucumbers)

13a (11b) Body with five movable, branched, and feathery arms; stalked or free-swimming; mouth and anus on oral surface, which is directed upward . . . . . **Class Crinoidea** (sea lilies, feather stars)

13b   Body with unbranched arms (except in some brittle stars); oral surface directed downward . . . . Go to 14

14a (13b) Arms not sharply set off from central disc; ambulacral grooves with tube feet on ventral side of each arm . . . . . **Class Asteroidea** (sea stars)

continued

Key to the Major Animal Taxa—*continued*

14b   Arms sharply set off from central disc; no ambulacral grooves. . . . . **Class Ophiuroidea** (brittle stars)

15a (5b)   No gill slits on pharynx, no internal skeleton (skull or vertebrae) of cartilage or bone . . . . Go to 16

15b   Lateral gill slits in pharynx present, or body with internal skeleton of cartilage or bone, or both; nerve cord dorsal and single . . . Go to 38

16a (15a)   Body slender, wormlike or leaflike, with no body segments, lateral appendages or fins, or shell . . . . Go to 17

16b   Body not as above; if body wormlike, then having appendages or segments or both . . . . Go to 23

17a (16a)   Body flat and soft, rarely cylindrical . . . Go to 18

17b   Body narrowly cylindrical with hard or tough cuticle . . . . . Go to 21

18a (17a)   Body flattened dorsoventrally; no anus, no proboscis . . . . . **Phylum Platyhelminthes** (flatworms) . . . . Go to 19

18b   Body long, soft, and highly contractile; long eversible proboscis present above mouth; digestive tract with anus; mostly free-living . . . **Phylum Nemertea** (ribbon worms)

19a (18a)   Mouth and digestive tract present . . . Go to 20

19b   No mouth or digestive tract; body of scolex and usually numerous pseudosegments (proglottids); body increasing in size posteriorly, usually long and ribbonlike . . . . . **Class Cestoda** (tapeworms)

20a (19a)   No attachment organs, no suckers around mouth; ciliated epidermis . . . **Class Turbellaria** (planarians)

20b   Attachment organs present, some with hooks at posterior end; suckers present around mouth; nonciliated epidermis; body leaflike or slender in shape . . . . . **Class Trematoda** (digenetic flukes)

21a (17b)   Lateral lines present . . . . **Phylum Nematoda** (roundworms)

21b   Lateral lines absent . . . . Go to 22

22a (21b)   Body extremely slender and elongate . . . . . . **Phylum Nematomorpha** (horsehair worms)

22b   Anterior retractile proboscis armed with spines . . . . . **Phylum Acanthocephala** (spiny-headed worms)

23a (16b)   Body slender and torpedo-shaped, with lateral and caudal fins; mouth with bristles or spines; planktonic . . . . . **Phylum Chaetognatha** (arrow worms)

23b   Body not as above. . . . . Go to 24

24a (23b)   Body soft and unsegmented; body enclosed in shell, or body with ventral muscular foot or both; some with fleshy arms or tentacles . . . . Go to 25

24b   Body segmented (often wormlike), with jointed appendages, or both . . . . . . Go to 30

25a (24a)   Shell of two valves arranged in dorsal and ventral position to each other; ventral shell usually larger than dorsal; stalk or peduncle for attachment . . . . . **Phylum Brachiopoda** (lamp shells)

25b   Shell single, or of two lateral valves, or of eight dorsal plates; or shell absent or reduced and internal; ventral muscular foot or fleshy arms or tentacles present; body unsegmented . . . . . . **Phylum Mollusca** (molluscs) . . . . . Go to 26

26a (25b)   No prehensile arms with suckers; eyes small or absent. Go to 27

26b   Head large and well developed with two large eyes and foot modified into 8 or 10 prehensile arms with suckers . . . . . **Class Cephalopoda** (nautiluses, squids, cuttlefishes, and octopuses)

27a (26a)   Shell of eight dorsal plates; radula present . . . **Class Polyplacophora** (chitons)

27b   Single shell, or shell of two valves, or shell absent or internal . . . . Go to 28

28a (27b)   Shell of two lateral valves with ligamentous hinge; muscular foot present; head reduced; no tentacles or radula . . . . **Class Bivalvia** (bivalves)

28b   Shell of one piece or absent or internal . . . . Go to 29

29a (28b)   Shell tubular and open at both ends; head absent; mouth with tentacles and radula . . . . **Class Scaphopoda** (tooth shells)

29b   Shell usually coiled or spiraled (uncoiled or absent in some); head with radula, one or two pairs of tentacles, and one pair of eyes . . . **Class Gastropoda** (snails, slugs, nudibranchs, tectibranchs)

30a (24b)   Body wormlike and segmented throughout; setae, parapodia, or both often present; no jointed appendages . . . . . **Phylum Annelida** (segmented worms). . . . . Go to 31

continued

Key to the Major Animal Taxa—*continued*

30b   Segmented body encased in firm exoskeleton of chitin; jointed appendages . . . . . . **Phylum Arthropoda** (arthropods) . . . . Go to 33

31a (30a)   Setae present on each segment; segments distinct; suckers absent . . . . Go to 32

31b   Setae absent; no parapodia; segments indistinct and with many annuli; clitellum present; suckers present anteriorly and posteriorly or posteriorly only . . . . . **Class Hirudinida** (leeches)

32a (31a)   Many setae on each segment; parapodia or fleshy lateral appendages present (may be reduced); clitellum absent . . . . . **Class Polychaeta** (marine segmented worms)

32b   Few setae per segment; no parapodia; clitellum present . . . . . **Class Oligochaeta** (earthworms; freshwater segmented worms)

33a (30b)   Paired antennae present . . . . Go to 34

33b   Antennae absent; body of cephalothorax and abdomen; segmentation often obscured . . . . Go to 37

34a (33a)   Two pairs of antennae; appendages mostly biramous and specialized for different functions; many with gills; head, thorax, and abdomen present but head and at least part of thorax fused . . . . . **Subphylum Crustacea** (crustaceans)

34b   One pair of antennae . . . . Go to 35

35a (34b)   Head, thorax, and abdomen distinct; three pairs of legs on thorax; one or two pairs of wings often present . . . . . **Subphylum Hexapoda, Class Insecta** (insects)

35b   Body elongate, 15 or more pairs of jointed legs . . . . **Subphylum Myriapoda** . . . . . Go to 36

36a (35b)   Each segment with one pair of legs; dorsoventrally flattened . . . . **Class Chilopoda** (centipedes)

36b   Each segment with two pairs of legs; subcylindrical body . . . . . . **Class Diplopoda** (millipedes)

37a (33b)   Four pairs of walking legs; head completely fused with thorax; no wings . . . . **Class Arachnida** (spiders, scorpions, ticks)

37b   Five pairs of walking legs; lateral compound eyes present . . . . **Class Merostomata** (horseshoe crabs)

38a (15b)   Short and wormlike; body divided into proboscis, collar, and trunk . . . . **Phylum Hemichordata** (acorn worms)

38b   Body not wormlike . . . . . **Phylum Chordata** Go to 39

39a (38b)   Cranium and brain absent . . . . Go to 40

39b   Cranium and brain present . . . . . **Subphylum Vertebrata** . . . . Go to 41

40a (39a)   Adults saclike and sedentary; body covered with a test and with two siphons at one end . . . **Subphylum Urochordata** (tunicates)

40b   Body lance-shaped; lateral musculature in conspicuous V-shaped segments; notochord and dorsal nerve cord extend length of body . . . . **Subphylum Cephalochordata** (amphioxus)

41a (39b)   Body fishlike . . . . Go to 42

41b   Body not fishlike . . . . Go to 45

42a (41a)   Without true fins, scales, or paired fins . . . **Superclass Agnatha** (hagfishes, lampreys) . . . . Go to 43

42b   With jaws and (usually) paired appendages; notochord replaced by vertebrae . . . . **Superclass Gnathostomata** (jawed fishes, all tetrapods) . . . . Go to 44

43a (42a)   Suctorial mouth with horny teeth; seven pairs of gill pouches . . . . . **Class Petromyzontida** (lampreys)

43b   Terminal mouth with 4 pairs of tentacles; 5 to 15 pairs of gill pouches . . . . **Class Myxini** (hagfishes)

44a (42b)   Skeleton cartilaginous; ventral mouth; placoid scales or no scales; five to seven pairs of gills, each in a separate pharyngeal cleft . . . . **Class Chondrichthyes** (sharks, rays, skates)

44b   Skeleton mostly bony; body primarily fusiform but variously modified; cycloid, ganoid, or ctenoid scales; terminal mouth with many teeth . . . . . **Class Actinopterygii** (ray-finned fishes)

45a (41b)   Skin with horny epidermal scales, sometimes with bony plates; paired limbs, usually with five toes, or limbs absent; no gills . . . . . **Class Reptilia** (non-avian reptiles: snakes, turtles, lizards, crocodilians)

45b   Skin without scales . . . . . Go to 46

46a (45b)   Epidermis with feathers on body and scales on legs; forelimbs (wings) adapted for flying; toothless horny beak . . . **Class Reptilia, Subclass Aves** (birds)

46b   No feathers, wings, or beak . . . . . Go to 47

47a (46b)   Skin naked, often moist, and sometimes warty . . . . **Class Amphibia** (frogs, toads, salamanders, caecilians)

47b   Body covered with hair (reduced in some); integument with sweat, sebaceous, and mammary glands . . . . . **Class Mammalia** (mammals)

Name_____

Date_____

Section_____

Use of a Key for Animal Identification

In each box, print the number of a specimen and taxon (phylum or class) or, if other animal keys are used, the species (genus name and species epithet) as required by the instructor. Enter the complete reference consulted to verify the identification.

A _____   _____

Reference

B _____   _____

Reference

C _____   _____

Reference

D _____   _____

Reference

E _____   _____

Reference

F _____   _____

Reference

LAB REPORT 2A

G _____ _____

Reference

H _____ _____

Reference

I _____ _____

Reference

J _____ _____

Reference

K _____ _____

Reference

L _____ _____

Reference

PART TWO

The Diversity of Animal Life

The Microscope

For a biologist, the compound microscope is probably the most important tool ever invented. It is indispensable not only in biology but also in medicine, biochemistry, and geology; in industry; in crime detection; and even in many hobbies. But although the microscope is one of the most common tools in the biologist's laboratory, too frequently it is used without any effective understanding of its construction and operation. The results may be poor illumination, badly focused optics, and misleading interpretations of what is (barely) seen.

EXERCISE 3A
Compound Light Microscope

The compound light microscope may be either monocular or binocular, and the oculars may be either vertical or inclined.

☞ Use both hands to carry a microscope. Grasp it firmly by the **arm** with one hand, and support the **base** with the other. Carry it in a fully upright position.

Core Study

Understanding the Parts and Operation of the Microscope

If you are not familiar with the parts of the microscope, please study Figures 3.1 and 3.2.

The image-forming optics consist of (1) a set of **objectives** screwed into a **revolving nosepiece,** and (2) a **body tube,** or head, with one or two **oculars (eyepieces).**

Each objective is a complex set of tiny lenses that provide most of the magnification. Your microscope may have two, three, or four objectives, each with its magnification, or power, engraved on the side. For example, if the objective magnifies an object 10 times, the magnification is said to be 10 diameters and is commonly written simply as 10×. Most microscopes include a **scanning objective** (3.5× or 4.5×), a **low-power objective** (10×), and a **high-power objective** (40×, 43×, or 45×). Some microscopes also carry an **oil-immersion objective** (95×, 97×, or 100×), which must always be used with a drop of oil to form a liquid bridge between itself and the surface of the slide being viewed.

☞ Rotate the nosepiece and note the clicking sound when an objective swings into place under the tube. Enter here the magnifications of the objectives on your microscope: _____ _____ _____ Does your microscope have an oil-immersion lens? _____ If so, what is its magnification? _____

Lenses in the ocular further magnify the image formed by the objective. The ocular most often used is the 10×. Often a pointer is mounted into the ocular. Enter here the magnification of the ocular on your microscope: _____

☞ If there is a pointer in your microscope, rotate the ocular and note the movement of the pointer. Remember that if you look through a binocular scope with only one eye you may miss the pointer. Always use both eyes.

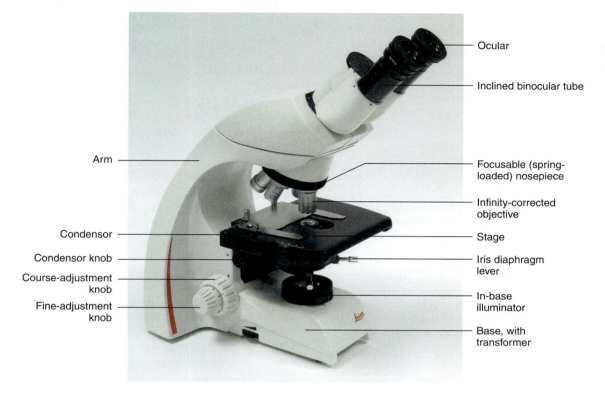

Figure 3.1

Optical and mechanical features of a compound microscope.

Directly beneath the **stage** of most microscopes is a **substage condenser,** a system of enclosed lenses that concentrates the light on the specimen above. The condenser may have a knob that allows you to move the condenser up and down.

Beneath the condenser, many microscopes have a built-in, low-voltage **substage illuminator.** Microscopes lacking a substage illuminator employ an adjustable **reflecting mirror** that has two surfaces. The **concave surface** of the mirror is used to reflect natural light or light from a separate microscope lamp when there is no condenser on the microscope. The **plane (flat) surface** of the mirror is used with a substage condenser.

☞ Turn on the substage illuminator. If your microscope lacks one, use a microscope lamp, or position your microscope and mirror to take advantage of natural light. With the low-power objective in place, adjust the mirror to bring a bright, evenly distributed circle of light through the lens.

An adjustable iris diaphragm under the stage is used to reduce glare caused by stray light from the illuminator.

☞ Raise or lower the substage condenser to near its uppermost limit. Then close down the iris diaphragm until all glare is gone (but don't close so far that dark halos appear around objects).[1]

On a straight microscope, you raise and lower the body tube with two sets of adjustment knobs. Use the **coarse-adjustment knob** for low-power work and for initial focusing. Use the **fine-adjustment knob** for final adjustment.

On a microscope with an inclined tube, you focus by *moving the stage* rather than the body tube. If you are using an inclined microscope, read "lower the stage" when the directions say "raise the objective." In either case, the distance between objective lens and the object is increased.

☞ Turn the coarse-adjustment knob and note how it moves the body tube (or stage). Find out which way to turn the knob to raise and lower the tube.

Never use the coarse-adjustment knob when a high-power objective is in place. Turn the fine-adjustment

---

[1]Correct adjustment of illumination, called Koehler illumination, is described in more detail in Appendix A (p. 278).

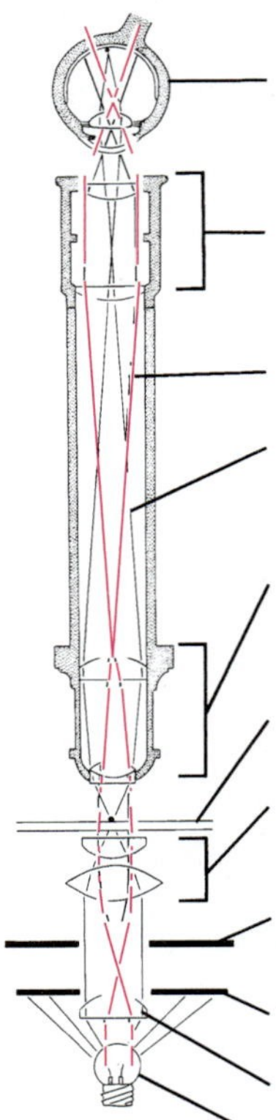

Your eye

Ocular (eyepiece) normally 10× magnification; may contain pointer to assist in locating objects in field of view.

Illuminating rays

Image-forming rays

Objective—two, three, or four objectives of different magnification are mounted in a rotating nosepiece: objective magnification times ocular magnification gives total magnification

Specimen stage

Condenser lenses concentrate light on specimen; correct position is at or just below its uppermost position

Condenser iris diaphragm controls amount of light passing through the specimen

Lamp diaphragm (also called the field stop) controls amount of light reaching condenser

Lens of lamp

Low-voltage lamp

Figure 3.2
Optical path of light through a microscope.

knob. This moves the tube so slightly that you cannot detect it unless you are examining an object through the ocular. The fine-adjustment knob works the same as the coarse-adjustment knob. To focus downward, turn the knob in the same direction as you would to focus down with the coarse adjustment. Practice this. **Always use the fine adjustment when the high-power objective is in place.**

Getting Acquainted with Your Microscope

You need not wear glasses when using the microscope unless they correct a severe astigmatism. Nearsightedness and farsightedness can be corrected by adjusting the microscope.

Clean the lenses of both the ocular and the objectives, when necessary, by wiping them gently with a clean sheet of lens paper. **Never touch lenses with anything except clean lens paper.**

Keep both eyes open while using a monocular microscope. If this seems difficult at first, hold a piece of paper over one eye while viewing the object with the other. Should one eye become tired, shift to the other one.

How to Focus with Low Power

Turn the low-power (10×) objective until it clicks in place over the aperture. Adjust the condenser and iris diaphragm for optimal illumination as already described.

Obtain a slide containing the letter *e* (or *a, h,* or *k*). Place it, coverslip up, on the stage with the letter centered under the objective lens. ***While watching from the side,*** lower the objective with the coarse adjustment until it is close to the slide surface.

While looking through the ocular, slowly raise the objective by turning the coarse adjustment toward you until the object on the slide is in sharp focus. Is the image upside down? _____ Is it reversed—that is, does the left side of the letter appear on the right, and the right side on the left? _____ On a separate sheet of paper, draw the letter as it appears.

Shift the slide very slightly to the right while viewing it through the ocular. In what direction does the image move? _____ Move the slide away from you. What happens to the image? Turn the fine-adjustment knob toward and then away from you and observe the effect on the image. To gain experience with interpreting depth, obtain a slide with three different-colored threads, and focus with low power. Which colored thread is on top? _____ In the middle? _____ On the bottom? _____

How to Focus with High Power

Focus the object first with low power; then slowly rotate the high-power objective into position. If the microscope lenses have been constructed in a particular way by the manufacturer, the object in focus with low power will be nearly in focus under high power. Such lenses are said to be **parfocal** with respect to each other. Now turn the **fine-adjustment knob** to bring the specimen into sharp focus.

Never use the coarse adjustment while looking at an object under high power; you may drive the objective into the slide, damaging the slide or ruining the lens of the objective.

If the microscope is not parfocal (your instructor will tell you), focus first with low power and then swing the high-power objective into place, raising the tube so that the objective clears the slide. Now, **still watching the high-power objective from the side,** lower it slowly to about 1 mm from the coverslip. Then, **looking through the ocular,** raise the tube with the fine adjustment until the object is in focus. Do this several times to acquire skill in focusing.

Because light decreases when you switch to high power, you must adjust the light with a mirror and/or the iris diaphragm.

While viewing the object, keep your hand on the fine adjustment and constantly focus up and down. This enables you to see detail throughout the depth of the object.

How to Use the Oil-Immersion Objective

Occasionally a project or demonstration requires using an oil-immersion objective. This is an objective in the 90× to 100× range. Because the resolving power of this objective is so great, you must use oil to bridge the separation between slide and objective.

To use oil immersion, bring the specimen into focus first with the low-power and then with the high-power objective. Carefully center the point of interest in the field of view; then rotate the nosepiece to move the high-power objective off to one side. Place a single drop of immersion oil on the coverslip at the point where the objective will come into position. Now move the oil-immersion objective carefully into position, **watching from the side** to be certain that the lens clears the coverslip. The oil should now form a bridge between lens and coverslip. **Carefully** adjust the fine focus to bring the specimen into focus. Adjust the iris diaphragm or substage condenser to increase light as required.

When finished, clean the lens with a lens tissue wetted with xylol and then with a lens tissue wetted with Kodak Lens Cleaner. **Never** use alcohol, which will dissolve the cement around the lens system. If the lens is to be used again soon (within a day or two), it is best not to clean the lens face. Residual oil will not harm the lens unless it is allowed to harden over a long period without use.

Taking Control of Your Microscope

1. **Proper lighting** is the first requirement for happy microscopy. Too much light is as bad as too little (beginners usually tend to use too much). Transparent objects are often clearer in reduced light. Reduce light by closing down the iris diaphragm, **not** by lowering the substage condenser.

2. **Focus with eyes relaxed.** The image appears to your eye as though it were about 10 inches away, but the eye should be relaxed as though it were viewing an image in the distance. Look up periodically and train your eyes on something across the room. Then, if you keep your eyes relaxed, you should not have to readjust your focus very much when looking through the microscope. If you get a headache, chances are you are trying to look *into* the microscope rather than *through* it.

3. **If using a binocular microscope,** it is important (a) to adjust the distance between the microscope's ocular to match the distance between your own pupils, and (b) to adjust the oculars for a sharp focus. If the focus is not sharp, you may have to focus each eye separately. The microscope will have one fixed and one adjustable ocular. Adjust focus for the fixed ocular first to suit that eye, and then adjust the other ocular (usually by rotation) until the focus is sharp for both eyes.

4. **Find the correct eye distance** from the oculars, one that affords a full view of the field. Stay relaxed, hold your head steady, and enjoy the view.

5. **Where's the dirt?** If spots or smudges appear in the field of vision, there may be dirt on the ocular, on the slide, or on the objective. To find out which, first rotate the ocular; if the spots move, the ocular needs to be cleaned (in laboratory, avoid using eye makeup, which may smear on the ocular surface). Move the slide; if the spots move with it, clean the slide. If, after cleaning the ocular and the slide, the spots persist, they probably signify a dirty objective lens. Use only special lens paper to clean lenses. The slide may be cleaned very gently with a soft damp cloth or damp cleansing tissue. If, after cleaning everything, you still see spots or smudges, try moving the condenser up or down a little. Still not satisfied? You may need help from the assistant. Some people see "floaters" that drift across the field of vision while viewing a brightly illuminated object. These are defraction images of red blood cell "ghosts" in the vitreous humor of the eye. Although they are annoying at first, one can usually learn to ignore them.

6. **Keep one hand on the fine adjustment** and constantly focus up and down. This is the only way to see everything.

7. **If the fine adjustment refuses to turn,** the knob has reached its range limit. To correct this, give the fine-adjustment knob several turns in the opposite direction, and then refocus with the coarse adjustment.

8. **Be friendly to your microscope.** It must be kept dust free, so return it **carefully** to its box or cupboard when finished with it. Before putting it away, put the low-power objective in place and raise the tube a little. Be sure not to leave a slide on the stage.

Prepared slides you will be using in the laboratory were made at the cost of great skill and patience and are quite expensive. They are fragile and should be handled with great care. Your instructor may provide a demonstration of how slides are made.

Magnification in the Microscope

How much your microscope will magnify depends on the power of the combination of lenses you are using. Your microscope is probably equipped with a 10× ocular, which magnifies the object 10 times in diameter. Other oculars may magnify 2×, 5×, or 20×. The objectives may be designated, respectively, 3.5× (scanning objective), 10× (low-power objective), and 45× (high-power objective). The total magnifying power is determined by multiplying the power of the objective by the power of the ocular. Here are examples of the magnification of certain combinations:

Ocular	Objective	Magnification
5×	3.5×	17.5 diameters
5×	10×	50 diameters
10×	3.5×	35 diameters
10×	10×	100 diameters
10×	45×	450 diameters
10×	90×	900 diameters

For handy reference, enter in Table 3.1 the total magnifying power for the lens combinations on *your* microscope.

How to Measure the Size of Microscopic Objects

It is often important to know the size of an organism or object you are viewing through the microscope because size may be a diagnostic characteristic. Or, if you are looking for a particular species of protozoan in a mixed culture and know that the species is usually about 400 µm long, it saves time to know just how large 400 µm will appear at either low or high power. Alternative methods for measuring object sizes are described here.

Measuring objects with transparent ruler calibration. With this simple method, the viewer measures the diameter of the field of view and then estimates the proportion of the field occupied by the object. This method is not as accurate as the alternative described in the next section, but often an approximate size is all that the viewer requires.

☞ With the scanning objective in position, place a transparent ruler on the microscope stage and focus on its edge so that you can see the scale. Move the ruler right or left so that one of the vertical millimeter lines is just visible at the edge of the circular field of view. Count the number of millimeter lines spanning the field; you will probably have to estimate the last decimal fraction of a millimeter. Enter the diameter in mm here: _____ Now convert this figure to micrometers by multiplying by 1000. Enter this value here: _____ This is the diameter in micrometers of the field for the scanning lens.

Diameters of the fields of view for your low-power and high-power objectives cannot be measured directly with the transparent ruler because of the high magnification. To calculate the field diameter of the low-power objective, multiply the diameter of the scanning lens field by the magnifying power of the scanning objective and divide by the magnifying power of the low-power objective:

$$\frac{\text{Magnification of scanning objective}}{\text{Magnification of low-power objective}}$$
$$\times \text{ Diameter of scanning objective field}$$
$$= \text{ Diameter of low-power field}$$

Make this calculation for your microscope, and enter the value here: _____ It usually ranges between 1.5 and 1.6 mm (1500 and 1600 µm).

Similarly, the diameter of the high-power field is determined as:

$$\frac{\text{Magnification of the low-power objective}}{\text{Magnification of the high-power objective}}$$
$$\times \text{ Diameter of low-power field}$$
$$= \text{ Diameter of high-power field}$$

Table 3.1 Magnifying Powers for Microscope Number _____						
Magnification For:	**Ocular**		**Objective**		**Magnification**	
Scanning lens	_____	×	_____	=	_____	diameters
Low-power lens	_____	×	_____	=	_____	diameters
High-power lens	_____	×	_____	=	_____	diameters
Oil-immersion lens	_____	×	_____	=	_____	diameters

Make this calculation for your microscope, and enter the value here: _____ It usually ranges between 0.36 and 0.42 mm (360 and 420 µm). Now enter these values in Table 3.2 for quick reference.

☞ Place a hair from your head and another from your eyebrow on a slide. Examine them with low and high power and measure their diameters. Write the diameters in micrometers here: eyebrow _____ head _____ Are they the same? Compare diameters with others in the class.

Measuring objects with ocular and stage micrometer calibration. An **ocular micrometer** can be fitted into the microscope's ocular. It is a disc on which is engraved a scale of (usually) either 50 or 100 units. These units are arbitrary values that always appear the same distance apart no matter which objective is used in combination with the eyepiece. Therefore, the ocular micrometer cannot be used to measure objects until it has been calibrated with a **stage micrometer.**

The stage micrometer resembles an ordinary microscope slide but bears an engraved scale on its upper surface, usually 1 or 2 mm long, apportioned into 0.1 and 0.01 mm divisions.

Place the stage micrometer on the microscope stage and focus on the engraved scale with the low-power objective. Both scales should appear sharply defined. Rotate the eyepiece until the two scales are parallel. Now move the stage micrometer to bring the 0 line of the stage scale in exact alignment with the 0 marking of the ocular scale. The scales should be slightly superimposed (Figure 3.3).

To calibrate the ocular scale for this objective, use the longest portion of the ocular scale that can be seen to coincide precisely with a line on the stage scale. For example, suppose that 70 units on the ocular scale equal 0.24 mm on the stage scale. Then:

$$70 \text{ ocular units} = 0.24 \text{ mm, and}$$

$$1 \text{ ocular unit} = \frac{0.24}{70} = 0.0034 \text{ mm, and}$$

$$1 \text{ mm} = 1000 \text{ µm}$$

Therefore:

$$1 \text{ ocular unit} = 0.0034 \text{ mm} = 3.4 \text{ µm}$$

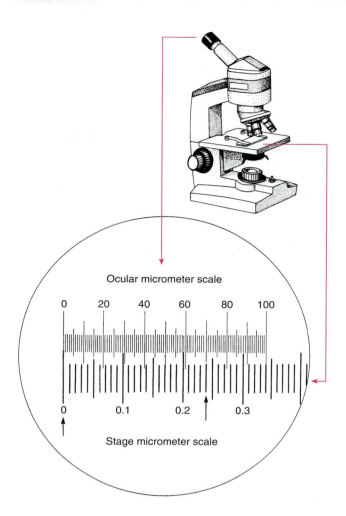

Figure 3.3
Calibrating an ocular micrometer.

Repeat the calibration procedure for the scanning lens and the high-power lens. For handy reference, enter in Table 3.3 the micrometer values you have calculated.

To measure any object available to you in the laboratory, it is only necessary to multiply the number of divisions covered by the specimen or part thereof by the micrometer value you have determined (3.4 µm in this example). Note that the micrometer value applies only to the objective with which the calibration was made.

Table 3.2 Diameters of Fields of View for Microscope Number _____	
Scanning lens:	_____ µm
Low-power lens:	_____ µm
High-power lens:	_____ µm

Table 3.3 Micrometer Values for Microscope Number _____		
Scanning lens, 1 ocular unit	=	_____ µm
Low-power lens, 1 ocular unit	=	_____ µm
High-power lens, 1 ocular unit	=	_____ µm

Exercises with the Compound Light Microscope

Try these or any similar exercises with the compound microscope.

1. Prepare a temporary wet mount (Figure 3.4). Select a clean slide and coverslip. Mount a piece of insect wing or appendage, a bit of feather, or some similar object on the slide. Add a drop of distilled water with a pipette. Hold the coverslip at an angle with one edge on the slide and the other held up with a dissecting needle. Slowly lower the raised edge onto the drop of water. This helps prevent air bubbles from forming. Examine with low power, and sketch what you see in the field of vision. Switch to high power without moving the slide, and draw the portion now visible.

2. Make a wet mount using a drop of water from a solution in which gum arabic has been dissolved and shaken up. With low or scanning power, look for large and small air bubbles, focusing up and down on them until you are sure you will always recognize an air bubble. This may later prevent you from interpreting a bubble as a small organism or other small object. What is the characteristic outline of an air bubble?

3. Mount a drop of pond water, cover, and look for living organisms in it. As you watch the swimming animals, remember that movement as well as size is magnified by the microscope. Sketch some of the organisms on a separate paper. On your drawing, state the estimated size of each animal and the magnification of your drawing.

EXERCISE 3B
Core Study

Stereoscopic Dissecting Microscope

The stereoscopic dissecting microscope (Figure 3.5) is as indispensable to the laboratory as is the compound microscope. It enables us to study objects too large or too thick for the compound microscope. It furnishes a three-dimensional view of objects at a very low power (5× to 50×, depending on the microscope). The image is not inverted, and there is ample space for manipulation and dissection under the lens. The microscope stage can be illuminated either by **transmitted light** (light passing through the object from below) or by **reflected light** (light illuminating the object from above and being reflected by the object into the microscope). Which type of lighting does the compound microscope use? _____ Some dissecting microscopes have a substage mirror or a substage lamp. Focusing is done with very little adjustment.

Place an object on the stage and illuminate with reflected light. Looking through the oculars, adjust them

Figure 3.4
Preparation of a wet mount.

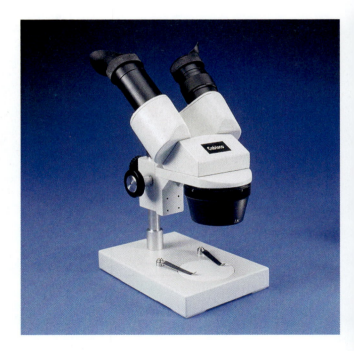

Figure 3.5
Stereoscopic dissecting microscope.

to fit the distance between your eyes so that you can look comfortably through both oculars at once. You should see a single field of vision. If the focus is not sharp, you may have to focus for each eye separately. The microscope will have one fixed and one adjustable ocular. Adjust focus for the fixed ocular first to suit that eye, and then adjust the other ocular (usually by rotation) until the focus is sharp for both eyes.

Move the object on the stage away from you. Which way does the image move? _____

Move the object to one side. Which way does the image move? _____ How does this compare with the compound microscope?

Further Study

Exercises with the Dissecting Microscope

Try these or any similar exercises with the dissecting microscope, and **keep a record** of the results of different lighting and background effects with different types of material. This will save you time in later studies.

1. Examine some pond water. How many kinds of organisms can you see? _____ Are they more distinct on a white or a black background _____?

2. Examine a prepared slide of the whole mount of a fluke or a tapeworm or a similar slide. Observe it with reflected lighting, moving the light source about for the best effects. Then try transmitted lighting. If the microscope lacks substage lighting, try placing the slide over a small microscope lamp with a piece of writing paper between the slide and the lamp. Which method gives better illumination?

3. Examine the surface of a preserved sea star or the gills of a crayfish. Study the material first without water; then study it submerged in a finger bowl or glass of water. In which preparation do you see the most detail? Why? Observe the gills both with top lighting and with transmitted light. Can both methods be used with the sea star? Why?

Be sure to make good use of the dissecting microscopes that are available in the laboratory. You will find them invaluable.

Protozoan Groups
Phylum Amoebozoa
Phyla Euglenozoa and Plantae
Phylum Apicomplexa
Phylum Ciliophora

The protozoa are a diverse assemblage of unicellular eukaryotic organisms having at least two animal-like properties: (1) absence of a cell wall and (2) presence of at least one motile stage in the life cycle. All the life functions of a protozoan are performed within the limits of a single plasma membrane. Although protozoans have no organs or tissues, there is division of labor within the cytoplasm, where various complex organelles are specialized to carry out specific tasks. These organelles function as skeletons, locomotory systems, sensory systems, conduction mechanisms, defense mechanisms, and contractile systems.

Although the protozoa are often called "simple" organisms, there are no simple organisms. In fact, many protozoans are exceedingly complex. Ciliates, for example, are not only the most complex of protozoans but also the most elaborately organized of all known cells. Protozoans are widespread ecologically, being found in fresh, marine, and brackish water and in moist soils. Some are free-living; others live as parasites or in some other symbiotic relationship.

See pp. 50–51 for a classification of protozoans.

EXERCISE 4A[1]
Phylum Amoebozoa—*Amoeba* and Others

Core Study

Amoeba
Higher-level taxonomic status uncertain
Order Euamoebida
Species *Amoeba proteus*

---

[1]An engaging initial exploration of protozoan diversity that may be implemented before beginning Exercise 4A is described in Projects and Demonstrations, p. 38.

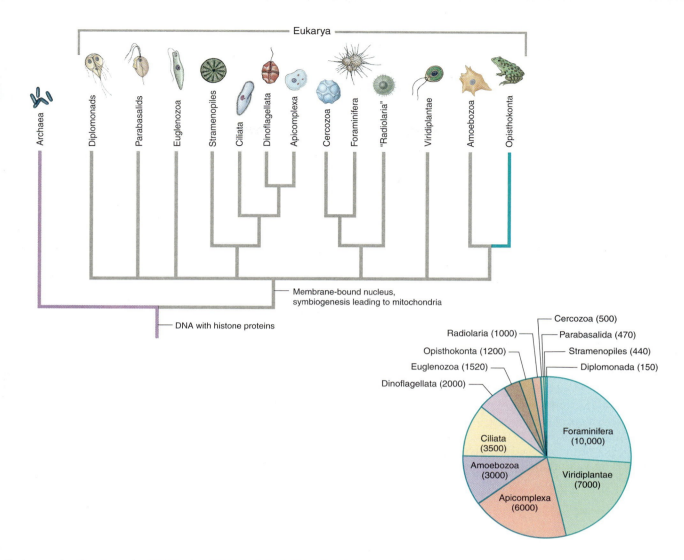

Cladogram showing two major prokaryotic branches (Bacteria and Archaea) and diversification of eukaryotes. Major eukaryotic clades containing protozoans are shown but several clades of amebas and other forms are not shown. The order of branching remains to be determined in most clades. The Opisthokonta clade contains all multicellular animals.

Where Found

The amebas[2] may be naked or enclosed in a shell. Naked amebas, which include the genera *Amoeba* and *Pelomyxa,* live in both fresh water and seawater and in soil. They are bottom-dwellers and must have a substrate on which to glide. *A. proteus*[3] is a freshwater species that is usually found in slow-moving or still-water ponds, and often on the underside of lily pads and other water plants. They feed on algae, bacteria, protozoa, rotifers, and other microscopic organisms.

Study of Live Specimens

☞ In the center of a clean slide, the instructor will place a drop of culture drawn from the bottom of the undisturbed culture bottle. Adjust the iris diaphragm of the microscope to provide *subdued light.* With low power, explore the contents of the slide, at first *without a coverslip.*

You may see some masses of brownish or greenish plant matter, and there will probably be some small ciliates or flagellates moving about. The ameba, in contrast, is gray, rather transparent, irregularly shaped, and finely

---

[2]The term "ameba" is used by many authors as a common name for any of the naked or shelled amebas. When referring to a specific genus, the name is latinized and italicized, such as *Amoeba, Pelomyxa,* and *Entamoeba.*

[3]The genus name (*Amoeba;* from the Greek, meaning "change") may be abbreviated when used in a context in which it is understood. Recall, however, that the species is binomial (*Amoeba proteus*) and that the species epithet (*proteus,* Gr. *Proteus,* sea god that could change shape at will) has no taxonomic meaning by itself.

granular in appearance (Figure 4.1). If it is not apparently moving, watch it a moment to see if the granules are in motion or if the shape is slowly changing.

☞ After you have found specimens and observed their locomotion, carefully cover with a coverslip. You will need to support the coverslip to avoid crushing the amebas as the water evaporates. Place two short lengths of hair or thread or two pieces of broken coverslip, one on each side of the drop of culture, before adding the coverslip. *Never use high power without a coverslip.* From time to time, add a drop of culture water to the edge of the coverslip to replace water lost by evaporation.

Some types of substage microscope lamps will warm the stage enough to kill the ameba. Turn off the lamp when not actually viewing the specimen.

General Features

The outer cell membrane is the **plasmalemma** (Gr. *plasma,* form, + *lemma,* skin). Electron micrographs show that the plasmalemma is fringed with fine, hairlike projections (too small to be seen with the light microscope), which are thought to be involved in adhesion of the cell surface to the substrate or to nutrient particles and to aid in the capture and intake (ingestion) of food.

The cytoplasm enclosed by the plasmalemma is differentiated into a thin, peripheral rim of stiff **ectoplasm** and an inner, more fluid **endoplasm.** The ectoplasm looks clear, even glassy, because it lacks the subcellular organelles present in abundance in the endoplasm (Figure 4.1B). As a result, the clear ectoplasm is often referred to as a **hyaline cortex,** or "glassy rind."

Locomotion. An ameba moves and changes shape by thrusting out **pseudopodia** ("false feet"), which are extensions of the cell body. The ameba creeps forward when a protruding pseudopodium grasps a surface and a motile force pulls the rest of the cell forward. The advancing end of the pseudopodium is called the **hyaline cap.** As endoplasm flows into the hyaline cap, it fountains out into the cortex and is converted into stiff ectoplasm, thus extending the sides of the pseudopodium much like a tube or sleeve. As the pseudopodium lengthens, ectoplasm at the temporary "tail end" converts again into streaming endoplasm to replenish the forward flow. At any time, the action can be reversed—the endoplasm streaming back, the tube shortening, and another pseudopodium forming elsewhere.

Within the pseudopodium is a fibrous network of actin, one of the major proteins of muscle of more derived organisms. Actin molecules form a rigid polymer meshwork in the clear ectoplasm that gives the pseudopodium its strength and form. The alternate rapid assembly and disassembly of this actin meshwork are responsible for the propulsive force of ameboid movement, although the molecular details are not yet fully understood.

Observe the formation of pseudopodia. Does the ameba have a permanent anterior and posterior end?

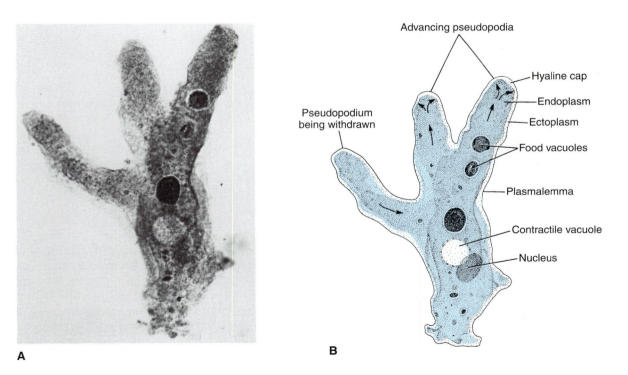

A

B

Figure 4.1
Ameba. **A,** Whole mount of a living specimen. **B,** Interpretive drawing.

_____ Can you observe the change from endoplasm to ectoplasm, and vice versa? Does the ameba move steadily in one direction? _____ Does more than one pseudopodium ever start at once? _____ How is a pseudopodium withdrawn?
_____

What happens when the ameba meets an obstruction? _____ Tap the coverslip gently with the tip of a pencil or probe. Does the ameba respond? _____ Do pseudopodia ever extend vertically as well as laterally? _____

Drawings

On separate paper, make a series of five or six outline sketches, each about 2 cm in diameter, to show changes in shape that occur in an ameba in a period of 1 to 10 minutes. Sketch rapidly and use arrows freely to indicate the direction of flow of the cytoplasm. Indicate the magnification of the drawings. In the space next to your sketches, briefly explain how this ameboid movement is achieved.

Feeding. Note the **food vacuoles,** which are particles of food surrounded by water and enclosed in a membrane. Cells and unicellular animals that engulf foreign particles are called **phagocytes,** and this type of ingestion is known as **phagocytosis** (Gr. *phagein,* to eat, + *kytos,* hollow vessel) (Figure 4.2). Phagocytosis involves encircling the prey by pseudopodia to form a food cup. Subsequently, the prey, along with a quantity of water, becomes completely enclosed by cytoplasm to form a food vacuole. The contents are digested by hydrolytic enzymes secreted into the food vacuoles by tiny, membrane-bound organelles called **lysosomes.** Undigested end products can be eliminated (egested) at any point along the plasmalemma.

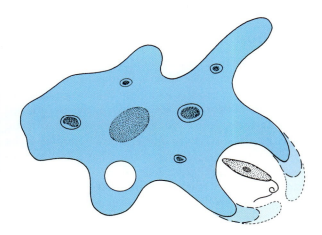

Figure 4.2
Feeding by an ameba. Successive positions of the pseudopodia are shown as the food cup forms and closes around the prey organism.

Cells also take in fluid droplets and minute food particles by a process of channel formation called **pinocytosis** (Gr. *pinein,* to drink, + *kytos,* hollow vessel). You will look for this in the exercise on p. 34.

Osmoregulation. Look for the **contractile vacuole,** a clear bubble containing no particles. Note that this bubble gradually increases in size by accumulation of fluid and then ruptures and disappears. This organelle rids the ameba of the excess water that has been taken in along with food vacuoles or acquired by osmosis. The contractile vacuole is surrounded by a network of continuous membranous channels that are populated by proton pumps. These pumps create an osmotic gradient that moves water into the vacuole. The vacuole is then pushed against the plasmalemma, where it ruptures and empties to the outside. Note that the vacuole, as it enlarges, tends to be located in the temporary posterior end of the moving ameba.

Over a period of several minutes, time the appearance and disappearance of the contractile vacuole of an ameba in the culture medium. Then place a drop of distilled water at the edge of the coverslip and draw the water through to the other side, using a piece of absorbent paper. Repeat. The culture fluid now has been largely replaced with distilled water. After a minute or so, make another recording of vacuole discharge and compare it with the first. Rate (discharges/min) in the culture medium: _____ Rate (discharges/min) in distilled water: _____ What is your conclusion? _____
_____

How do you explain any observed difference in discharge rate? _____
_____

Nucleus. Locate the **nucleus.** It is disc-shaped, often indented, finely granulated, and somewhat refractive to light. You can distinguish it from the contractile vacuole because the latter is perfectly spherical, increases in size, and finally disappears. The nucleus is usually carried along in the cytoplasm and is often found near the center of the animal. As the nucleus turns over and over, it sometimes appears oval and sometimes round.

Reproduction. It is possible, although not too likely, that you will find a specimen in division. If you do, call it to the attention of your instructor. Amebas reproduce asexually by a type of mitotic cell division known as **binary fission.** The life cycles of some amebas are much more complex than that of *Amoeba.*

Written Report

Answer the questions in the lab report on p. 34.

Further Study

Study of a Stained Slide

☞ On a stained slide, locate the structural features that you studied in the living specimen. Note especially the **nucleus, plasmalemma, ectoplasm, endoplasm, contractile vacuole,** and **food vacuoles.**

Other Amebas

☞ Examine stained slides or specimens from living cultures of as many of the following amebas as are available in your laboratory.

Parasitic Amoebozoans

A number of species of *Entamoeba* are found in humans and other vertebrates. *E. gingivalis* lives in the mouth and feeds on bacteria around the base of the teeth. *E. histolytica* lives chiefly in the large intestine, where it causes dysentery by feeding on body tissues and red blood cells (Figure 4.3). *E. histolytica* exists in two phases—a trophozoite (trōf'uh-zō'īt; *trophe,* food, + *zōon,* animal), or active feeding phase, and an encysted phase, in which nuclear divisions occur. The mature cyst usually contains four small nuclei. After ingestion by a suitable host, the multinucleated ameba emerges from its cyst and undergoes a series of fissions, resulting in uninucleate daughter amebas.

Endamoeba is a similar genus found in cockroaches and termites.

Shelled Amoebozoans

Protozoa employ many kinds of materials for building their shells, or tests. *Arcella,* which occurs in bogs or swamps where much vegetation exists, has a hemispherical, hat-shaped skeleton (up to 260 μm in diameter) made up of siliceous or chitinous material set in a base of polymerized proteins. Small, finger-like pseudopodia extend through an opening in the flat underside of the skeleton (Figure 4.4).

Difflugia is often found in leaf-choked puddles or in delicate aquatic vegetation. This protozoan has an inverted, flask-shaped skeleton (up to 500 μm long) made up of sand grains cemented together with a polymerized protein base, sometimes with the addition of diatom shells or sponge spicules (Figure 4.4). In living cultures from biological supply houses, they are often seen feeding on the single-celled filamentous algae *Spirogyra.*

Foraminiferans

Foraminiferans, such as *Globigerina* (Figure 4.4), are marine amebas that secrete a skeleton of one or more

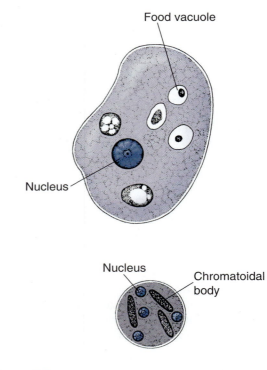

Figure 4.3

Entamoeba histolytica, the cause of amebic dysentery in humans. The trophozoite *(above)* is the actively moving and feeding form. The cyst *(below)* can tolerate conditions outside the body and is infective to the new host. Note the four small nuclei and the four chromatoidal bodies.

chambers, usually calcareous and sometimes incorporating silica, sand, or sponge spicules. Long, delicate feeding pseudopodia extend through pores in the skeleton. As the foraminiferan grows, it adds new chambers to the shell. Upon death, the skeletons join the ooze of the ocean floor. Enormous limestone deposits were formed over millions of years by the skeletons of these animals.

Stramenopiles and Cercozoans

Spherical amebas with long, slender feeding pseudopodia radiating out in all directions in a sunburst arrangement belong to several groups including Stramenopiles and Cercozoa. In *Clathrulina,* a Cercozoan, the pseudopodia project from perforations in a lattice-like organic capsule (Figure 4.4). *Actinophrys,* a Stramenopile, lacks a capsule, but its cytoplasmic surface has a frothy, vacuolated appearance (Figure 4.4).

Radiolarians Marine actinopod amebas that secrete about themselves a transparent skeleton of silica and then thrust out slender pseudopodia through pores in the skeleton are radiolarians. They float in surface plankton; when they die, their skeletons become a part of the ocean bottom ooze. Examine a prepared slide showing a variety of shell types.

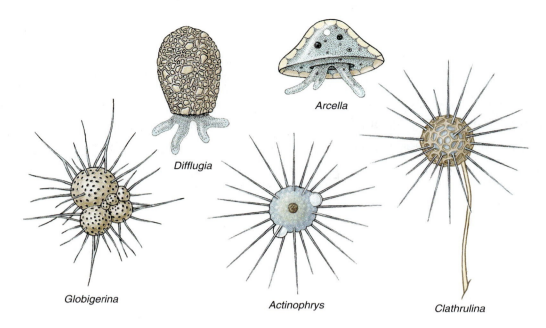

Figure 4.4

Some ameboid protozoa. *Arcella* and *Difflugia* are shelled amebas with lobopodia in Amoebozoa. *Globigerina* is a foraminiferan. *Actinophrys* is an actinopod ameba with long, slender feeding pseudopodia called axopodia; it is a Stramenopile. *Clathrulina* also has axopodia, but it is placed in Cercozoa.

Drawings

On separate paper, sketch any of the amebas or their skeletons.

Projects and Demonstrations

1. *Initial exploration of protozoa.* This exercise can introduce you to protozoan diversity, using a sample of natural pond water or a commercially available culture (for example, Carolina "Protozoa Survey Mixture," 13-1940). The Carolina culture comes with a simple dichotomous key that enables students to identify protozoa to genus. Place a drop of water taken from the bottom of a mixed protozoa culture sample on the center of a clean microscope depression slide, and examine at low power with your microscope. Add a coverslip to the depression slide to avoid evaporation.

   You may find that you have sampled masses of algae and debris. Look carefully on and around the debris to locate the organisms in your sample. How many types of organisms can you find? Are they all the same size? If not, what differences do you observe?

   _____

   _____

   _____

   Are they all the same color? What differences in color do you observe? Do they all move in the same way? If you have used a dichotomous key to identify the protozoans to genus, list the genera here: _____

   _____

   _____

   _____

   What characteristics seem most useful in identifying the unknowns?

   _____

   _____

2. *Demonstrating the contractile vacuole (water-expulsion vesicle).* Add a drop of 10% nigrosine to a drop of *Amoeba* culture on a slide, add a coverslip, and examine with high power.

Amebas

1. What is the average size of the specimens on your slide? _____

2. Do amebas react to stimuli? _____ What reactions have you observed? _____
   _____
   _____

3. What is the main function of the contractile vacuole? _____

4. What was the vacuole's discharge rate with the ameba in culture medium? _____
   _____ Discharge rate in distilled water? _____ Explain any
   observed difference in discharge rate:
   _____
   _____
   _____

5. How does the ameba respire? _____
   _____
   _____

6. How does the ameba reproduce? _____
   _____

7. If you observed ingestion or egestion occurring, or if you could identify the contents of any of the food
   vacuoles, describe what you saw.
   _____
   _____
   _____
   _____
   _____

8. Distinguish between phagocytosis and pinocytosis.
   _____
   _____
   _____
   _____

9. How would you describe and explain ameboid movement? (Include answers to the questions on p. 31.)
   _____
   _____
   _____
   _____
   _____
   _____
   _____

EXERCISE 4B

Phyla Euglenozoa and Plantae, Class Viridiplantae—*Euglena, Trypanosoma,* and *Volvox*

Core Study

Members of these phyla include protozoa having one or more flagella—undulating, whiplike organelles that move these protozoa efficiently through their fluid environment. Flagellates belong to several different groups including Euglenozoa, Parabasalids, Diplomonads, and Dinoflagellata. Many of these groups contain pigmented and non-pigmented species. Pigmented species, such as *Euglena* and *Volvox,* possess chlorophyll and consequently can synthesize carbohydrates from carbon dioxide and water as does any green plant. They are **autotrophic** (although they may also gather nutrients by pinocytosis).

Non-pigmented species obtain their nutrients by absorbing them through the plasma membrane or by engulfing prey in food vacuoles. Thus, they are all **heterotrophic.** Many are free-living, but some, such as *Trypanosoma,* parasitize humans and other animals.

Euglena

Phylum Euglenozoa
   Subphylum Euglenida
      Class Euglenoidea
         Genus *Euglena*
            Species *Euglena gracilis* (or *Euglena viridis*)

Where Found

Euglenoids are most common in still pools and ponds, where they often give a greenish color to the water. Ornamental lily ponds are excellent sources. *Euglena gracilis* and *Euglena viridis* are commonly studied species.

Study of Live Specimens

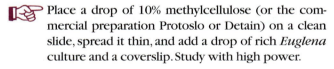 Place a drop of 10% methylcellulose (or the commercial preparation Protoslo or Detain) on a clean slide, spread it thin, and add a drop of rich *Euglena* culture and a coverslip. Study with high power.

Members of *E. gracilis,* which range from 35 to 65 μm in length, are spindle-shaped and greenish in color. The color is caused by the presence of **chloroplasts,** which contain **chlorophyll** and are scattered through the cytoplasm (Figure 4.5). A nucleus is located centrally but is difficult to see in living specimens.

Locomotion. The blunt anterior end bears a whip-like **flagellum** that you may be able to see with reduced light when the specimen has slowed down considerably.[4]

---
[4]Dilute methyl violet or Noland's stain may help make the flagella easier to see.

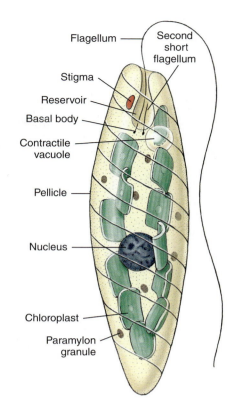

Figure 4.5
Euglena. Features shown are a combination of those visible in living and stained preparations.

The flagellum emerges from the **reservoir,** a clear, flask-shaped space in the anterior end. In some flagellates, such as *Peranema,* the flagellum extends forward, but in *Euglena* it is directed backward along the side of the body. Movement involves generation of waves originating at the base of the flagellum and transmitted along its length to the tip (Figure 4.6). The flagellum beats at the rate of about 12 beats per second. It not only moves the organism forward but also rotates it and pushes it to one side, causing it to follow a corkscrew path, rotating about once every second as it goes.

Euglenoid Movement. When a *Euglena* stops swimming, watch how it changes shape. Keep notes on your observations. These will be used in writing your lab report on p. 41. These peculiar, wormlike contractions are referred to as "euglenoid movements," and authorities believe that the movements are made possible by microtubules—tiny, hollow fibrils about 20 nm in diameter, lying just beneath the pellicle (Figure 4.7).

Body Covering. The body is covered with a protective but flexible **pellicle** secreted by the clear **ectoplasm** that surrounds the **endoplasm.** The **stigma,** or "eyespot," is a reddish pigment spot that shades a swollen basal area of the flagellum that is thought to be light-sensitive. Early microscopists, however, believed the red stigma itself was sensitive to light, so they named the genus *Euglena,*

The figure labels (Figure 4.5) include: Flagellum, Second short flagellum, Stigma, Reservoir, Basal body, Contractile vacuole, Pellicle, Nucleus, Chloroplast, Paramylon granule.

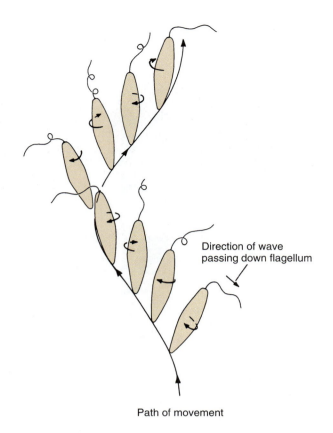

Direction of wave
passing down flagellum

Path of movement

Figure 4.6

Locomotion in *Euglena.* Base-to-tip waves curl down the flagellum, forcing *Euglena* forward. The organism rotates as it swims, and the anterior end follows a corkscrew path. The posterior end "drags," following a straighter path.

Source: Adapted from T. L. Jahn and E. C. Bovee, The motile behavior to protozoa, in Research in Protozoology, vol. 1, edited by T. Chen. Pergamon Press, New York, 1967.

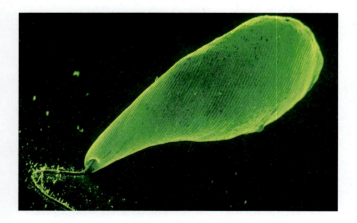

Figure 4.7

Scanning electron micrograph of *Euglena,* showing "candy-striped" ridges the pellicle. The odd spiraling pattern of microtubules is thought to provide the flexibility required for the twisting and contraction of euglenoid movement. (About ×1180.)

meaning "true eyeball." Why is light sensitivity important to a *Euglena?* _____

Osmoregulation. A large contractile vacuole empties into the reservoir. It is fed by smaller vesicles around it.

Feeding. The *Euglena* takes in no solid food; the opening of the reservoir is merely for elimination of waste and excess water discharged by the contractile vacuole. Nutrition is **holophytic** (Gr. *holo,* whole, + *phyt,* plant), making use of photosynthesis—the manufacture of food from carbon dioxide and water with the aid of chlorophyll and sunlight. The chlorophyll is carried within shield-shaped **chloroplasts** in the cytoplasm. Carbohydrates synthesized within the chloroplasts are stored as starch granules and as **paramylon** (a carbohydrate similar to starch) concentrated in the chloroplast centers. In older cultures of *Euglena,* you may see dispersed in the cytoplasm small red, brown, or orange droplets containing lipids and pigments.

Not all photosynthetic forms are holophytic. Some are also saprozoic, others are holozoic, and still others use a combination of methods.[5]

Reproduction. *Euglena* reproduces by **longitudinal fission,** which may occur when the organism is free or when it is in the encysted state. Fission begins at the anterior end and proceeds posteriorly. During encystment, *Euglena* rounds up and forms a gelatinous wall around itself. In this state, it can withstand drought and other harsh conditions. A number of longitudinal divisions may occur during encystment, and many separate individuals may emerge later from a single cyst.

Stained Slide of *Euglena*

☞ Select a good specimen on the slide. On separate paper, sketch the specimen and label the flagellum, stigma (eyespot), chloroplasts, pellicle, reservoir, and nucleus. What is the function of each of the labeled structures?

Written Report and Drawings

🖋 In the lab report on p. 41, describe and explain the manner in which your specimens move about (locomotion) and alter their shape (euglenoid movement). Can you see any flagella? Record any other interesting observations. Also on p. 41, make a series of 2- to 3-cm-long drawings showing the shapes your specimen assumes in its euglenoid movement. Do *Euglena*

[5]Saprozoic (Gr. *sapros,* rotten, + *zōon,* animal) nutrition is accomplished by absorption of dissolved salts and simple organic nutrients from the surrounding medium. Holozoic (Gr. *holo,* whole, + *zōon,* animal) nutrition involves ingestion of solid or liquid organic food particles. Thus, all major forms of nutrition—holophytic, holozoic, and saprozoic—are represented among the phytoflagellates.

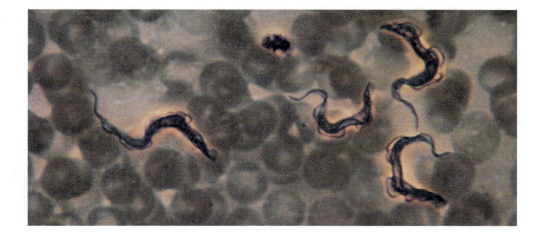

Figure 4.8
Trypanosoma brucei gambiense in the blood of a human.

push or pull themselves with their flagella? Can they move without using their flagella?

Further Study

Trypanosoma

Phylum Euglenozoa
  Subphylum Kinetoplasta
    Class Trypanosomatidea
      Genus *Trypanosoma*

Trypanosoma (Gr. *trypanon,* auger, + *soma,* body) are parasites living in blood or other tissues of all classes of vertebrates (Figures 4.8 and 4.9). Trypanosomes are **obligate heterotrophs;** that is, they lack chlorophyll and obtain their organic and inorganic requirements from the environment. Because a trypanosome's environment is the body fluid of its host, its nutrition is **osmotrophic** meaning that it absorbs its nutrients directly from the surrounding blood or other body fluid.

Some trypanosomes are nonpathogenic, but others produce severe disease, especially **sleeping sickness.** *T. brucei gambiense* and *T. brucei rhodesiense* cause African sleeping sickness in humans, and *T. brucei brucei* causes a related disease in domestic animals (note that the organisms are all subspecies of the same species, *T. brucei). T. b. rhodesiense,* the more virulent of the sleeping sickness trypanosomes, and *T. b. brucei* have natural reservoirs (antelope and other wild animals) that apparently are not harmed by the parasites. Trypanosomes of African sleeping sickness are transmitted by tsetse flies of the genus *Glossina.*

T. cruzi causes Chagas disease in humans in Central and South America. It is transmitted by the bite of the "kissing bug" *(Triatoma).* Many believe that Charles Darwin contracted Chagas disease in Chile during the voyage of the *Beagle,* resulting in his renowned ill health later in life.

☞ Using oil immersion, examine a slide containing a stained blood smear from an infected animal. Note the numerous purple-stained trypanosomes (see Figure 4.8). The body of the organism is fusiform in shape and slightly twisted. Are the two ends alike? The flagellum runs closely opposed to the body surface together with a fold of the plasma membrane; these together (fold and

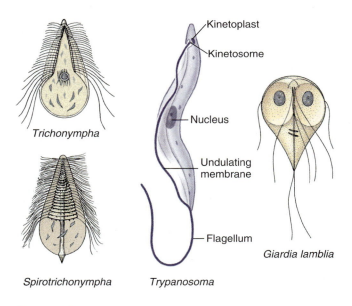

Figure 4.9
Some flagellate protozoa. *Trypanosoma* is a blood parasite of various animals; some cause serious disease in humans and domestic animals. *Trichonympha* and *Spirotrichonympha* are parabasalids commonly found in the gut of termites and roaches, where they help digest cellulose in wood eaten by the insects. *Giardia* is a diplomonad that causes a highly contagious infection sometimes seen in hikers and campers who drink untreated water. Art is not to the same scale.

flagellum) constitute the **undulating membrane** so characteristic of the genus.

The two subspecies of *T. brucei* (*rhodesiense* and *gambiense*) cannot be distinguished morphologically, but they differ in pathogenesis and growth rate. *T. b. rhodesiense* causes the most virulent form of sleeping sickness in central and east central Africa. *T. b. gambiense* produces the "classical" chronic form of sleeping sickness, with protracted somnambulism (sleepwalking) and nervous disorders.

Drawings

On separate paper, sketch one or more trypanosomes from your slide, paying attention to scale relative to other cells on the slide.

Written Report

Answer the questions relating to *Trypanosoma* in the lab report on p. 42.

Volvox

Phylum Plantae
  Class Viridiplantae
    Order Volvocida
      Family Volvocidae
        Genus *Volvox*
          Species *Volvox globator*

Where Found

Volvox (L. *volvere*, to roll) is a beautiful, large, spherical genus with colonial and multicellular members often found with other protozoa in stagnant pools and ponds. It is an important component of summer plankton "blooms" that appear in nitrogen-rich ponds, pools, and ditches.

General Features

Colonies of *Volvox* may reach a diameter of 2 mm (most are smaller) and are easily visible to the unaided eye. *Volvox* is especially interesting because we see the beginning of **cell differentiation,** resulting in the division of labor among cells. In multicellular organisms, as opposed to colonial organisms, only certain cells reproduce sexually. Thus, some species of *Volvox* are multicellular. This group also illustrates an important stage in the **development of gender,** where large gametes are designated female and small gametes are designated male.

Study living or preserved colonies and stained slides. For living specimens, use a pipette to transfer a colony (visible to the unaided eye) to a ringed or depression slide and cover with a coverslip. Study with *low power* and focus up and down to get all the details. Be especially careful if you use high power to avoid damaging the slides or objective.

Each spherical colony is composed of a variable number of one-celled individuals called **zooids.** There may be a few hundred or many thousands of cells in a colony, arranged on the surface of a gelatinous ball and connected to each other by fine **protoplasmic strands.** The cells are differentiated into **somatic cells** and a smaller number of **reproductive cells.** The somatic cells are quite similar to other flagellate animals and make up most of the colony. They handle nutrition, locomotion, and response to stimuli for the entire colony. Each somatic cell contains **chloroplasts** (which give the colony its green color), a **stigma** for light sensitivity, and a pair of **flagella** for locomotion.

Locomotion. If you have living colonies, observe locomotion. Do you notice that one end usually goes foremost? This is the anterior pole, and the anterior cells are so highly specialized that they are unable to reproduce. Some of the colonies on your slide will contain smaller **daughter colonies** revolving about inside the gelatinous center of the mother colony (Figure 4.10).

Reproduction. *Volvox* reproduces both sexually and asexually. During spring and summer, *Volvox* reproduces asexually by repeated division of the cells, which form miniature daughter colonies. These eventually escape by rupture of the mother colony. In the fall, the asexual colonies develop **sex cells.** Some cells enlarge to form **macrogametes** (eggs). Other cells divide repeatedly to form packets of spindle-shaped **microgametes** (sperm). The sperm escape and fertilize the eggs, which become **zygotes.** Each zygote secretes a spiny cyst wall around itself, providing protection during winter after the mother colony dies. In spring, the zygotes break out of the cysts to give rise by cell division to new asexual colonies (Figure 4.10).[6]

Written Report and Drawings

Answer the questions on p. 42, and on separate paper, sketch one of the colonies for your report.

Projects and Demonstrations

1. Demonstration of Living *Trichonympha,* a Symbiont of *Termites*

The parabasalid *Trichonympha* (Figure 4.9) belongs to an order of flagellates that have many flagella and live in the digestive tracts of roaches and termites. *T. campanula,* which lives in the

---

[6]Several invertebrate groups use the strategy of employing asexual reproduction during the warm months of the year when rapid reproduction and colonizing new habitats are more important than the genetic variability furnished by sexual reproduction. Sexual reproduction is then used during the unstable conditions of winter.

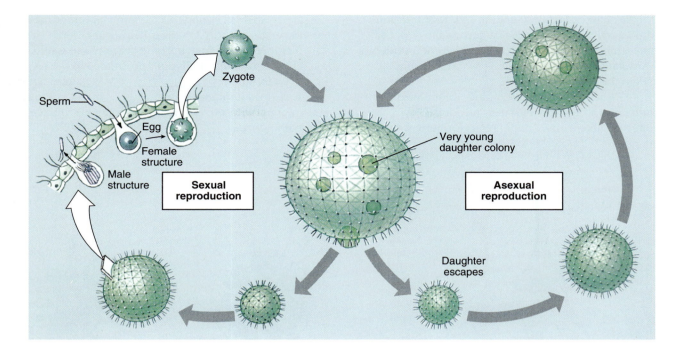

Figure 4.10
Life cycle of *Volvox,* a colonial flagellate, phylum Chlorophyta.

gut of American termites, is a good example of **symbiotic mutualism.** Neither the termite nor the flagellate can live without the other. Most flagellates harbor bacteria in their gut or in their cytoplasm that secrete the enzyme cellulase, which digests the wood (cellulose) that termites eat. Also, recent evidence suggests that at least some termite species secrete their own cellulase.

Squeezing the posterior abdomen of the termite to extrude a drop of fluid from the anus may provide enough specimens on a slide for study. Or hold the termite with forceps and, with another pair of forceps, pull off the posterior segment. Pull the intestine out of the body and tease it apart on a slide. Add a drop of 0.6% NaCl solution and cover. Why is *Trichonympha* important to termites? _____ In what ways does *Trichonympha* resemble *Trypanosoma?* _____ In what ways does it differ? _____

2. Dinoflagellates

Dinoflagellates, a group of flagellates of the phylum Dinoflagellata, are found mostly in marine plankton. They have a peculiar arrangement of two flagella—one extends backward to propel the organism; the other encircles the body in a transverse groove and causes it to rotate. Some

dinoflagellates have chloroplasts of various colors; others are colorless. Some, such as *Noctiluca* (Figure 4.11), possess luminescent granules in the cytoplasm, which are largely responsible for the phosphorescence observed in some coastal waters. Some occasionally appear in enormous numbers and cause "red tides" off the coasts of Florida *(Ptychodiscus)* and California and New England *(Gonyaulax). Ceratium,* with long spines and green chloroplasts, is found in both marine and fresh water (Figure 4.11). Why are red tides of particular concern to humans?

3. *Giardia intestinalis (lamblia)*

Giardiasis is a highly contagious infection caused by the diplomonad *Giardia* (see Figure 4.9). It enters drinking water contaminated with raw sewage and causes severe diarrhea with dehydration, intestinal pain, and weight loss. If slides of the feeding stage (trophozoite) are available, note the characteristic shape, rounded anteriorly, pointed posteriorly, and bearing several flagella. *Giardia* lives in the intestine and can swim rapidly using its flagella. The protozoan also divides rapidly and can quickly build up enormous numbers; a single diarrheic stool can contain 14 billion parasites. As the protozoans pass into the large intestine, they encyst; the cysts transmit the infection to new hosts. When swallowed, they pass through the stomach unharmed, then undergo excystation in the intestine, grow flagella, and begin their nefarious work.

Giardia is rapidly becoming ubiquitous in untreated water, having been introduced by infected hikers and campers into formerly pristine areas, such as springs and streams in remote areas of western North America. Beaver also serve as a major reservoir host in some areas of the world. How is *Giardia* similar in appearance to *Trypanosoma*? _____ In what way does it differ? _____ How might you test to determine if a person is infected with *Giardia*? _____

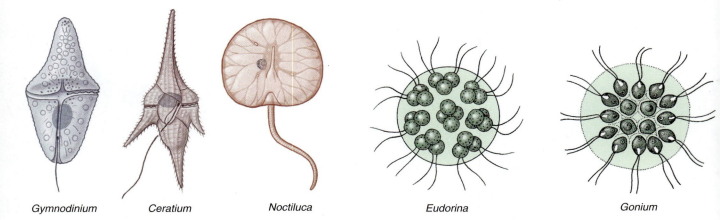

Gymnodinium   Ceratium   Noctiluca   Eudorina   Gonium

Figure 4.11
Some flagellate protozoa. *Gymnodinium, Ceratium,* and *Noctiluca* are dinoflagellates. *Eudorina* and *Gonium* are members of Viridiplantae, a class within Phylum Plantae.

Phylum _____

Subphylum _____

Euglena

1. What is the average length of *Euglena* specimens in your culture? _____

2. Describe the different forms of movement and locomotion in *Euglena*. _____
   _____
   _____
   _____

3. How do *Euglena* satisfy their nutritional requirements? _____
   _____
   _____

4. What function might the stigma of *Euglena* serve? _____
   _____

5. Following the survival strategy of encystment, are the *Euglena* that emerge from a cyst genetically similar to or different from organisms of previous generations? _____
   _____

6. Given that cyst formation is probably adaptive to conditions in *Euglena*'s habitat, in what types of habitat do you think *Euglena* is found? _____
   _____

Observations on *Euglena* Locomotion

Trypanosoma

1. Using the average diameter of the mammalian red blood cell (7.5 μm) as a guide, what do you estimate the length of *Trypanosoma* to be? _____

2. How must *Trypanosoma* organisms obtain their nutritional requirements? _____
_____

3. A doctor suspects that you might be infected with *Trypanosoma*. Name two places you might recently have visited that would help decide what species of *Trypanosoma* you are carrying.
_____

4. What type of tissue sample will the doctor collect to confirm his or her diagnosis? _____
_____

5. What key characteristic is possessed by all species of this genus? _____

6. What characteristics possessed by *Trypanosoma* organisms would tell you that they belong to the phylum Euglenozoa? _____

7. Would you expect *Trypanosoma* to have a system for digesting food like that of *Amoeba?* Why or why not?
_____

Volvox

1. In what ways is *Volvox* similar to *Euglena?* _____
_____

   In what ways does it differ from *Euglena?* _____

2. Were the colonies you examined reproducing asexually or sexually? _____ Give reasons for your answer. _____
_____
_____
_____

3. How does *Volvox* move? _____
_____
_____

4. How do somatic and reproductive cells of *Volvox* differ in function? _____
_____

5. Which type of cell is more abundant in the colony and why? _____
_____

6. Why might sexual reproduction in this group be more common in the fall, whereas asexual reproduction tends to dominate in the spring? _____
_____

EXERCISE 4C
Phylum Apicomplexa—
Gregarina and *Plasmodium*

Core Study

Apicomplexans are entirely endoparasitic. On the whole, they lack special locomotor organelles, although some can move by gliding or by changing body shape, similar to euglenoid movement, and some species have flagellated gametes.

Apicomplexans are distributed via resistant sporelike stages. The process of **sporogony** (spōr-ahg′uh-nē; *spores,* seed, + *gonos,* progeny), or spore formation, follows the formation of the zygote in most apicomplexans, so that sexual reproduction leads into the production of spores.[7] Many apicomplexans also undergo multiple fission, or **schizogony** (ski-zahg′uh-nē; *schizein,* to split, + *gonos,* progeny), a means of spreading the population within the host, and this is followed by sporogony.

Apicomplexans have no mouth, and their nutrition is osmotrophic, meaning that they absorb nutritives from the immediate environment.

Common apicomplexans are the gregarines (class Gregarinea), which are parasitic in such invertebrates as annelids, echinoderms, and ascidians, and the coccidians (class Coccidea), which are endoparasites in both invertebrates and vertebrates. Among the Coccidea are *Eimeria,* which causes coccidiosis in domestic rabbits and chickens, and *Plasmodium,* which causes malaria in humans.

Further Study

Plasmodium

Phylum Apicomplexa
   Class Coccidea
      Order Eucoccidia
         Family Plasmodiidae
            Genus *Plasmodium*

Plasmodium (*plasma,* molded, image, + *eidos,* form) is the genus of coccidean parasites that causes malaria. Malaria ("bad air") is the most important disease in the world today in terms of lives lost and impact on global economy. Malaria sickens more than 300 million people every year and kills more than 1 million. It has always outranked war as a source of human suffering. *Plasmodium* requires two hosts—one a vertebrate, the other an invertebrate. It is

---

[7]The term "spore" is unfortunate, because protozoan spores are not homologous to the spores produced by plants, and confusing, because the term refers to a variety of propagating, infective, and resistant stages in the life cycles of many different species. In *Plasmodium,* the "spore" is the **oocyst** (Gr. *oon,* egg, + *kystis,* bladder), a cystlike stage in which numerous daughter spores, called **sporozoites,** are produced by sporogony.

transmitted to a human by a female *Anopheles* mosquito that has had a blood meal from a malaria-infected human. The mosquito, as it feeds, also injects into the human bloodstream some of its salivary juice, which contains infective sporozoites (Figures 4.12 and 4.13).

In humans, the sporozoites leave the bloodstream and penetrate liver cells, where they undergo **schizogony** (asexual cleavage multiplication) and produce **merozoites** (mer-uh-zō′īts; *meros,* part, + *zōon,* animal). These may either infect other liver cells or be released into the blood to enter red blood cells (erythrocytes). In either case, they undergo further multiplication.

In red blood cells, merozoites become trophozoites (adult stage). The cytoplasm of the trophozoite becomes nucleated and takes on a ringlike appearance, called the "signet-ring stage" (Figures 4.12 and 4.13).

☞ Examine a slide containing a stained blood smear with various infective stages. Find a red blood cell containing a signet-ring stage. Slides stained with blood-differentiating dyes usually show the cytoplasm of the trophozoite blue and the nucleus reddish pink.

The signet-ring trophozoite now develops into a **schizont** (Gr. *schizō,* cleave, + *ontos,* a being) (Figure 4.12). The nucleus divides repeatedly, producing 8 to 32 merozoites. The host red blood cell ruptures, releasing merozoites as well as metabolic wastes that are largely responsible for the characteristic symptoms of malaria: recurring episodes of fever with intense shivering, nausea, vomiting, pain, and delirium. Many merozoites enter and infect more red blood cells, with the cycle repeating until an enormous number of host cells have become parasitized.

After several such asexual generations, some of the merozoites enter red blood cells to become sexual forms called **macrogametocytes** and **microgametocytes** (Figures 4.12 and 4.13). To which gamete (egg or sperm) in the human life cycle is each of these analogous? Macrogamete = _____ Microgamete = _____ The gametocytes may be picked up by a feeding *Anopheles* female. Gametocytes give rise within the mosquito to gametes, which unite to become zygotes. The zygotes divide to produce sporozoites, which migrate to the salivary glands, from which they may be injected along with the saliva into the blood of a human, perhaps to cause another malarial infection.

Written Report

✍ Prepare a sketch of your own version of the life history of this organism, using words to indicate stages and arrows to show events such as reproduction, growth, and release of stages. Be sure to note in which host the parts of the cycle occur.

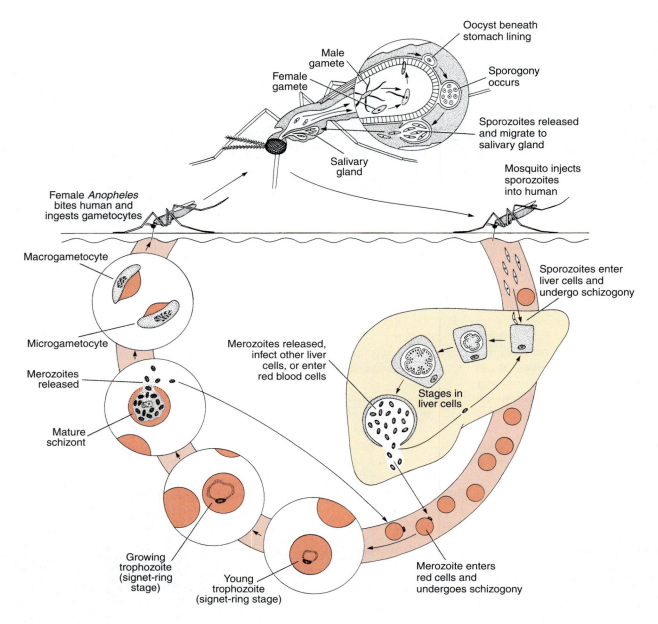

Figure 4.12

Plasmodium falciparum life cycle.

Answer the following questions in your report, referring to your textbook when necessary. When does meiosis occur in the life cycle of this organism? In which host do the gametes come together?

Gregarina

Phylum Apicomplexa
  Class Gregarinea
    Order Eugregarinida
      Family Gregarinidae
        Genus *Gregarina*

Various species of *Gregarina* (L. *gregarius*, belonging to a herd or flock) can be found in the guts of cockroaches (such as *Blatta* and *Periplaneta*), mealworms (the larvae of *Tenebrio*), and grasshoppers. Some species are surprisingly large for unicells and at one point were placed among worms by nineteenth-century zoologists.

☞ Using a live cockroach or mealworm, cut the anus free from the body wall with scissors. Remove the complete digestive tract by holding the animal and pulling off the head and attached gut with forceps. Place the gut in a watch glass, and cover with a

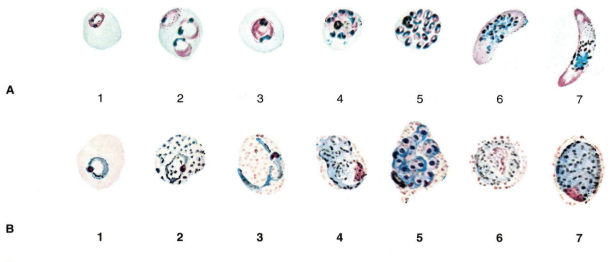

A 1 2 3 4 5 6 7

B 1 2 3 4 5 6 7

Figure 4.13

Stages in the life cycle of the malaria parasite. **A,** *Plasmodium flaciparum.* 1, Young trophozoite in signet-ring form. 2, Three medium trophozoites in one red blood cell. 3, Mature trophozoite showing clumped pigment. 4, Developing schizont containing several merozoites. 5, Mature schizont filled with merozoites. 6, Mature, crescent-shaped microgametocyte (length about 1.5 × diameter of red blood cell). 7, Mature macrogametocyte (about same size as microgametocyte). **B,** *Plasmodium vivax.* 1, Early ring-form trophozoite. 2, Young trophozoite with heavy chromatin dots. 3, Developing trophozoite. 4, Mature trophozoite with large mass of chomatin. 5, Schizont containing merozoites. 6, Microgametocyte almost filling red blood cell. 7, Macrogametocyte, similar to microgametocyte but with cytoplasm staining darker blue.

little insect saline solution.[8] Tease into small bits; then transfer a little of the contents in a drop of the saline to a slide for examination. Add a coverslip.

If adult feeding stages, called **trophozoites** (trōf′uh-zō′its; *trophe,* food, + *zōon,* animal), are present, they appear as large, distinctly shaped forms ranging from nearly round to elongate and wormlike, depending on the species. In general, the body is constricted into two unequal parts: a small anterior compartment (the protomerite), which bears a holdfast device (but which usually becomes detached during transfer of the gut and its contents to the slide), and a much larger posterior compartment (the deutomerite), which contains the nucleus (Figure 4.14).

☞ Did you find trophozoites of gregarines in your sample? If so, describe them in your notebook.

The entire life cycle takes place within the midgut of a cockroach or mealworm. Two trophozoites join together, encyst, and produce gametes, which, after fertilization, divide into **sporozoites** (spō′ruh-zō′ īts; *sporos,* seed, + *zōon,* animal). These pass out with the feces. When ingested by another cockroach or mealworm, the sporozoites develop into trophozoites—the adult stage.

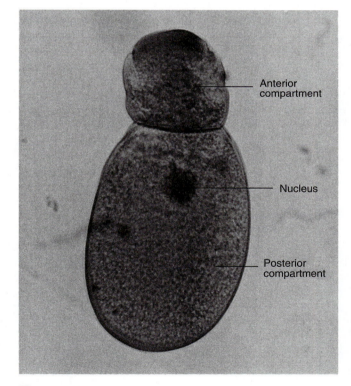

Anterior compartment

Nucleus

Posterior compartment

Figure 4.14

Trophozoite of *Gregarina,* a parasitic apicomplexan that lives in the gut of the cockroach or mealworm. The body is divided into two compartments.

[8]See Appendix A, p. 279, for preparation of insect saline.

EXERCISE 4D
Phylum Ciliophora— *Paramecium* and Other Ciliates

Core Study

Paramecium
Phylum Ciliophora
  Class Oligohymenophorea
    Subclass Hymenostomatia
      Genus *Paramecium*

Where Found

Paramecium is an active ciliate protozoan common in most fresh water that contains vegetation and decayed organic matter. Pond scum and even cesspools are also good sources. Some of the most commonly studied forms are *P. multimicronucleatum, P. caudatum, P. aurelia,* and *P. bursaria.*

Study of Live Specimens

Locomotion and Behavior.

☞ Spread out a *very few* fibers of absorbent cotton[9] on a slide, and add a drop or two of *Paramecium* culture. Cover with a coverslip. Examine first with a scanning lens to locate the paramecia; then switch to low power for study. Add water as necessary to prevent drying out.

A paramecium is slipper-shaped, rather transparent and colorless, and very active. It can swim at the rate of 1 to 3 mm per second. Watch its swimming habits. Does it swim in a straight line, a circle, or a zigzag? _____ Does it keep one side uppermost or revolve on an axis? _____ Does it have a definite anterior end? _____ Can it reverse its direction? _____ Does it seem to be contractile, as *Euglena* is? _____ Why is the term "spiral movements" used to describe its swimming habits?

What does a paramecium do when it encounters a barrier? _____ How does it go about finding an opening? _____ Is its body flexible enough to bend or to squeeze through tight places?

_____

Written Report

✍ Describe the locomotion of *Paramecium*. Use the preceding questions to guide your description, answering as many as you can.

_____

[9]Strands of cotton fibers serve as barriers to slow movement. Other methods for slowing *Paramecium* for study are given in Appendix A (p. 280).

Describe the attempts of *Paramecium* to avoid or to pass under or around a cotton fiber barrier. Use diagrams and arrows if you wish.

What characteristics possessed by *Paramecium* tell you it belongs in phylum Ciliophora?

General Structure and Function

☞ Locate a *Paramecium* that has been stopped by a cotton fiber, and study its structure with both low and high power.

Note the **oral groove** that extends obliquely from the anterior end to about the middle of the body. At the posterior end of the groove is the **mouth (cytostome).** The oral groove extends from the mouth into the body as a little canal, the **gullet (cytopharynx).** The groove and gullet are lined with strong cilia that are used to draw in food (Figure 4.15).

Pellicle. The cytoplasm is made up of two zones, a clear outer **ectoplasm** and an inner **endoplasm.** Are there any granules in the endoplasm? Outside the ectoplasm is a complex living **pellicle.** A delicate **plasma membrane** lies just underneath the pellicle. The pellicle and plasma membrane are more easily seen in stained preparations.

Osmoregulation. A contractile vacuole (water-expulsion vesicle) is usually located in each end of the body. *P. multimicronucleatum* may have more than two.

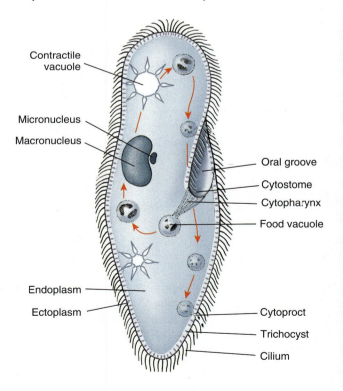

Figure 4.15
Anatomy of *Paramecium*.

Observe pulsations of the vacuoles caused by their alternate filling and emptying. After one empties, note the starlike **radiating canals** that appear. These radiating, or nephridial, canals collect liquid from a network of minute tubules and empty into the vacuoles, which rupture when filled, thus expelling fluid to the outside.

Do the two vacuoles empty at the same time or alternately? _____ How much time (in seconds) is there between vacuole discharges? _____ In normal pond water, a vacuole may discharge once every 6 to 10 seconds. How does this animal's rate compare (faster, slower, about the same)? _____

The pulsation period depends on the temperature and osmotic pressure of the medium. As the temperature increases, the pulsation rate increases. The pulsation rate can be slowed by adding drops of 0.25% solution of sodium chloride to the slide. Knowing this, how do you think the rate in marine ciliates would compare with that of freshwater ciliates? Why? _____

_____

Written Report

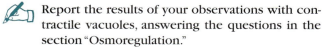 Report the results of your observations with contractile vacuoles, answering the questions in the section "Osmoregulation."

Ciliary Action. Cilia perform very much like the oars of a boat; that is, they have an effective stroke that propels the animal forward and a recovery stroke that offers little resistance (Figure 4.16). For a discussion of ciliary action, see your text.

 Find a quiet specimen, cut down the light, and look closely with high power at the margin of a paramecium. Study the action of the **cilia**.

Nuclei. The nuclei may be difficult to see in living paramecia.

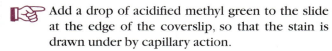 Add a drop of acidified methyl green to the slide at the edge of the coverslip, so that the stain is drawn under by capillary action.

This will kill the organisms but should give the cytoplasm a bluish tinge and stain the nuclei green. Note the presence of two nuclei: a large macronucleus that regulates metabolism of the cell (controlling cellular functions, such as feeding and digestion) and a small micronucleus that is reproductive and contains the animal's genome. The micronucleus is sometimes obscured by the macronucleus.

Trichocysts. If you reduce your light properly, you can, with careful focusing, observe the many small, spindle-shaped **trichocysts** lying in the ectoplasm just under the pellicle and perpendicular to the surface. Under certain kinds of stimulation, the trichocysts explode, each releasing a liquid

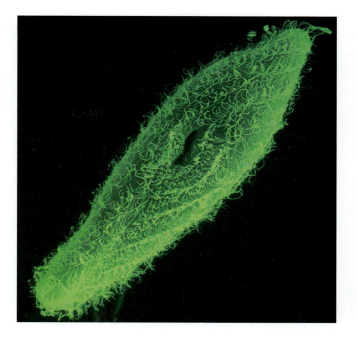

Figure 4.16
Scanning electron micrograph of *Paramecium,* showing how the cilia beat in waves. The cytostome is visible at the center. (× 400.)

that hardens in water to form a long, slender, threadlike filament. The tangle of filaments is believed to have some protective function. In some paramecia discharged trichocysts may be used to anchor the animal while feeding. Trichocysts are absent in some ciliates.

Perhaps the acidified methyl green you just used to demonstrate the nuclei has caused explosion of trichocysts on your slide. Alternatively, trichocysts may be made to discharge in live paramecia by adding a drop of dilute picric acid to the slide at the side of the coverslip and drawing the acid through with filter paper touched to the opposite side of the coverslip. As the fluid reaches the paramecia, you may be able to observe the discharged trichocysts by using subdued light and a high-power lens. How does the length of discharged trichocysts compare with the length of cilia?

Written Report

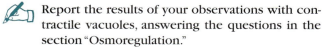 Report the results of your observations, including a sketch of a *Paramecium* with discharged trichocysts.

Further Study

Feeding and Response to Stimuli

Feeding. Nutrition in ciliates is holozoic. A paramecium is a particulate feeder; that is, it feeds on small particles, such as bacteria, which it moves toward its cytostome by action of cilia in the oral groove.

☞ On a clean slide, spread out a few cotton fibers and add a drop or two of *Paramecium* culture. Dip a toothpick into a preparation of yeast stained with Congo red, and transfer a *very small* quantity to the culture on the slide. Mix gently with the toothpick. The mixture should be light pink. If too much yeast is added, the protozoans will be obscured. Carefully apply a coverslip.

Do not allow the culture to dry out during your observation.

Find a specimen trapped by cotton fibers. Note the currents created by cilia in its **oral groove.** Watch yeast particles pass into the groove and through the **cytostome,** or cell mouth, into a passageway called the **cytopharynx.** Watch the formation of a **food vacuole,** which is a membranous sac containing water and suspended food particles. When it reaches a certain size, the vacuole breaks away and another forms. In what direction are food vacuoles carried by streaming endoplasm (cyclosis)? _____

Follow the course of a food vacuole. Does it vary in size during its trip? _____ If so, how? _____ Congo red, which is red in weak acid to alkaline solutions (pH 5.0 or above), turns blue in stronger acid solutions (pH 3.0 or below). Can you observe any changes in color in the vacuoles that might indicate a change in the condition of vacuole contents? _____ Is there any subsequent color change as vacuoles near the anal pore? _____ How might this be explained? _____ Are vacuoles any smaller as they complete their circuit than when first formed? _____

An **anal pore (cytoproct)** is found between the mouth and posterior end of the body. It is a temporary opening where indigestible food is discharged, and it is seen only at that time. Have you noticed such a discharge of material in one of your specimens?

Response to Stimuli. The following simple experiments are designed to determine the type of response (taxis) paramecia make to selected types of stimuli. A **taxis** (pl., **taxes**) is a directed reaction and orientation of the body to a specific stimulus. A movement toward a source is a positive taxis; a movement away from a source is a negative taxis. In these experiments, the responses are **chemotactic.** Perhaps you can design and perform other experiments to determine responses to other types of stimuli, such as **phototaxis** (response to light rays), **thigmotaxis** (to contact), or **geotaxis** (to gravity).

☞ 1. Place a drop of culture on a clean slide with no coverslip. Use a hand lens, scanning lens, or

dissecting microscope. (Some of you will be able to see paramecia with the naked eye.) Place a drop of *weak* acetic acid on the slide *near* the culture but *not touching it.* Locate the specimens; while observing them, draw a line with the point of a pin or toothpick from the acid to the culture.

Describe the reaction of the animals to the approach of the acid. Are they positively or negatively chemotactic to weak acid? _____ Do they move to the area where the acid is strongest, weakest, or in between? _____

☞ 2. Place a drop of culture on a clean slide. Place a few grains of salt on the slide near the culture but not touching it. While observing the animals, draw a grain or two of salt into the side of the culture drop.

Describe the reaction of the animals. Do they choose an area nearest the salt, farthest from the salt, or in between? _____

Study of Stained Slides

☞ With both low and high power, study a paramecium on a stained slide.

Focus up and down on the body and note that its entire surface is covered with **cilia.** Examine the **pellicle.** Specially prepared slides will show the peculiar pattern of hexagonal areas. Look especially for features difficult to see in the living unstained specimens, such as the **macronucleus,** one or more **micronuclei, trichocysts, oral groove, cytostome, cytopharynx,** and **contractile vacuoles.**

Binary Fission.

☞ Study a stained slide of paramecia undergoing binary fission.

Note the constriction across the middle of the body. What is happening to the macronucleus and the micronucleus? At the end of the process, the halves produced by the constriction will be separate daughter animals. How does this fission process compare with that of *Trypanosoma* (see *Exercise 4B*)?

Conjugation.

☞ Study also a stained slide of conjugation. Look for paired individuals lying with oral grooves attached.

In this position, they exchange micronuclear material. Consult your textbook for details of the process of conjugation.

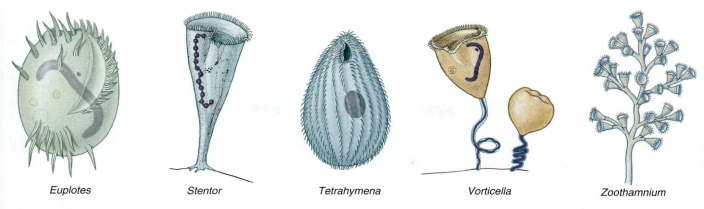

| *Euplotes* | *Stentor* | *Tetrahymena* | *Vorticella* | *Zoothamnium* |

Figure 4.17
Representative ciliates. *Euplotes* has stiff cirri used for crawling about. Contractile myonemes in the ectoplasm of *Stentor* and in the stalks of *Vorticella* allow great expansion and contraction. Note the macronuclei, which are long and curved in *Euplotes* and *Vorticella* and shaped like a string of beads in *Stentor*.

Drawings

Sketch a paramecium in the process of binary fission and a pair of paramecia in the process of conjugation. If you can find a live specimen dividing, watch it over a period of time and make a series of sketches to illustrate the process.

Written Report

On separate paper, summarize your observations and describe asexual reproduction in paramecia. On what axis do they divide? What is happening to the macro- and micronuclei? How would you recognize when two paramecia are in the process of sexual reproduction (conjugation)? Which nucleus is involved in sexual reproduction? Explain what happens in this process.

Further Study

Other Ciliates

Vorticella

Vorticella (Figures 4.17 and 4.18) is a solitary sessile ciliate. It clings to aquatic vegetation in stagnant ponds and streams. The blanket algae of ponds and small lakes is a favorite place for this animal.

Study both living specimens and the stained slides.

What is the color of the living animal? _____
The vorticellid is attached by a long, slender **stalk** that can contract into a spiral spring shape when it is disturbed. The **body** is bell-shaped with a flaring rim, the **peristome,** at its distal end. Within the peristome is

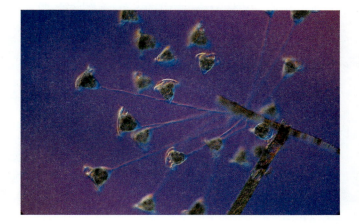

Figure 4.18
Group of living *Vorticella,* a solitary, sessile ciliate.

a circular **oral disc.** Note the **cilia** on the edges of the peristome and the oral disc. Note the beating of the cilia. What is their function? _____
The **cytostome** (mouth) is found between the peristome and the oral disc. From the cytostome, a short tube, the **cytopharynx,** leads into the interior. Food particles are swept into the cytopharynx by the action of the cilia, and **food vacuoles** are formed as they are in other protozoa. Note that the **nucleus** is made up of an elongated, U-shaped body, the **macronucleus,** and a much smaller **micronucleus.** Does the animal have a **contractile vacuole?** _____ *Vorticella* has a surrounding **pellicle,** which helps maintain the shape of the body. Reproduction is mostly by longitudinal binary fission, but **budding** also occurs.

Sketch a specimen in your notebook, and annotate your sketch with descriptions of the behavior of the specimens you observed.

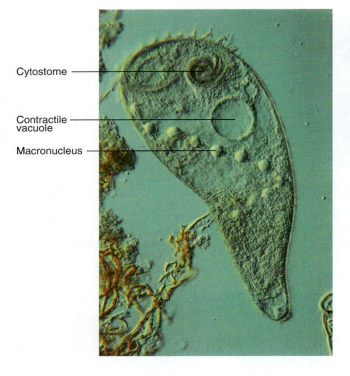

Cytostome

Contractile vacuole

Macronucleus

Figure 4.19
Living *Stentor* specimen.

Zoothamnium (Figure 4.17) is a stalked ciliate similar to *Vorticella*, but it is colonial (*Vorticella* is solitary) and has noncontractile stalks.

Stentor

☞ **Spread methylcellulose thinly on the slide before adding *Stentor* culture and a coverslip.**

Stentor (Figures 4.17 and 4.19) is a large ciliate with many of the same characteristics as *Paramecium* and *Vorticella*. How long is it when it is expanded? _____ When it is contracted? _____ How would you describe its shape? _____ See how many structures of this ciliate you can identify from your knowledge of paramecia. Can you locate the cytostome, cytopharynx, and contractile vacuole? _____ Is there an oral groove? _____ Can you see food vacuoles? _____ Are the cilia uniform in length? _____ Observe *stentor*'s striped appearance, which is caused by longitudinal bands of pigmentation. The blue pigment stentorin causes the blue-green color of *S. coeruleus*. Note the large **macronucleus,** stretched out like a string of beads. Small dots nearby are the micronuclei.

☞ **In your notebook, draw and label an extended specimen. Sketch its shape when the animal is swimming. Estimate its size.**

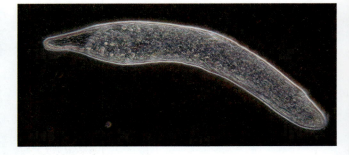

Figure 4.20
Spirostomum, an unusually large ciliate.

Spirostomum

Spirostomum (Figure 4.20) is one of the largest common freshwater protozoa. Estimate its length: _____ Different species range from 50 µm to 3 mm long. Can you see why it might be mistaken for a worm? Locate the large contractile vacuole posteriorly. It is fed by a long canal. The macronucleus is long and beadlike. Locate the cytostome and long oral groove. Focus on the surface and see the arrangement of cilia and trichocysts. Describe its swimming or other movements. Do you think it has myonemes?

☞ **In your notebook, draw and label an extended specimen. Outline a contracted one.**

Project: Microaquariums

A fascinating study of microorganisms and their relationships to each other may be carried out in the laboratory at no expense and with only a few minutes of time during each laboratory period. Use a series of small, clean, empty jars, such as those that may have held jelly or mayonnaise. Fill each jar two-thirds full with water from the tap, a pond, a ditch, or any other source. Add to each a teaspoonful of some source material: rich soil, plants (dry grass, leaves, hay, water plants, moss, or rotting leaf mold), pond scum, sludge from a sewage plant, or any other source. Label each jar. To retard evaporation, cover with a piece of glass, plastic wrap, or the jar lid placed loosely. Examine weekly—or more often, if you like—for several weeks or months. Each jar becomes a community that provides an interesting variety and an ever-changing cycle of life. In addition to protozoans, you may find a variety of crustaceans, flatworms, rotifers, gastropods, annelids, hydras, algae, and diatoms. Keep a weekly record of what you found in each jar.

Classification: Protozoa

The protozoa do not form a monophyletic group; they are an artificial assemblage of organisms placed together for convenience. Phylogenetic studies have shown that the protozoa represent numerous clades. Both the flagellate and ameboid bodies are likely to have evolved several times, and thus occur across clades.

Following is a condensed classification of the protozoan groups as presently recognized:

Phylum Amoebozoa (uh-mee′bo-zo′a) (Gr. *amoibē*, to change, + *zōon*, animal, living thing). Naked and shelled amebas, many with flagellated stages in the life cycle; mitochondria, when present, have tubular and branched cristae. Free-living and parasitic. Examples: *Entamoeba, Dictyostelium, Chaos, Arcella, Amoeba, Difflugia*.

Phylum Foraminifera (for′a-min-if′er-a) (L. *foramin*, hole, + *fero*, to bear). Shelled amebas bearing slender pseudopodia that extend through many openings in the test, forming a net that ensnares prey. Examples: *Vertebralina, Globigerina*.

Phylum Stramenopiles (stra-men′o-piles) (L. *stramen*, straw, + *pile*, hair). Flagellates with two different flagella, one long and one short; mitochondria with tubular cristae; free-living and parasitic plant-like and animal-like forms. Examples: *Actinosphaerium, Actinophyrs, Phytophthora infestans*.

Phylum Cercozoa (ser-ko-zo′a) (Gr. *kerkos*, tail, + *zōon*, animal). Diverse group of amebas, heterogeneous in morphology and lifestyle, but monophyly strongly supported by genetic studies. Most are free-living, some parasitic. Examples: *Euglypha, Clathrulina, Chlamydophrys*.

Phylum "Radiolaria" (ra′de-a-la′re-a) (L. *radiolus*, small sunbeam). Marine amebas with silica skeletons, forming beautiful test. Examples: *Tetrapyle, Pterocarys*.

Phylum Euglenozoa (yu-glen-a-zo′a; Gr. *eu-*, good, true, + *glēnē*, cavity, socket, + *zōon*, animal). Movement by flagella; cortical microtubules. Examples: *Euglena, Trypanosoma*.

Phylum Plantae, Class Viridiplantae Gr. *chlōros*, green, + *phyton*, plant). Unicellular and multicellular algae; photosynthetic chlorophyll pigments; flagella of equal length and smooth; mostly free-living photoautotrophs. Example: *Volvox, Chlamydomonas*.

Phylum Parabasala (par′a-bas′a-la) (Gr. *para*, beside, + *basis*, base). With a stiffening rod, the axostyle, along longitudinal axis of body; modified region of Golgi apparatus, the parabasal body; all parasitic or endosymbiotic. Examples: *Dientamoeba, Trichomonus, Trichonympha, Spirotrichonympha*.

Phylum Dinoflagellata (dy′no-fla-jel-at′a; Gr. *dinos*, whirling, + *flagellum*, little whip). Typically with two flagella, one transverse, one trailing; body usually grooved transversely and longitudinally, each groove containing a flagellum; chromoplasts bearing chlorophyll; free-living, planktonic, parasitic, or mutualistic. Examples: *Noctiluca, Ceratium, Gonyaulax*.

Phylum Diplomonada (di′plo-mon′a-da) (Gr. *diploos*, double, + L. *monas*, unit). One or two karyomastigonts (group of kinetosomes with a nucleus); individual organisms with one to four flagella; mitotic spindle within nucleus; cysts present; free-living or parasitic. Example: *Giardia*.

Phylum Apicomplexa (a′pi-com-plex′a; L. *apex*, tip, + *complex*, twisted around, + *a*, suffix). Characteristic set of organelles (apical complex) at anterior end in some stages; cilia and flagella usually absent; all species parasitic.

    Class Gregarinea (gre-ga-ryn′e-a; L. *gregarius*, belonging to a herd or flock). Mature, gamete-producing individuals; large, extracellular; gametes usually alike in shape and size; parasites of digestive tract or body cavity of invertebrates; life cycle with one host. Examples: *Monocystis, Gregarina*.

    Class Coccidea (kok-sid′e-a; Gr. *kokkos*, kernel, grain). Mature, gamete-producing individuals; small, typically intracellular; parasites mostly of vertebrates. Examples: *Plasmodium, Toxoplasma, Eimeria*.

Phylum Ciliophora (sil-i-of′o-ra; L. *cilium*, eyelash, + Gr. *phora*, bearing). Cilia or ciliary organelles present in at least one stage of life cycle; usually two types of nuclei; binary fission across rows of cilia; budding and multiple fission also occur; sexuality involving conjugation, autogamy, and cytogamy; heterotrophic nutrition; mostly free-living; contractile vacuole typically present. (This is a very large group, now divided into three classes and numerous orders.) Examples: *Paramecium, Colpoda, Tetrahymena, Stentor, Blepharisma, Epidinium, Vorticella, Euplotes, Didinium*.

EXPERIMENTING IN ZOOLOGY
Effect of Temperature
on the Locomotor Activity of *Stentor*

In this simple experiment, you will measure the effect of temperature on the biological activity of the large ciliate *Stentor coeruleus*. *Stentor*, like most other protozoans, has a body temperature that is passively decided by the temperature of its surroundings. Most people would call such organisms "cold-blooded," but biologists try to avoid this term (as well as its counterpart, "warm-blooded") because they find it hopelessly subjective. For one thing, *Stentor* does not have blood. For another, "cold-blooded" organisms living in warm environments are not cold. We call such organisms **poikilotherms** (meaning "variable temperature") or, alternatively, **ectotherms** (meaning the body temperature is determined by the temperature of the environment). Humans and the rest of the mammals, as well as the birds, are among the few members of the animal kingdom that have developed some orderly (and rather complicated) ways for maintaining a stable body temperature. We humans are **homeotherms** ("same temperature") and **endotherms** (meaning that internally generated heat is used to elevate the body temperature).

Because biological activity is based on chemical reactions that are much influenced by temperature, protozoans such as *Stentor* find that everything they do is paced with water temperature. In cold water, all processes slow down: swimming speed, metabolism, rate of food digestion, and so on. In warm water, it all speeds up. You might suspect that, in cold water, *Stentor* would have trouble catching its prey, but its prey have slowed down too, as have its potential predators.

Getting Ready[10]

Work with a partner for this experiment. First, with cellophane tape, tape a 2 cm square piece of finely divided graph paper to the underside of a depression slide (Figure 4.21). Press the cellophane tape firmly to the slide so that it will not come loose when it gets wet. Place two or three *Stentor* in a drop of water into the depression of the slide. Apply petroleum jelly in a neat ring around the depression. The ring must be complete to prevent water from leaking in (if tap water containing chlorine gets in, the *Stentor* will die). Carefully lower a coverslip over the depression to avoid trapping bubbles beneath the coverslip. Place the slide in the bottom of a dish half-filled with cold water, place the

dish on the dissecting microscope stage, and position the slide so that you can see the *Stentor*. Now carefully add crushed ice to bring the temperature down to 5°C. Stir the water with a laboratory thermometer to speed cooling. *Stentor* will be seen swimming over the graph paper grid—at this temperature, rather slowly.

How to Proceed

Choose a *Stentor* that looks active. Count the number of times this *Stentor* crosses a line in 1 minute. Your partner watches time and temperature while you watch *Stentor*. Your partner should also stir the bathwater gently with the thermometer to keep the water isothermal and add bits of crushed ice as necessary to keep the temperature steady at 5°C.

After making three separate counts at 5°C, raise the water temperature to 15°C by adding a little warm water to the bath. Hold the temperature steady at 15°C for

Figure 4.21

Measuring the effect of temperature on the locomotor activity of *Stentor*.

---

[10] Additional notes for implementing this exercise are found in Appendix A, pp. 280–281.

several minutes while measuring the swimming rate three times.

Biologists often express the effect of temperature on a biological process with the temperature coefficient Q_{10}:

$$Q_{10} = \frac{k_t + 10}{k_t},$$

where k_t is the velocity constant at temperature t and $k_t + 10$ is the velocity constant at 10° higher. Velocity constants are lines crossed per minute. The Q_{10} is an expression of temperature sensitivity of a biological function over a 10°C temperature range. For temperature intervals of *precisely* 10°C, you can use the following simple equation:

$$Q_{10} = SS_{(t+10)} / SS_t,$$

where SS_t is the swimming speed (lines crossed per minute) at the lower temperature, and $SS_{(t+10)}$ is the swimming speed at the higher temperature.

In general, chemical reactions have Q_{10} of about 2 to 3, while purely physical processes (such as diffusion) have much lower Q_{10} values, usually closer to 1.0.

Stentor's True Swimming Speed

How fast is *Stentor* actually swimming? The swimming speeds you have measured are "lines crossed per minute"; this is an expression, or index, of swimming speed but not a *true* swimming speed, which is distance swum per unit of time. If the graph paper grid beneath the depression slide is a square millimeter rule, you can easily determine *Stentor's* swimming speed in millimeters per second. To get an accurate measurement, stabilize the water temperature at 20°C. Time *Stentor* with a stopwatch only when it is moving straight across or straight up the grid. Take an average of several timings. Dividing the number of lines crossed by the seconds elapsed gives swimming speed in millimeters per second.

Now, estimate *Stentor's* body length by comparing it with the millimeter grid. Calculate how many body lengths per second *Stentor* is swimming at 20°C. Many fish (trout, for example) swim maximally at 10 body lengths per second. Humans (practiced swimmers) can swim about 1 body length per second. How does *Stentor* compare?

Written Report

On separate paper, summarize your findings in a short paragraph. Attach a graph of swimming speed plotted against temperature.

Questions for Independent Investigation

1. At what temperature does *Stentor* reach top swimming speed? To determine this, measure swimming speed at 5° intervals, from 5° to 25°C; then plot the number of lines crossed per minute on the Y-axis of graph paper against water temperature on the X-axis.

2. Calculate Q_{10} for three different temperature increments (5° to 15°, 10° to 20°, and 15° to 25°). Are the values constant over the entire temperature range?

3. Does the temperature of acclimation affect *Stentor* swimming speeds? Compare, for example, the swimming speed at 10°C of *Stentor* from two cultures, one held for at least 3 days at room temperature and another held for 3 days at 10°C.

References

Jennings, H. S. 1904. Contributions to the study of the behavior of lower organisms. Carnegie Institution of Washington, Pub. No. 16. Experiments on the reactions of *Stentor* to light are described on pp. 31–48.

Jennings, H. S. 1915. Behavior of lower organisms. New York, Columbia University Press.

The Sponges
Phylum Porifera

M embers of phylum Porifera are considered the simplest metazoans—little more than loose aggregations of cells, with little or no tissue organization. They show the **cellular level of organization.** There is division of labor among their cells, but sponges have no organs, no systems, no mouth or digestive tract, and only very rudimentary nervous integration. Because adult sponges have no germ layers, sponges are neither diploblastic nor triploblastic. Adult sponges are all sessile in form. Some have no regular form or symmetry; others have a characteristic shape and radial symmetry. They may be either solitary or colonial.

Chief characteristics of sponges are their **pores** and **canal systems;** the flagellated sponge feeding cells, called **choanocytes,** which line their cavities and create currents of water; and their peculiar internal skeletons composed of **spicules** or organic fibers **(spongin).** They also have some form of internal cavity **(spongocoel)** that opens to the outside by an **osculum.**

EXERCISE 5
Class Calcarea—_Sycon_
   Sycon, a Syconoid Sponge
   Other Types of Sponge Structure
Classification: Phylum Porifera

EXERCISE 5
Class Calcarea—_Sycon_

Core Study

Sycon, a Syconoid Sponge
Phylum Porifera
   Class Calcarea
      Order Heterocoela
         Genus _Sycon_ (=_Scypha, Grantia_)

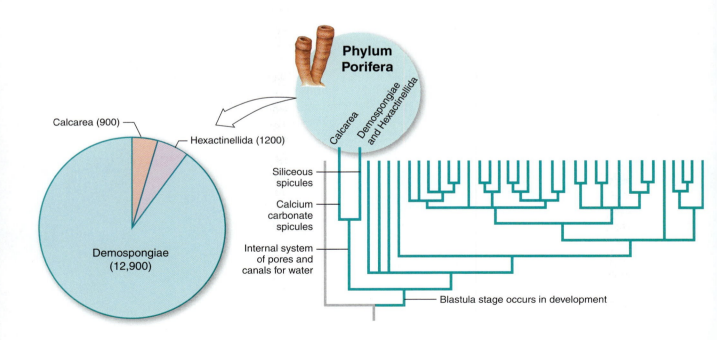

Where Found

Most sponges are marine, but there are a few freshwater species. Freshwater forms are found in small, slimy masses attached to sticks, leaves, or other objects in quiet ponds and streams.

Sycon is strictly a marine form, living in clusters in shallow water, usually attached to rocks, pilings, or shells. *Sycon* is chiefly a North Atlantic form. *Rhabdodermella* is a somewhat similar Pacific intertidal form, also belonging to class Calcarea.

Gross Structure

☞ Place a preserved specimen in a watch glass and cover with water. Examine with a hand lens or dissecting microscope.

Sycon (Gr., like a fig) is a **syconoid** type of sponge (Figure 5.1). What is the shape of the sponge? _____ The body wall is made up of a system of tiny, interconnected, dead-end canals whose flagellated cells draw in water from the outside through minute pores, take from it the necessary food particles and oxygen, and then empty it into a large central cavity for exit to the outside. What is the name of this central cavity? _____ All sponges have some variation of this general theme of canals and pores on which they depend for a constant flow of water.

External Structure. Is the base of the sponge open or closed? _____ The opening at the other end is the **osculum** (L., a little mouth), surrounded by a fringe of stiff, rodlike **spicules.** The external surface appears bristly when examined under magnification. Why? _____

Note that the body wall seems to be made up of innumerable, fingerlike processes pointing outward (Figure 5.1). Inside each of these processes is a **radial canal,** which is closed at the outer end but opens into a central cavity, the **spongocoel** (Gr. *spongos,* sponge, + *koilos,* hollow). External spaces between these enclosed canals are **incurrent canals,** which open to the outside but end blindly at the inner end. What is the name of the openings, or pores, to the outside of the sponge? _____

Water enters the incurrent canals and passes through minute openings called **prosopyles** (Gr. *prosō,* forward, + *pylē,* gate) into radial canals and then to the spongocoel and out through the osculum. There is no mouth, anus, or digestive system. What kind of symmetry does this sponge have? _____

Spongocoel. To study the spongocoel, do the following:

☞ Make a longitudinal cut through the midline of the body from osculum to base with a sharp razor blade. Place the two halves in a watch glass and cover with water.

Find the small pores, called **apopyles** (Gr. *apo,* away from, + *pylē,* gate), that open from the radial canals into the spongocoel (Figure 5.1). Can you distinguish the tiny canals in the cut edge of the sponge wall?

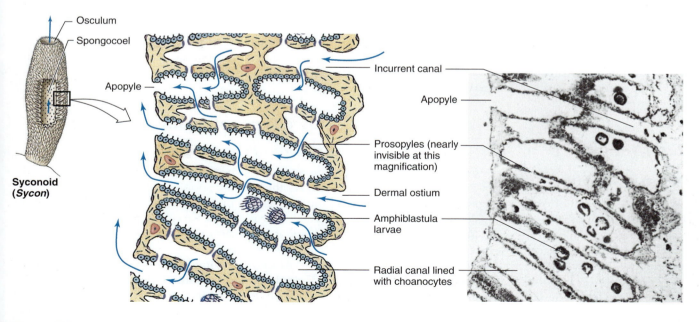

Figure 5.1

Structure of *Sycon,* showing a section through the body wall. *Sycon* is syconoid, one of three types of sponge structures. Sponges with simpler (asconoid) and more complex (leuconoid) canal system organization than *Sycon* are shown in Figures 5.3 and 5.4.

_____ Which direction does the water move through these canals? _____

Study of Prepared Slide

Transverse sections of sponge are difficult to prepare for slides because the spicules make it impossible to cut sections thin enough for studying the cells. Therefore, the spicules have been dissolved away for slide preparation.

☞ On a prepared slide of a cross section of *Sycon,* examine the entire section with low power to get an idea of its general relations.

Note the **spongocoel** in the middle of the section (Figure 5.1). Study the canal system. Find the **radial canals,** which open into the spongocoel by way of the **apopyles.** Are the apopyle openings smaller or larger in diameter than the radial canals? _____ Some apopyle openings will be absent in this section, and some of the radial canals will appear closed at the inner end. Follow the radial canals outward. Do they open to the outside or end blindly? _____ The radial canals may contain young flagellate larvae. Identify the **incurrent canals,** which open to the exterior by the **dermal ostia.** Follow these canals inward and note that they also end blindly. Water passes from incurrent canals into radial canals through a number of tiny pores, or **prosopyles,** which will not be evident on the slides.

Drawings and Report for Core Study

✎ On p. 61, draw: 1) an external view of *Sycon,* showing its gross structure, and 2) a longitudinal section through the sponge showing spongocoel and internal ostia. Use arrows to show direction of water flow. Answer the questions in the laboratory report.

Further Study

Reproduction

In **sexual reproduction,** most sponges are monoecious, having both male and female sex cells in the same individual. Sperm are shed into the water and carried to other sponges, where eggs are fertilized internally. The zygotes typically are retained within the sponge to develop into free-swimming, flagellated larvae. These leave the parent through the osculum to soon settle down on a substrate where they begin development into sessile adults. What is the advantage

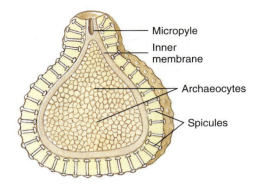

Figure 5.2
Gemmule of freshwater sponge.

to a sessile animal of producing free-swimming larvae? _____

Many sponges continue to grow almost indefinitely by budding and branching. They also bud off new individuals that become detached and carried away to form new sponges, a form of **asexual reproduction** in which the genotype of the parent sponge is copied in new offspring. What would be the disadvantage if this were the sole means of reproduction? _____ Is there a bud on your specimen?

Freshwater sponges and some marine Demospongiae reproduce asexually by means of **gemmules,** made up of clusters of amebocytes. Gemmules (L. *gemma,* bud, + *ula,* dim.) of freshwater sponges are enclosed in hard shells (Figure 5.2) and can withstand adverse conditions that would kill an adult sponge. In spring, cells in gemmules escape through the micropyle and develop into young sponges. Marine gemmules give rise to flagellated larvae.

Other Types of Sponge Structure

Asconoid Type of Canal System

The asconoid canal system is best seen in *Leucosolenia,* another marine sponge of the class Calcarea. *Leucosolenia* grows in a cluster, or colony (Figure 5.3), of tubular individuals in varying stages of growth. Large individuals may carry one or more buds.

☞ After observing the external structure of a submerged specimen, cut it in half longitudinally, place it on a slide with a little water, and cover. Study with low power or use a prepared slide.

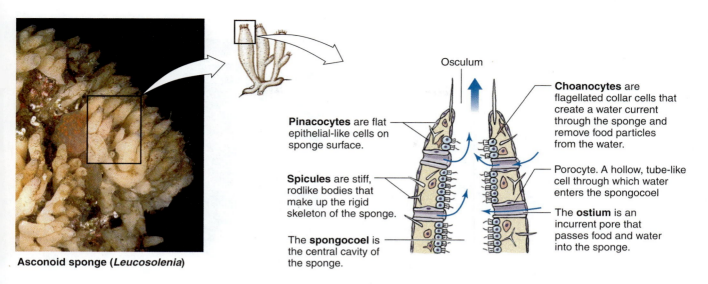

Asconoid sponge (*Leucosolenia*)

Osculum

Choanocytes are flagellated collar cells that create a water current through the sponge and remove food particles from the water.

Pinacocytes are flat epithelial-like cells on sponge surface.

Porocyte. A hollow, tube-like cell through which water enters the spongocoel

Spicules are stiff, rodlike bodies that make up the rigid skeleton of the sponge.

The **ostium** is an incurrent pore that passes food and water into the sponge.

The **spongocoel** is the central cavity of the sponge.

Figure 5.3

Asconoid type sponge *(Leucosolenia)*. Asconoid sponges have the simplest organization, with water drawn directly into the spongocoel and out the osculum. Because choanocytes are limited to capturing food from water passing adjacent to the spongocoel wall, asconoid sponges are all small and tubular. All live in dense colonies.

The body wall is covered with pinacocytes on the outside and filled with mesohyl that contains amebocytes and spicules. Incurrent pores extend from the external surface directly to the spongocoel, which is lined with flagellated choanocytes. On living specimens, you may be able to see some flagellar activity in the spongocoel.

Leuconoid Type of Canal System

Most sponges are of the leuconoid type, and most leuconoids belong to the class Demospongiae (Figure 5.4). Leuconoid sponges have clusters of flagellated chambers lined with choanocytes, and water enters and leaves the chambers by systems of incurrent and excurrent canals. Water from the excurrent canals is collected into spongocoels and emptied through the oscula. In large sponges, there may be many oscula. Many marine sponges, such as *Halichondria, Microciona, Cliona,* and *Haliclona*—all belonging to Demospongiae—are of the leuconoid type.

☞ Examine any such sponges available, in both external view and cut sections, to see this type of canal system.

Cellular Structure

The "connective tissue" of sponges is the **mesohyl** (Gr. *mesos,* middle, + *hyle,* wood) (also called mesenchyme).

It holds together the various types of ameboid cells, skeletal elements, and fibrils that make up the sponge body.

Choanocytes

☞ With high power, observe the "collar cells," or choanocytes (Gr. *choanē,* funnel, + *kytos,* hollow vessel), that line the radial canals (Figure 5.5).

Although they are flagellated, you probably will not see the flagella. What is the function of the choanocytes?

_____

Pinacocytes. Dermal pinacocytes (Gr. *pinax,* tablet, + *kytos,* hollow vessel) may be seen as extremely thin (squamous) cells lining the incurrent canals and spongocoel and covering the outer surface (Figure 5.5). What is their function? _____

Amebocytes. In the jellylike **mesohyl** that lies in the wall between the pinacocytes and choanocytes, look for large, wandering amebocytes of various functions (Figure 5.5). Some may differentiate into spicule-forming cells, some form sex cells, and some secrete spongin or spicules, serve as contractile cells, or aid in digestion.

☞ If living sponges are available, tease a bit of tissue on a slide with a drop of seawater and look for the various types of cells.

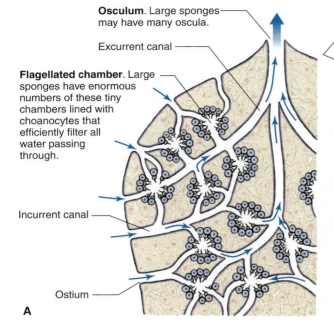

Osculum. Large sponges may have many oscula.

Excurrent canal

Flagellated chamber. Large sponges have enormous numbers of these tiny chambers lined with choanocytes that efficiently filter all water passing through.

Incurrent canal

Ostium

A

B

C

Figure 5.4

A, Leuconoid type sponge. **B,** *Aplysina lacunose,* globular sponge, each with single osculum; **C,** *Aplysina* sp., an encrusting sponge with numerous oscula. The difference in complexity between simple asconoid and complex leuconoid involves the water canals and skeletal systems, accompanied by infolding and branching of the collar cell layer. Leuconoid is the principal body plan for sponges because it permits large size and more efficient water circulation.

 On p. 62, draw a pie-shaped segment of a transverse section through *Sycon,* showing a few canals and some of the cellular details of their structure.

Skeleton

Place a small bit of the sponge on a clean microscope slide; add a drop of commercial chlorine bleach, such as Clorox (sodium hypochlorite); and set aside for a few minutes to allow the cellular matter to dissolve. Break up the piece with dissecting needles, if necessary. Add a coverslip and examine under the microscope.

Look for **short monaxons** (short and pointed at both ends), **long monaxons** (long and pointed) (Figure 5.6A), **triradiates** (Y-shaped with three prongs), and **polyaxons** (T-shaped). These spicules of crystalline calcium carbonate ($CaCO_3$) form a sort of network in the walls of the animal. What is the advantage of spicules to a loosely constructed animal such as *Sycon?*

_____

_____

Spicule types are used in the classification of sponges, along with the types of canal system. Demospongiae have

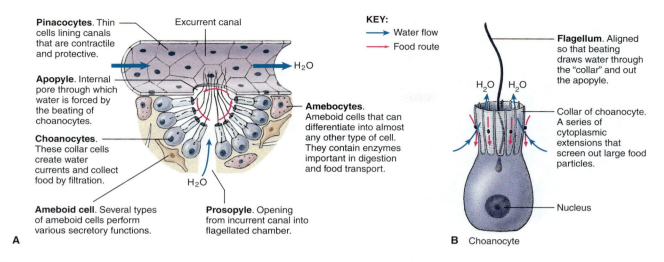

Pinacocytes. Thin cells lining canals that are contractile and protective.

Excurrent canal

Apopyle. Internal pore through which water is forced by the beating of choanocytes.

Choanocytes. These collar cells create water currents and collect food by filtration.

Ameboid cell. Several types of ameboid cells perform various secretory functions.

H_2O

H_2O

Amebocytes. Ameboid cells that can differentiate into almost any other type of cell. They contain enzymes important in digestion and food transport.

Prosopyle. Opening from incurrent canal into flagellated chamber.

A

KEY:
→ Water flow
→ Food route

H_2O H_2O

Flagellum. Aligned so that beating draws water through the "collar" and out the apopyle.

Collar of choanocyte. A series of cytoplasmic extensions that screen out large food particles.

Nucleus

B   Choanocyte

Figure 5.5
A, Section through wall of leuconoid sponge, showing flagellated chamber. **B,** Choanocyte as a food-catching cell.

A

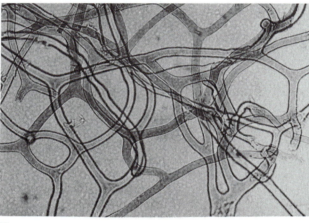

B

Figure 5.6
Skeletal elements, **A,** Spicules, mostly monaxons, from the cut and dehydrated surface of *Sycon*. SEM about × 480. **B,** Spongin fibers found in Demospongiae (greatly enlarged).

siliceous (mainly $H_2Si_3O_7$) spicules, spongin fibers (composed of an insoluble scleroprotein that is resistant to protein-digesting enzymes) (Figure 5.6B), or a combination of both. Their spicules are either straight or curved monaxons or tetraxons, but never six-rayed. The glass sponges (Figure 5.7) have siliceous triaxon (six-rayed) spicules.

See Appendix A, p. 281, for projects on the preparation of spicule samples and a study of sponge gemmules, including the preparation of permanent mounts of gemmules.

Classification: Phylum Porifera

Class Calcarea (cal-ca're-a; Gr. *calcis,* limy). Sponges with spicules of calcium carbonate, needle-shaped or three-rayed or four-rayed; canal systems asconoid, syconoid, or leuconoid; all marine. Examples: *Sycon, Leucosolenia.*

Class Hexactinellida (hex-ak-tin-el'i-da; Gr. *hex,* six, + *aktis,* ray). Sponges with three-dimensional, six-rayed siliceous spicules; spicules often united to form network; body often cylindrical or funnel-shaped;

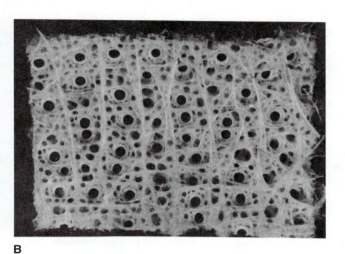

A

B

Figure 5.7
A, Beautiful skeleton of a glass sponge, *Euplectella*, of the class Hexactinellida. *Euplectella*, known as Venus's flower basket, like all glass sponges, is a deep-water form collected by dredging at depths of 1000 m or more in the western Pacific. **B,** Portion of wall of *Euplectella* with six-rayed spicules fused into a rigid latticework (about natural size).

canal systems syconoid or leuconoid; all marine, mostly deep water. Examples: *Euplectella* (Venus' flower basket), *Hyalonema.*

Class Demospongiae (de-mo-spun′je-e; Gr. *demos,* people, + *spongos,* sponge). Sponges with siliceous spicules (not six-rayed), spongin, or both; canal systems leuconoid; one family freshwater, all others marine. Examples: *Spongilla* (freshwater sponge), *Spongia* (commercial bath sponge), *Cliona* (a boring sponge). Most sponges belong to this class.

Core Study

Phylum _____

Subphylum _____

Genus _____

Name _____

Date _____

Section _____

Sycon

External View   **Longitudinal Section**
                                                    Internal View
_____   _____

Sycon

1. Describe the pathway of water through *Sycon,* naming all canals and openings through which water passes from entrance to exit._____

   _____

2. What drives the flow of water through a sponge? _____

3. To what level of organization do sponges belong? _____

4. In what way(s) does a sponge show evolutionary advancement, as compared with a colonial protozoan, such as *Volvox*? _____

   _____

   _____

Further Study

A Segment of a Transverse Section of *Sycon*

1. Explain how the two forms of reproduction in sponges, sexual and asexual, differ from each other.
    _____
    _____

2. What is the advantage to a sessile animal of producing free-swimming larvae? _____
    _____. What would be the disadvantage to asexual
    budding being the sole means of reproduction? _____
    _____

3. To what class of sponges do most marine sponges belong? _____

4. What is (are) the function(s) of choanocytes? _____

5. Explain how the skeletons of the three classes of sponges differ. _____
    _____
    _____

Characteristic	Characteristics of the Classes of Sponges		
	Calcarea	Hexactinellida	Demospongia
Common name(s)			
Spicule composition			
Type of canal system			
Habitat			
Representative genus			

5-9

EXERCISE 6

The Radiate Animals
Phylum Cnidaria

he two radiate phyla, Cnidaria (ny-dar′e-a) (= Coelenterata) and Ctenophora (te-nof′o-ra), are the most primitive of the metazoans—true multicellular animals. They are called radiates because all are radially (or biradially) symmetrical, a form of symmetry in which the body parts are arranged concentrically around the oral-aboral axis. They are the simplest animals having a **tissue level of organization** in which similar cells become aggregated into definite patterns or layers. With minor exceptions, however, the tissues of radiates are not organized into organs having specialized functions, and thus lack a feature characteristic of all the more complex metazoans.

Still, for all their simplicity, the radiates include several successful groups familiar to most people, such as sea anemones, jellyfish, and corals, as well as groups not so familiar, such as hydroids, zoanthids, and comb jellies. Many are brilliantly colored, and one group, the

EXERCISE 6A
Class Hydrozoa—*Hydra, Obelia,* and *Gonionemus*
Hydra, a Solitary Hydroid
Obelia, a Colonial Hydroid
Gonionemus, a Hydromedusa
Projects and Demonstrations

EXERCISE 6B
Class Scyphozoa—*Aurelia,* a "True" Jellyfish
Aurelia
Demonstrations

EXERCISE 6C
Class Anthozoa—*Metridium* and *Astrangia*
Metridium, a Sea Anemone
Astrangia, a Stony Coral
Projects and Demonstrations

Classification: Phylum Cnidaria

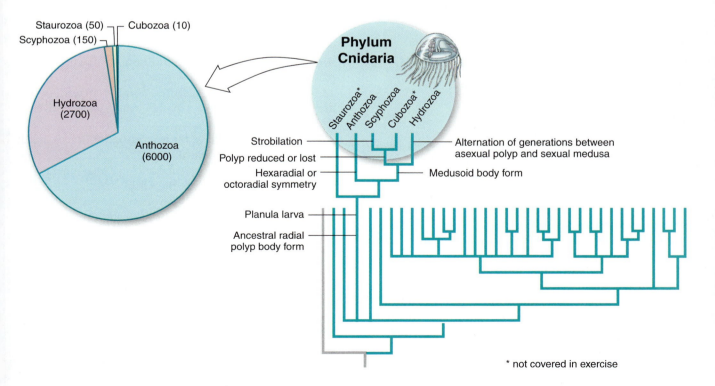

corals, form great tropical coral reefs that harbor a diversity of life rivaled only by tropical rain forests.

Two important metazoan features shared by all radiates are (1) two embryological primary **germ layers** (an ectoderm and an endoderm) that are homologous to those of higher metazoans, and (2) an internal space for digestion, the **gastrovascular cavity,** that lies along the polar axis and opens to the outside by a mouth.

Some of the cnidarians have a skeleton (coral, for example), but in most radiates, fluid in the gastrovascular cavity serves as a simple form of **hydrostatic skeleton.**

Although both cnidarians and their sister phylum, the Ctenophora (comb jellies, not covered in this exercise), are grouped together as radiate phyla, they differ in important ways. Cnidarians have characteristic stinging organelles (cnidocytes), which are usually absent in ctenophores. **Dimorphism**—the presence in a species of more than one morphological kind of individual—is common in cnidarians but absent in ctenophores.

There are two main types of body form in the cnidarians—the **polyp** (hydroid) form, which is often sessile, and the **medusa** (jellyfish) form, which is free-swimming. In some groups of cnidarians, both polyp and medusa stages occur in the life cycle; this is the dimorphic condition. In others, such as sea anemones and corals, there is no medusa; and in still others, such as the scyphozoans, or "true" jellyfish, the polyp stage is reduced or absent. In life cycles having both polyps and medusae, the juvenile polyp stage gives rise asexually to the medusa, which reproduces sexually. Both polyp and medusa have the diploid number of chromosomes, but the gametes are haploid.

EXERCISE 6A
Class Hydrozoa—*Hydra, Obelia,* and *Gonionemus*

Core Study

Three hydrozoan forms are traditionally used as examples of class Hydrozoa. Freshwater hydras are easily available, are conveniently large (2 to 25 mm in length), and can be studied alive. Hydras are solitary polyp forms, but they are atypical of the class because they have no medusa stage.

Obelia is a marine colonial hydroid that is more plantlike than animal-like in appearance. Its hydroid colonies are 2 to 20 cm tall, depending on the species, but its medusae are minute (1 to 2 mm in diameter). *Gonionemus,* another marine form, has a minute, solitary polyp stage that produces beautiful little medusae about 2 cm in diameter. For convenience, we combine the *Obelia* hydroid and the *Gonionemus* medusa for a life-history study. The hydroid stage, considered a

juvenile stage, produces medusae by asexual budding. The medusae are sexual adults.

Hydra, a Solitary Hydroid

Phylum Cnidaria
    Class Hydrozoa
        Order Hydroida
            Genus *Hydra, Pelmatohydra,* or *Chlorohydra*[1]

Where Found

Hydra (Gr., a mythical, nine-headed monster slain by Hercules) is the common name applied to any of about 16 species known to occur in North America. Hydras are found in pools, quiet streams, and spring ponds, usually on the underside of leaves of aquatic vegetation, especially lily pads. To collect hydras, bring in some aquatic plants with plenty of pond water and place in a clean jar or an aquarium. In a day or so, hydras, if present, may be seen attached to the plants or to sides of the jar.

Behavior and Structure

☞ Place a live hydra in a drop of culture water on a depression slide or watch glass. Examine with a hand lens or dissecting microscope. Answer the questions in this section.

The hydra may be contracted at first. Watch it as it recovers from the shock of transfer. Does it have great powers of contraction and expansion? _____. How long is it when fully extended? _____. Note the cylindrical body. A conical **hypostome** is at the oral end and bears a **mouth** surrounded by **tentacles.** How many tentacles does it have? _____. Compare with other specimens at your table. Touch a tentacle with the tip of a dissecting needle or fine artist's brush and note its reaction. What parts of the animal are most sensitive to touch? _____. After a quiet period, when the animal is extended, tap the watch glass and note what happens _____.

Does your hydra have **buds** or **gonads?** Does it attach itself to the glass by its **basal disc?** _____. The basal disc secretes a sticky substance for attachment. Can the hydra move about in the dish? _____. How? _____. Can you make out the outline of the **gastrovascular** cavity by focusing up and down? _____. Is there an anal opening? _____.

The warty appearance of the tentacles is caused by clusters of special cells, or **cnidocytes** (nī′dō-sīt; Gr. *knidē,*

---

[1]See p. 281 for notes on geographic distribution of hydra and selection of hydra species for class use. Species commonly provided for zoology classes are brown hydras, *Hydra littoralis* (eastern United States); false brown hydra, *Pelmatohydra pseudoligactis* (central North America); and green hydra, *Chlorohydra viridissima* (widely distributed in North America).

nettle, + *kytos,* hollow vessel), which contain stinging organelles, called **nematocysts** (ne-mat′o-cyst; Gr. *nēma,* thread, + *kystis,* bladder). Each nematocyst is a tiny capsule containing a coiled, threadlike filament that can be everted (Figure 6.1). If you are using a dissecting microscope, focus on a battery of cnidocytes on the edge of a tentacle. Note the tiny, projecting hairlike **cnidocils** (nī′dō-sil), which are involved in the discharge of the nematocyst. However, chemical stimulation (as from food) is necessary to lower the threshold before a cnidocil can be stimulated by contact. A hydra may discharge a quarter of its nematocysts during food capture. But within about two days, all are replaced by interstitial cells that develop into new cnidocytes.

Feeding and Digestion. To observe the feeding reaction of hydra, do the following:

☞ Add to the hydra culture on your slide a drop of water containing *Artemia* larvae (thoroughly washed to remove salt; see p. 281) or other suitable food organisms, such as *Daphnia* or enchytreid worms.

How does the hydra react to the presence of food? _____. How does it capture prey? _____. Does the prey struggle to escape? _____. When does the prey stop moving—before it is eaten or after? _____. What

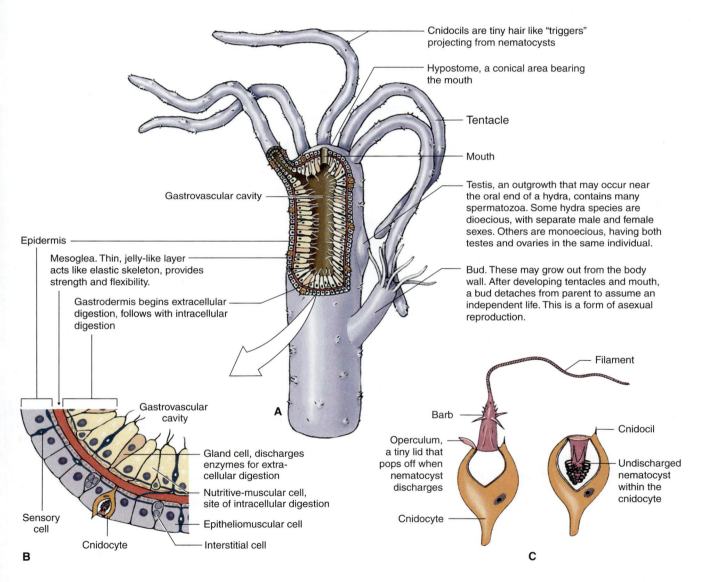

Figure 6.1

A, Structure of hydra. Both bud and developing gonad are shown, but in early life they rarely develop simultaneously. **B,** Cnidocytes with discharged (*left*) and undischarged (*right*) nematocysts. **C,** Diagrammatic cross section of a portion of the body wall.

is the reaction of the hypostome? _____.
How long does the feeding reaction take? _____.
How does the hydra act when its appetite has been
satisfied? _____ .

If food is not available, or if the hydra will not eat,
your instructor may want to demonstrate the feeding
reaction by adding a little reduced **glutathione** (0.03 g/l)
to a small dish containing some hydras. Glutathione is
found in living cells. It is released, in certain of the
hydra's prey, from the wounds made by nematocysts. It
stimulates the hydra to open its mouth and secrete
mucus to aid in the swallowing process.

Some digestion occurs within the **gastrovascular**
cavity, into which gland cells secrete digestive enzymes
(extracellular digestion). Food particles are then
engulfed by cells of the gastrodermis, in which digestion is completed **(intracellular digestion).** Indigestible materials must be regurgitated because there is
no anus.

Written Report

On separate paper record any feeding reactions
you have observed, using notes and sketches.

Further Study

Cross Section, Stained Slide

☞ Study a prepared stained slide of a cross section
of the body. Examine under both low and high
power.

Note that the body wall is made up of two layers
of cells, an outer **epidermis** (derived from what germ
layer? _____) and an inner **gastrodermis**
(derived from what germ layer? _____),
separated by a thin, noncellular layer, the **mesoglea**
(mez'o-glee'a; Gr. *mesos,* middle, + *glia,* glue)
(Figure 6.1B).

Epidermis. Most cells in the epidermal layer are
epitheliomuscular cells (Figure 6.1B). These are
medium-size cells with darkly stained nuclei that cover
the body and make possible rapid contraction of the
hydra's body and/or tentacles. Contractile fibers are
closely associated with the **nerve net,** which also lies
just beneath the epidermal layer. Now look for occasional cnidocytes containing spindlelike **nematocysts.**
At the bases of the epitheliomuscular cells may be found
some small, dark **interstitial cells.** These are embryonic
cells that can transform into the other kinds of cells
when needed. **Gland cells** secrete mucus onto the body
surface, particularly around the mouth and basal disc.

Mesoglea. This noncellular layer, lying between
epidermis and gastrodermis, extends over both body

and tentacles of hydra as an elastic "skeleton," providing increased flexibility to the animal. It is very thin in
hydroid polyps.

Gastrodermis. The gastrodermis layer is composed
principally of ciliated columnar **nutritive-muscular
cells** (Figure 6.1B). These cells perform several functions. Being muscular, they enable the hydra to change
shape (become longer and thinner). They also engulf
food (by phagocytosis), forming food vacuoles in which
food is digested intracellularly. They additionally discharge enzymes into the gastrovascular cavity, where
extracellular digestion occurs. Finally, their cilia create
currents that keep food particles constantly suspended
and circulating. What types of cells are common to both
layers of the body wall?_____
How does digestion in the hydra compare with that
in the sponge? _____
Does the sponge have intracellular or extracellular
digestion? _____

Sensory cells are found in both epidermis and
gastrodermis but would be difficult to identify.

Reproduction

Asexual. Budding is the asexual method of reproduction. A part of the body wall grows out as a hollow
outgrowth, or bud, that lengthens and develops tentacles and a mouth at its distal end. Eventually the bud
constricts at the basal end and breaks off from the parent. Buds may be found on live hydras.

☞ On a stained slide showing a budding hydra, study
the relation of the bud to the parent. Note that
the gastrovascular cavities of the two are continuous and that both layers of the parent wall extend
into the bud.

Sexual. Some species are **monoecious** (mow-
nee'shus; Gr. *monos,* single, + *oikos,* house), having both
testes and ovaries; other species have separate male and
female sexes are **dioecious** (di-ee'shus; Gr. *di,* two, +
oikos, house). Sex organs develop in the epidermis (from
the interstitial cells) of a localized region of the body column. Do sponges also have both asexual and sexual
reproduction? _____

☞ Examine stained slides showing testes (spermaries)
and ovaries.

Testes, small outgrowths containing many spermatozoa, are found toward the oral end; the single **ovary** is
a large, rounded elevation nearer the basal end. The
ovary produces a large, ripe **egg** (Figure 6.2), which
breaks out and lies free on the surface. **Spermatozoa**
break out of the testis wall, pass to the egg, and fertilize
it in position. The **zygote** so formed undergoes several
stages of development before dropping off the parent.

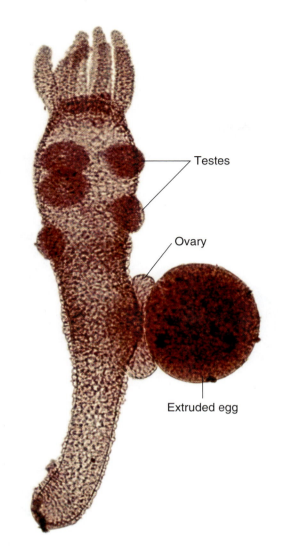

Testes

Ovary

Extruded egg

Figure 6.2
Hydra with five testes and large egg lying over the ovary from which
it was recently extruded.

Drawings

On separate paper, sketch a budding hydra and specimens that show testes (spermaries) and ovaries.

Core Study

Obelia, a Colonial Hydroid

Phylum Cnidaria
    Class Hydrozoa
        Order Hydroida
            Genus *Obelia*

Where Found

Obelia (Gr. *obelias,* round cake) is one of many colonial
hydroids found in marine waters and attached to

seaweeds, rocks, shells, and other objects. Its minute
medusae make up part of the marine plankton.

Life History and General Study

The life history of *Obelia* is dimorphic, having both polyp
(hydroid) and medusa (jellyfish) stages (Figure 6.3). A
hydroid colony arises from a free-swimming **planula** larva,
which settles and attaches to a substratum. Then, by budding, a colony is formed.

For study of a preserved colony, place a small piece
of marine algae with attached *Obelia* colonies[2] in a
small dish of water and observe with dissecting
microscope.

Colonies resemble tiny plants attached to the seaweed by
rootlike stolons. From the stolon arises a main stem that
gives rise to many lateral branches (Figure 6.3). The living
part of the colony is the **coenosarc,** which is encased in a
thin, transparent, protective **perisarc,** secreted by the epidermis. On the branches are two kinds of polyps (also
called zooids): nutritive polyps called **hydranths**
(Gr. *hydōr,* water, + *anthos,* flower) and reproductive
polyps called **gonangia** (sing. **gonangium;** N.L. *gonas,* primary sex organ, + *angeion,* dim. of vessel). Hydranths can
be recognized by their vase shape and the tentacles at their
free ends; gonangia, by their elongated club shape and lack
of tentacles. Are hydranths or gonangia more numerous?
_____ Medusa buds produced in the gonangia
break away to become free-swimming **medusae** (jellyfish).
Medusae are dioecious. When each sex is mature, the
medusa discharges its gametes into the surrounding water,
where fertilization occurs. The zygote develops into a planula larva, which attaches to a substratum and the cycle is
repeated. Thus, medusae give rise sexually to asexual
hydroid colonies, which in turn produce medusae.

Further Study

Study of a Stained Slide

Study a stained slide of *Obelia* and compare with
the preserved colony and with any living colony
you may have examined. Use low power of the
compound microscope.

Examine the main stem. Its inner protoplasmic part,
the **coenosarc** (sē′nō-sark; Gr. *koinos,* shared, + *sarx,*
flesh), is a hollow tube composed, like the hydra, of
epidermis, mesoglea, and **gastrodermis.** It encloses
a **gastrovascular cavity** that is continuous throughout
the colony. Surrounding this living part is the transparent, nonliving perisarc (Figure 6.3).

---

[2]If living *Obelia* colonies are preferred for this exercise, see p. 295 for
sources of living material.

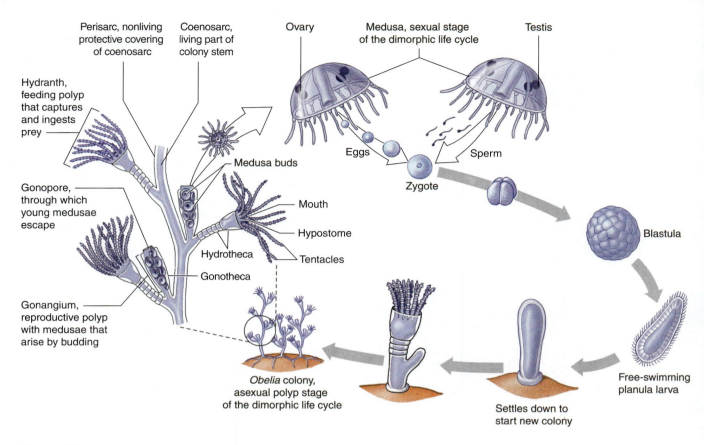

Figure 6.3
Dimorphic life cycle of *Obelia*, showing alternation of polyp (asexual) and medusa (sexual) stages. In *Obelia,* both polyps and stems are protected by continuations of the perisarc. In some hydroids, only the stems are so protected.

Each feeding polyp, or *hydranth,* is continuous with the coenosarc. A transparent extension of the perisarc forms a protective cup around the hydranth. Each hydranth has an elevated **hypostome** terminating in a **mouth** and bearing a circle of **tentacles** around the base. Each tentacle has in its epidermis rings of swellings caused by clusters of **cnidocytes,** which bear the **nematocysts.** Trace the continuous **gastrovascular cavity** from the hydranth through the branches and main stem of the colony. Food taken by the hydranths can thus pass to every part of the colony.

Club-shaped reproductive **gonangia** arise at the junction of the hydranth and coenosarc. A number of saucer-shaped **medusae buds** grow from a central stalk within the gonangium; these will develop into mature medusae. Where are the most mature buds located? ———————————————————— Young medusae escape through the opening, or **gonopore,** at the distal end.

Core Study

Gonionemus, a Hydromedusa

Phylum Cnidaria
    Class Hydrozoa
        Order Hydroida
            Genus *Gonionemus*

The medusa stage of *Obelia* is microscopic, too small to use for a continued study of the hydroid life cycle. However, another hydroid species with similar life history, *Gonionemus*, has an inconspicuous hydroid colony but a large medusa form, suitable for our continued study.

Where Found

Gonionemus[3] is a marine medusoid species found mainly in shallow protected coastal and bay areas along both coasts of the United States. Hydrozoan medusae are often called **hydromedusae,** as distinguished from the usually larger **scyphomedusae,** or jellyfish of the class Scyphozoa.

General Structure

☞ Place a preserved *Gonionemus* in a small dish filled with water. Examine with a hand lens. It is fragile, so do not grasp with forceps. Orient it by lifting or pushing with a blunt instrument.

———————————————————————————
[3]The name *Gonionemus*, meaning "angled thread," refers to a bend near the end of the slender tentacles where is located a tiny adhesive pad the animal uses to attach itself to seaweed (see Figure 6.4A).

The convex outer (aboral) surface is called the **exumbrella;** the concave (oral) surface is the **subumbrella.** Each of the **tentacles** around the margin of the bell bears rings of **cnidocytes,** an **adhesive pad** near its distal end, and a pigmented **tentacular bulb** at its base (Figure 6.4B). Tentacular bulbs make and store nematocysts, help in intracellular digestion, and act as sensory organs. Between the bases of the tentacles are tiny **statocysts** (Gr. *statos,* stationary, + *kystis,* bladder), considered to be organs of equilibrium. They are little sacs containing calcareous concretions. You will need low power of a compound microscope to see them.

Now look at the subumbrellar surface. Around the margin, find a circular, shelflike membrane, the **velum** (L., veil, covering), that aids in subumbrellar movements. Medusae move by a form of jet propulsion, forcing water out of the subumbrellar cavity by muscular contractions, thus propelling the animals in the opposite direction.

Note the **manubrium** (L., handle) suspended from the central surface of the subumbrellar cavity. At its distal end is the **mouth,** with four liplike **oral lobes** around it.

The **gastrovascular cavity** includes the **gullet,** the **stomach** at the base of the manubrium, four **radial canals** extending to the margin, and a **ring canal** around the margin.

Note the convoluted **gonads** suspended under each of the radial canals. The sexes look alike and can be determined only by microscopic inspection of the gonadal contents.

The medusa has the same cell layers as the hydroid or polyp form. All surface areas are covered with **epidermis,** the **gastrodermis** lines the entire gastrovascular cavity, and between these two layers is the jellylike **mesoglea,** which is much thicker in the medusa than in the hydroid form.

Drawings

On p. 77, label the diagrammatic drawing of an oral-aboral section through *Gonionemus.*

Further Study

Behavior

If living hydromedusae (freshwater *Craspedacusta,* marine *Gonionemus,* or others) are available, watch their movements.

Note the pulsating contractions that force water from the underside of the umbrella (subumbrella) and so propel the animal by a feeble "jet propulsion" (Figure 6.4A). Can they change direction? _____ How? _____ Does the animal swim continuously, or does it float sometimes? _____ What happens if a tentacle is touched? _____ Feeding reactions differ with different species. Some apparently depend on chance contact with food; others, such as *Gonionemus,* swim to the surface and then turn over and float downward with tentacles spread in search of food. *Gonionemus* is often found moving slowly about among marine algae and sea grass. If algae are present in the aquarium with *Gonionemus,* note how it uses its tentacles to move about. How does it use the adhesive pads on the tentacles? _____
_____

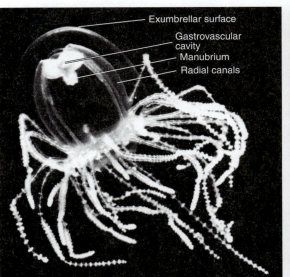

A

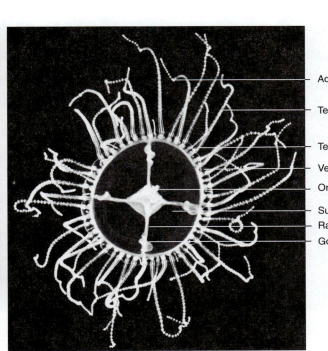

Adhesive pad

Tentacle

Tentacular bulb

Velum

Oral lobe

Subumbrella

Radial canal

Gonad

B

Figure 6.4
Gonionemus. **A,** Swimming. **B,** Oral view.

Written Report

Record your observations on separate paper.

Projects and Demonstrations

1. *Portuguese man-of-war* (*Physalia pelagica*). The Portuguese man-of-war illustrates the highest degree of **polymorphism** among cnidarians, for several types of individuals are found in the same colony. Study a preserved specimen. Note the **pneumatophore** (Gr. *pneuma*, air, + *pherein*, to bear), or bladder. What is its function? Suspended from the pneumatophore are many zooids (polyps) that have budded from it. Two kinds of polyp individuals are feeding polyps (gastrozooids) and fishing polyps (dactylozooids), which are equipped with long, stinging tentacles. Male and female gonophores are modified medusoid individuals that remain attached as buds, producing eggs and sperm.

2. *Obelia*. Examine some medusae of *Obelia*.

3. *Various colonial hydroids*. Examine forms of colonial hydroids, such as *Eudendrium, Tubularia, Pennaria, Sertularia,* or *Bougainvillia,* either preserved or on slides.

EXERCISE 6B
Class Scyphozoa—*Aurelia,* a "True" Jellyfish

Core Study

Aurelia

Phylum Cnidaria
   Class Scyphozoa
      Order Semaeostomeae
         Species *Aurelia aurita*

Where Found and Life History

Aurelia aurita, the "moon jelly," is common along both coasts of North America. It is a cosmopolitan species distributed from temperate and tropical to subpolar latitudes. *Aurelia* (L. *aurum,* gold) (Figure 6.5) is a scyphozoan (sy-fo-zo′an) medusa (often referred to as a scyphomedusa). Scyphomedusae are generally larger than hydrozoan medusae (hydromedusae); most of them range from 2 to 40 cm, but some reach as much as 2 m or more in diameter. The jelly layers (mesoglea) are thicker and contain cellular materials, giving scyphomedusae a firmer consistency than hydromedusae. Nevertheless, all jellyfish are largely water (94% to 96% water in marine species, such as *Aurelia,* and up to 99% water

Figure 6.5
Swimming *Aurelia aurita.*

in some freshwater hydromedusae) and active tissues are mostly epithelial.

Scyphozoans are often called "true" jellyfish. Scyphomedusae are constructed along a plan similar to that of hydromedusae, but they lack a velum (the velum of hydromedusae is described on p. 69). Their parts are arranged symmetrically around the oral-aboral axis, usually in fours or multiples of four, so they are said to have tetramerous radial symmetry. Their gastrovascular systems have more canals and more modifications than those of hydrozoans.

Sexes are separate in *Aurelia,* as they are in all scyphozoans. Sex cells are shed from the gonads into the gastrovascular cavity and are discharged through the mouth for external fertilization. Within folds of the oral arms, the young embryos develop into free-swimming **planula larvae** (Figure 6.6). These escape from the parent, attach to a substratum, and develop into tiny polyps called **scyphistomae** (sy-fis′to-mee; Gr. *skyphos,* cup, + *stoma,* mouth). The scyphistoma later becomes a **strobila,** which begins to bud off young medusae (**ephyrae** [Gr. *Ephyra,* Greek city, in reference to castle-like appearance]) in layers resembling a stack of saucers (Figure 6.6). This budding process is called **strobilation** (Gr. *strobilos,* pinecone).

The large size and fiery nematocysts of many jelly-fish make them disagreeable and sometimes dangerous to swimmers. One of these is *Cyanea capillata,* the "lion's

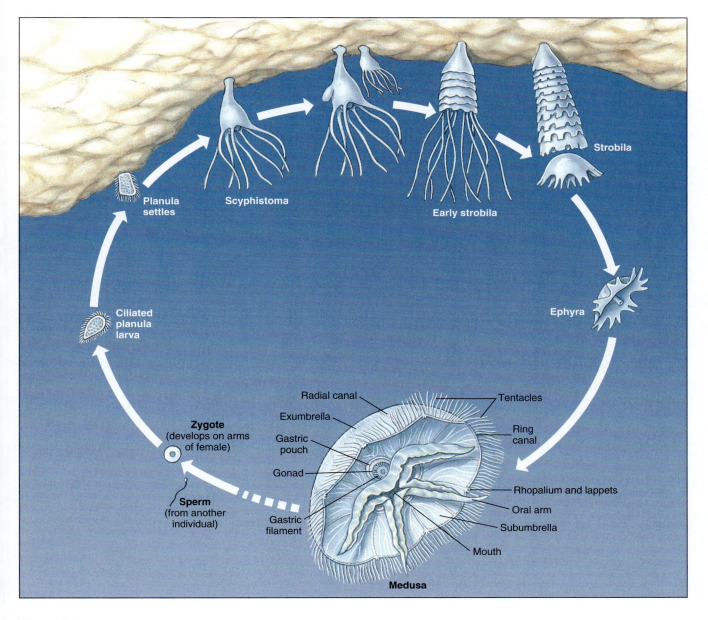

Figure 6.6
Life cycle of *Aurelia,* a marine scyphozoan medusa.

mane jellyfish" of the North Atlantic.[4] Even more dangerous is the cubomedusan *Chironex fleckeri,* sea wasp of the Australian region. This jellyfish has caused numerous fatalities in Australia; deaths occur rapidly from anaphylactic shock.

General Structure

☞ Using a ladle (the medusa is too fragile to be handled with a forceps), transfer a preserved specimen of *Aurelia* to a finger bowl of water and spread out flat.

Note that *Aurelia* is more discoidal and less cup-shaped than *Gonionemus.* When spread flat, the jellyfish shows a circular shape broken at eight regular intervals by marginal notches (Figure 6.5). Each marginal notch contains a **rhopalium** (ro-pay'li-um; N.L. from Gr. *rhopalon,* a club), a sense organ consisting of a statocyst and an ocellus. This is flanked on each side by a marginal extension, the **lappet** (Figures 6.5 and 6.6). What is the function of a statocyst? _____ An ocellus? _____

☞ Snip out a rhopalium[5] with scissors and examine under the higher power of a dissecting microscope.

In living medusae, the excision of all rhopalia would interfere with swimming, either slowing down the contractions or stopping them altogether.

Note the short **tentacles** that form a fringe around the animal's margin. Compare these tentacles with those of *Gonionemus* medusae. How do they differ?

_____

Gastrovascular System. In the center of the oral side are four long, troughlike **oral arms.** These are modifications of the manubrium. Note that the oral arms converge toward the center of the animal, where the square **mouth** is located. The mouth opens into a short **gullet,** which leads to the **stomach.** From the stomach, four **gastric pouches** extend. They can be identified by the horseshoe-shaped **gonads** that lie within them. Near the inner edges of the gonads are numerous thin processes, the **gastric filaments,** which are provided with **nematocysts.** What would be the use of stinging cells here? _____
On the oral side of each gastric pouch is a round aperture that leads into a blind depression, the **subgenital** pit. These pits have no connection with the gonads or gastric pouches and may be respiratory in function.

---

[4]The lion's mane jellyfish appears in a Sherlock Holmes episode, "The adventure of the lion's mane," in which a school professor becomes the victim of the jellyfish while swimming.

[5]The specific epithet *aurita* of the species *Aurelia aurita* means "eared," probably in reference to the lobelike rhopalia.

A complicated system of radiating canals runs from the gastric pouches to the **ring canal,** which follows the outer margin (Figure 6.6). This system of stomach and canals, resembling hub, spokes, and rim of a wheel, forms the medusoid gut, or gastrovascular cavity. Contrast this with the simple, saclike gut of polyp individuals such as hydra or *Obelia* polyps.

Jellyfish are carnivorous, most feeding on fish and a variety of marine invertebrates. *Aurelia,* however, is a suspension feeder that feeds largely on zooplankton. The tentacles are not used in food capture. Food organisms are caught in mucus secreted on the subumbrella and moved by cilia to the bell margin. Food is collected from the margin by the oral arms and is transferred by cilia to the stomach. Gastric filaments with nematocysts help pull in and subdue larger organisms and secrete digestive enzymes. Partially digested food and seawater are then circulated by cilia through the system of canals. In this way, nutrients and oxygen are carried to all parts of the body. The canals are lined with cells that complete digestion of the food. Thus, digestion is both extracellular and intracellular.

Further Study

Behavior

☞ If living *Aurelia* or other scyphozoan genera are available, observe their swimming movements (Figure 6.5).

How is movement achieved? _____ Are they strong swimmers? _____ How many times per minute does the medusa pulsate? _____ Does it swim horizontally or vertically? _____ Use a gentle touch with a small artist's brush to test reaction to touch. To test response to food chemicals, dip the brush into glucose solution, clam or oyster juice, or other food substances; small crustaceans or other small food organisms may be placed near their tentacles.

Written Report

✍ Record your observations on separate paper.

Demonstrations

1. *Slides.* Examine slides showing stages in the life cycle of *Aurelia*—scyphistoma, strobila, and ephyra.

2. *Scyphozoan jellyfish.* Examine various species of preserved scyphozoans or, if available, living specimens of the "upside-down" jellyfish *Cassiopeia* (see pp. 282 for additional information).

EXERCISE 6C
Class Anthozoa—*Metridium,*
a Sea Anemone, and *Astrangia,*
a Stony Coral
Core Study

Metridium, a Sea Anemone

Phylum Cnidaria
    Class Anthozoa
        Subclass Hexacorallia
            Species *Metridium senile*

Where Found

Metridium senile, a name from the Greek that means "ancient womb," is the most common sea anemone on the Atlantic Coast from Delaware north to the Arctic and on the Pacific Coast from Santa Catalina Island, California, north to the polar seas. It occurs from low intertidal, where it is commonly seen on rocks and pilings, to depths of perhaps 75 m. Most sea anemones are solitary sessile animals but some make clonal patches or colonies. Members of the class Anthozoa are all polyps in form; there are no medusae. There is a great variety in size, structure, and color among the sea anemones. All are marine.

Behavior

☞ If living anemones[6] are available, allow one to relax completely and then touch a tentacle lightly with a new (untouched) coverslip held in clean forceps.

Do the nematocysts discharge? _____ Will nematocysts discharge if the coverslip is touched more vigorously to the tentacle? _____ Now put some saliva (which contains protein) on a dry coverslip and again touch a tentacle. Is the response stronger? Why? _____ Discharged nematocysts should stick to the glass and may be examined under the microscope.

Using another relaxed anemone, drop bits of clean filter paper on the tentacles and time the type and speed of the response. Now test with bits of filter paper soaked in shrimp, clam, or mussel juice and compare the reactions. Test again with bits of clam or other sea food. What conclusions can you draw from this simple experiment? _____

Is response to food similar to response to touch? _____ Is the reaction of these animals a part of the normal feeding reaction? _____ If live sea stars are available, try touching an anemone with the arm of a star. What happens? Is this a feeding response or a defense reaction? _____ Some anemones react to certain predatory stars by detaching their pedal discs and moving away from the star.

☞ Use a glass rod to probe the pedal disc. What happens? _____ If you prod the animal vigorously, it will shoot out white, thread-like **acontia** (a-con'she-a; Gr. akontion, dart), which are filled with nematocysts used for defense.

Place an acontium thread on a slide, cover with a coverslip, and examine with a microscope. Do the acontia move? _____ Examine the edge of an acontium with high power. What do you see that might explain acontia movement? _____

For a dramatic demonstration of nematocysts in action, draw some methyl green under the coverslip using a piece of filter paper touched to the opposite edge of the coverslip. Viewed at high power, the undischarged nematocysts in the acontia look like grains of long rice. What happens when they discharge?

Written Report

✍ Record your observations on separate paper.

External Structure

☞ Place a preserved specimen in a dissecting pan. Note the sturdy nature of its body structures compared with those of other cnidarians you have studied.

The **body** is cylindrical (Figure 6.7), but in preserved specimens it may be somewhat wrinkled. Note that the body of the animal can be divided into three main regions: (1) **oral disc,** or free end, with numerous conical **tentacles** and **mouth;** (2) cylindrical column, forming the main body of the organism; and (3) **basal disc** (aboral end), by which during life the animal attaches itself to a solid object by means of its glandular secretions. Although it is called a sessile animal, a sea anemone can glide slowly on its basal disc.

Is there more than one row of tentacles? _____ Note that the inner surface of the mouth is lined with ridges and that a smooth-surfaced, ciliated groove, the **siphonoglyph** (sy-fun'o-glif; Gr. *siphōn,* tube, siphon, + *glyphē,* carving), is found at one side of the mouth. (In some specimens, there are two of these grooves.) Siphonoglyphs, aided by cilia, circulate water throughout the gastrovascular cavity. The mouth is separated from the nearest tentacles by a smooth space, the **peristome** (Gr. *peri,* around, + *stoma,* mouth).

---

[6]See p. 282 for notes on implementing behavior studies with sea anemones.

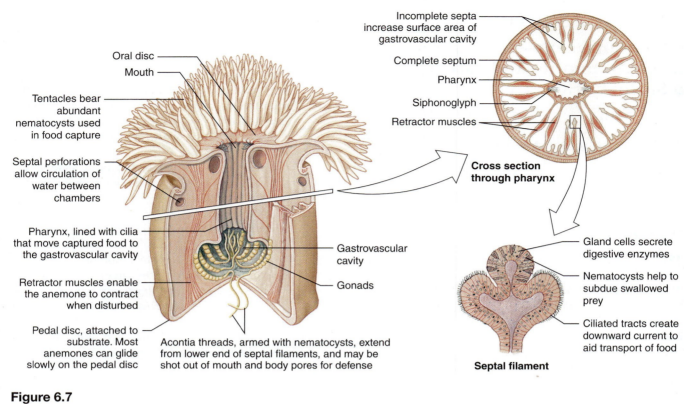

Figure 6.7
Structure of a sea anemone.

Note the tough outer covering **(epidermis)** of the specimen. Small pores on tiny papillae are scattered over the epidermis, but they are hard to find.

Further Study

Internal Structure

☞ Study of internal anatomy is best made by comparing two sections, one cut longitudinally through the animal and a second cut transversely. Study first one and then the other of these two sections to understand the general relations of the animal. Determine through what part of the animal the transverse section was made.

Look at the longitudinal sections and note that the **mouth** opens into a **pharynx** (fair'inks; Gr. *pharynx,* gullet) (Figure 6.7), which extends down only part way in the body to where it opens into the large **gastrovascular cavity.** Thus, the upper half of the body appears as a tube within a tube; the outer tube is the **body wall,** and the inner tube is the pharynx. Look at the cross section and determine these relations. Notice that not only is the gastrovascular cavity the space aboral to the pharynx, but it also extends upward to surround the pharynx.

The gastrovascular cavity is subdivided into six chambers by six vertical pairs of **primary** (complete) **septa** (Figure 6.7). These six chambers are partially subdivided by smaller pairs of **incomplete septa** with free edges expanded into **septal filaments** armed with nematocysts and bearing gland cells that secrete digestive enzymes. The septa greatly increase the digestive surface of the gastrovascular cavity. The nematocysts may help to subdue struggling prey and, by driving enzymes deep into the prey, also speed digestion.

The **gonads** are thickened bands resembling stacks of coins, often orange-red in color, that lie in the septa just peripheral to the septal filaments. Anemones are dioecious.

Astrangia, a Stony Coral

Phylum Cnidaria
   Class Anthozoa
      Subclass Hexacorallia
         Order Scleractinia
            Genus *Astrangia*
               Species *Astrangia danae*

Stony corals resemble small anemones but are usually colonial. Each polyp secretes a protective calcareous cup, into which the polyp partly withdraws when disturbed. Some corals form colonies consisting of millions of individuals, each new individual building its skeleton upon the skeletons of dead ones, thus forming, over many years, great coral reefs. Reef-building corals build only in tropical or subtropical waters, where the water temperature stays at or above 21° C.

Astrangia (Gr. *astron,* star, + *angeion,* vessel), known as the star coral, is our only shallow-water northern coral. It occurs on the Atlantic and Gulf coasts locally from Florida north to Cape Cod. It does not build reefs but forms small colonies of 5 to 30 individuals encrusted on rocks and shells. Its food consists of small organisms, such as protozoans, hydroids, worms, crustaceans, and various larval forms.

Behavior

The same sort of touching and feeding experiments as suggested for *Metridium* are applicable to corals, making allowance for the smaller size.

Structure

☞ Examine living or preserved coral polyps.

Note the delicate, transparent polyps extending from the circular skeletal cups, also called corallites (Figure 6.8A). The polyps resemble those of anemones, with a column, an oral disc, and a crown of tentacles. Two dozen or more simple tentacles, supplied with nematocysts, are arranged in three rings around the mouth. Siphonoglyphs are absent. The edges of the septa can usually be seen through the transparent polyp walls.

As in sea anemones, stony corals are built on a plan of six or multiples of six (hexamerous). Digestion in the gastrovascular cavity is similar to that of anemones.

Colonies usually arise by budding or division from a single polyp that has been sexually produced. The surface of the colony between skeletal cups is covered by a sheet of living tissue, which is an extension of the polyp walls. This tissue connects all members of the colony.

☞ Now study a piece of skeleton from which the polyps have been removed.

Each cup (Figure 6.8B) was the home of a polyp, which secreted it. The rim of the cup is called the **theca** (L. *theca,* box), and the base is the **basal plate.** The **sclerosepta,** or radial partitions within the theca, form the same hexamerous pattern as do the septa. They are laid down by the folds of the epidermis at the base of the polyp. The theca is secreted by the epidermis of the lower part of the column. As the coral polyp grows, the cup tends to fill up with calcium carbonate, so that the theca and sclerosepta are continually extended upward. The entire skeleton is outside the body of the polyp.

A

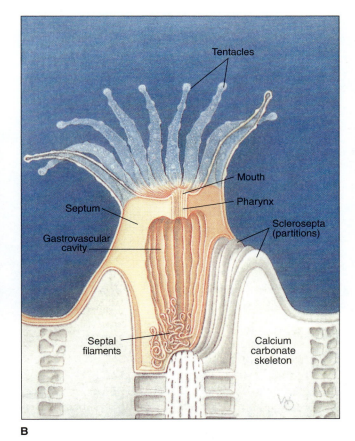

B

Figure 6.8
Stony corals. **A,** Living polyps of *Astrangia danae.* **B,** Structure of a coral polyp, diagrammatic.

Projects and Demonstrations

1. *Various hard coral skeletons.* Examine such skeletons as white corals; red coral, *Corallium;* staghorn coral, *Acropora;* organpipe coral, *Tubipora;* or brain coral, *Meandrina,* looking for the structures just described.

2. *Nematocysts.* Nematocysts may be collected from various live cnidarians. Zoantharian corals (soft corals) usually have fairly large nematocysts, and the little red colonial anemone *Corynactis* is good for this purpose. Wiping a clean coverslip across the oral disc is likely to attract both discharged and undischarged nematocysts, which can be studied by inverting the coverslip over a drop of seawater on a slide. A good-quality microscope is necessary.

   Collect and compare the nematocysts of various hydrozoans, scyphozoans, and anthozoans.

Classification

Phylum Cnidaria (ny-dar'e-a)

Class Hydrozoa (hy-dro-zo′a) (Gr. *hydra,* water serpent, + *zōon,* animal). Both polyp and medusa stages represented, although one type may be suppressed; medusa with a velum; found in fresh and marine water. The hydroids. Examples: *Hydra, Obelia, Gonionemus, Tubularia, Physalia.*

Class Scyphozoa (sy-fo-zo′a) (Gr. *skyphos,* cup, + *zōon,* animal). Solitary; medusa stage emphasized; polyp reduced or absent; enlarged mesoglea; medusa without a velum. The true jellyfish. Examples: *Aurelia, Rhizostoma, Cassiopeia.*

Class Staurozoa (sta′ro-zo′a) (Gr. *stauros,* a cross, + *zōon,* animal). Solitary, polyps only, no medusa; polyp surface extended into eight clusters of tentacles surrounding mouth; all marine. Examples: *Haliclystis, Lucernaria.*

Class Cubozoa (ku′bo-zo′a) (Gr. *kybos,* a cube, + *zōon,* animal). Solitary; polyp stage reduced; bell-shaped medusae square in cross section, with a tentacle or group of tentacles at each corner; margin without velum but with velarium; all marine. Examples: *Carybdea, Chironex.*

Class Anthozoa (an-tho-zo′a) (Gr. *anthos,* flower, + *zōon,* animal). All polyps, no medusae; gastrovascular cavity subdivided by mesenteries (septa).

Subclass Hexacorallia (hek-sa-ko-ral′e-a) (Gr. *hex,* six, + *korallion,* coral) **(Zoantharia).** Polyp with simple, unbranched tentacles; septal arrangement hexamerous; skeleton, when present, external. Sea anemones and stony corals. Examples: *Metridium, Tealia, Astrangia.*

Subclass Ceriantipatharia (se-re-an-tip′a-tha′ri-a) (N. L. combination of Ceriantharia and Antipatharia, from type genera). With simple, unbranched tentacles; mesenteries unpaired. Tube anemones and black or thorny corals. Examples: *Cerianthus, Antipathes.*

Subclass Octocorallia (ok′to-ko-ral′e-a) (L. *octo* + Gr. *korallion,* coral) **(Alcyonaria).** Polyp with eight pinnate tentacles; septal arrangement octamerous. Soft and horny corals. Examples: *Gorgonia, Renilla, Alcyonium.*

Phylum _____

Subphylum _____

Genus _____

Name _____

Date _____

Section _____

Hydra

1. What is meant by the term "polymorphism"? _____

    Is hydra polymorphic? _____

2. Describe feeding and digestion in hydras, explaining how extracellular digestion differs from intracellular.

    _____

    _____

    What is the function of nutritive-muscular cells? _____

3. How does hydra respire? _____

    How does hydra excrete nitrogenous wastes? _____

4. Cnidarians are "diploblastic." What does this term mean? _____

5. Describe and contrast sexual and asexual reproduction in hydra. _____

    _____

    Is the species of hydra you used for behavioral observations monoecious or dioecious? _____

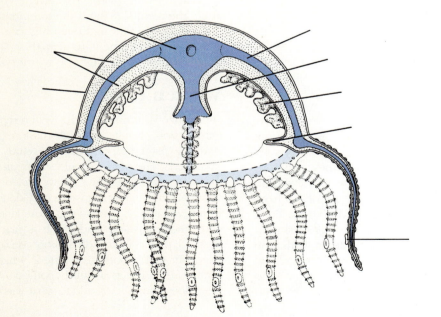

Gonionemus. Diagrammatic oral-aboral section—to be labeled.

EXPERIMENTING IN ZOOLOGY[7,8]
Predator Functional Response:
Feeding Rate in *Hydra*

An important property of the predator-prey relationship is that the behavior of predators often changes as the availability of prey in the environment changes. One aspect of such behavior is known as the **functional response,** in which a predator changes its feeding rate as prey density changes (Hurd and Rathet, 1986). This measure of predation efficiency is important not only to understanding the behavioral ecology of predators but also to evaluating the ability of predators to control pest populations in biological control programs. The cnidarian, *Hydra,* is unlikely to be a satisfactory candidate for pest control, but it is a good model predator with which to measure functional response.

The obvious null hypothesis to be tested here is the following: Functional response is not affected by changing prey density. What is the most likely alternative hypothesis?

The most convenient prey to use for this experiment is *Artemia,* the brine shrimp. Brown hydra *(Hydra littoralis)* readily feed on the first stage nauplii of brine shrimp but may have trouble capturing later developmental stages as the shrimp increase in size. Therefore, brine shrimp should be hatched no earlier than 1–2 days before they are needed. Functional response rate will be recorded as the number of prey eaten during a 10-minute period as prey density increases.

How to Proceed

Assemble the materials you will need for this experiment: a dissecting microscope, depression slides, disposable pipettes, a Petri dish, a clock or stopwatch, pond water, brown hydra, and *Artemia* nauplii.

Fill a depression slide 2/3 with pond water and place a living hydra in the depression. Do not use a coverslip. With a pipette, place a few brine shrimp nauplii in the Petri dish. Put in only enough to count accurately. When the hydra appears adjusted to its new environment and is fully extended, use a pipette to transfer an exact number of shrimp (1, 2, 3, 4, 6, 8, 10, or 12) from the Petri dish to the depression slide and note the time. At the end of 10 minutes, record the number of shrimp that have been killed by the hydra. Begin with one shrimp for the first trial. Then, with a new hydra, repeat the procedure with two shrimp, then with three, and so on through the series of prey densities, ending with

12 shrimp. *Use a new hydra with each trial.* Add your data to a master sheet for the whole class.

Results and Discussion

The class data represent replicates, so you can calculate the mean and standard deviation for each prey density. Plot x = prey density and y = mean predation rate. How does the functional response of hydra change with increasing prey density? Does the curve level off or continue increasing for each successive prey density? Does the functional response variability, as reflected in the standard deviation, change with changing prey density? Why would results be variable among replicates (different students, different hydra, etc.)? What is the value of replication in such experiments?

Extending the Experiment (Optional)

Other experimental variables to consider might be increased volume of water, which might be expected to slow predation rate by reducing spatial density of brine shrimp and making it easier for the shrimp to avoid being stung. Another variable is the size of the hydra. If budding hydra are available, test them against nonbudding specimens to see if feeding rate is influenced by asexual reproduction. Why might you expect this to affect the results? What if the temperature at which you ran the experiment had been higher or lower? Can you design an experiment to test this? Can you think of other variables that might be worth testing?

Written Report

Summarize your findings in a written report organized into introduction, results, and discussion. The introduction should explain the purpose of the study and state the null hypothesis. The results should summarize the data and tell the reader what the data show. Include a table with the class data and a graph showing prey density plotted against mean predation rate. Note that tables and graphs have legends and each table and graph is placed on a separate sheet of paper. In the discussion consider the following: Was your null hypothesis rejected? What does rejection of (or failure to reject) a null hypothesis mean biologically? How does your finding add to what was known?

Reference

Hurd, L. E., and I. H. Rathet. 1986. Functional response and success in juvenile mantids. *Ecology* **7:**163–167.

---

[7]Materials and method for implementing this exercise are found in Appendix A, p. 282–283.

[8]This exercise was contributed by Lawrence Hurd, Washington & Lee University.

The Flatworms
Phylum Platyhelminthes

The flatworms are members of the phylum Platy-helminthes (Gr. *platys,* flat, + *helmins,* worm). This large and economically important group contains both free-living forms, such as the common planarians, and parasitic forms, such as tapeworms and flukes. Flatworms are more complex in organization than the radiate animals in several ways: (1) They exhibit **bilateral symmetry,** with a distinct head and associated sense organs that allow forward, directed movement. (2) They have a **third germ layer—mesoderm**—that arises between the ectoderm and endoderm. This creates the triploblastic condition as distinguished from the two-germ-layer, or diploblastic, condition of the radiate animals. The mesoderm serves as a source for many tissues, organs, and systems. (3) Their **excretory system** is made up of specialized flame cells and tubules for the removal of nitrogenous wastes. (4) They possess a highly organized nervous system, with nervous tissue and sense organs concentrated in the anterior end (cephalization).

EXERCISE 7A
Class Turbellaria—Planarians
Dugesia
Projects and Demonstrations

EXERCISE 7B
Class Trematoda—Digenetic Flukes
Clonorchis, the Liver Fluke of Humans
Schistosoma, the Human Blood Fluke

EXERCISE 7C
Class Cestoda—Tapeworms
Taenia or *Dipylidium*
Projects and Demonstrations

Classification: Phylum Platyhelminthes

EXPERIMENTING IN ZOOLOGY
Planaria Regeneration Experiment

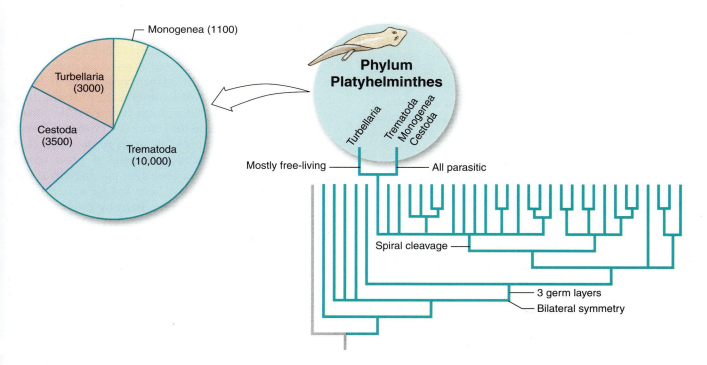

The flatworms are **acoelomate** (Gr. *a*, not, + *koilōma*, cavity) animals, meaning they have no coelom. A coelom is a cavity lying within the mesoderm. In flatworms, the mesodermal space is filled with muscle fibers and a loose tissue called parenchyma, rather than a cavity. The radiate animals, considered in Exercise 6, also lack a coelom but are not considered acoelomate because they do not possess mesoderm.

EXERCISE 7A
Class Turbellaria—Planarians

Core Study

Dugesia

Phylum Platyhelminthes
    Class Turbellaria
        Order Tricladida
            Family Planariidae
                Genus *Dugesia*

Where Found

Freshwater triclads,[1] or planarians, are found on the underside of stones or submerged leaves or sticks in freshwater springs, ponds, and streams. They are often confused with leeches, which they resemble in color and somewhat in shape. There are 9 genera and more than 30 species of planarians in North America. Common and widely distributed species are the brown planarians, *Dugesia*[2] *tigrina* and *Dugesia dorotocephala*, which are adapted to warm and standing (or slowly moving) water. Black planarians, *Dendrocoelopsis*[3] *vaginata*, are found west of the Continental Divide.

Observation of Live Planarians

☞ Using a small artist's brush, place a live planarian on a ringed slide, depression slide, or Syracuse dish in a drop of culture water. Replace water as it evaporates. Keep the surrounding glass dry to prevent the animal from wandering out of range. By holding the slide above a mirror, you can also observe its ventral side.

External Structure

☞ Observe the animal's **anterior, posterior, dorsal, and ventral** aspects.

---

[1] Triclads, members of the order Tricladida (Gr. *treis*, three, + *klados*, branched), are so named for the three-branched intestine characteristic of all planarians.

[2] The genus *Dugesia* was named to honor Antoine-Louis Dugès, a nineteenth-century French zoologist and physician.

[3] *Dendrocoelopsis* derives from the Greek *dendron*, stick, + *koilos*, hollow.

Note the triangular **head.** Its earlike **auricles** (Figure 7.1) bear many sensory cells, but they are tactile and olfactory, not auditory, in function. Are the **eyes** movable? _____ Do they have lenses? _____ Note the pigmented **skin.** Is it the same color on the underside? _____ Is its coloring protective? _____ Are the length and breadth of the worm constant? _____ Holding the slide for a moment over a bright light, can you locate the muscular **pharynx** along the midline? When the animal feeds, the pharynx can be protruded through a ventral **mouth** opening. Can you verify this by using the mirror?

☞ Use a hand lens, a dissecting microscope, or the low power of a compound microscope to further examine the eyes and body surface.

Locomotion. Observe the animal's gliding movement. Glands in its ciliated epidermis secrete a path of mucus on which the planarian propels itself by means of its cilia. Do you think cilia alone are responsible for its movement? _____ What do you think causes the waves of contraction along its body? _____ Does the animal ever leave the drop of water and travel on the dry glass? _____ Why? _____ How does it use the head and auricles? _____ Does it ever move backward? _____

Reactions to Stimuli

☞ Observe the responses of planarians to touch **(thigmotaxis),** food **(chemotaxis),** and light **(phototaxis)** by doing some of the following simple experiments.

1. *Response to touch.* *Very gently* touch the outer edges of the worm with a piece of lens paper or a soft brush. What parts of its body are most sensitive to touch? Are its reactions more localized or less localized than those of the hydra?
2. *Response to food.* To observe the pharynx, stick a *very small* bit of fresh beef liver (or cut-up mealworms) on the center of a coverslip. Invert the coverslip over a deep depression slide containing one or two planarians in pond water. Examine under a dissecting microscope. If the planarians have not been fed for several days, they will soon find the meat by gliding upside down across the coverslip; you will then be able to watch them extend the pharynx to feed.
3. *Response to directional illumination.* Direct a strong beam of light at the planarians from one side of the dish and observe their movements. Move the light 90° around the dish and direct it again at the planarians. Do you think their movements indicate a "trial and error" response or a directed response? _____

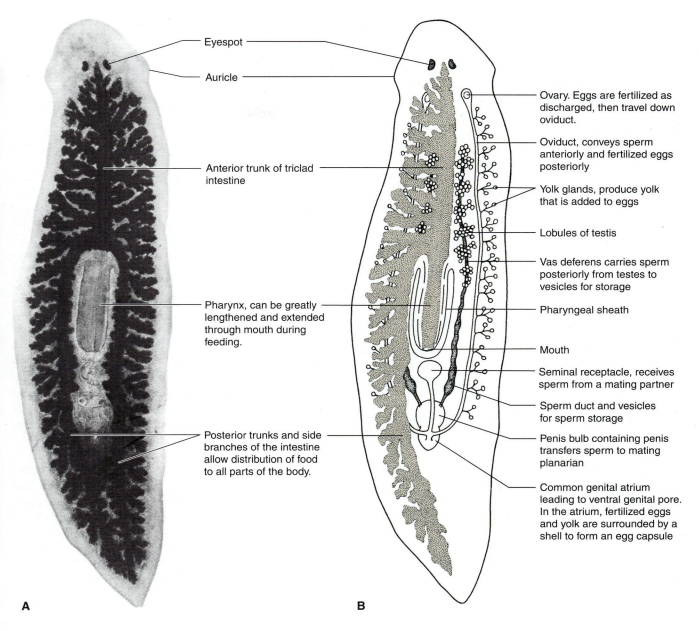

Eyespot

Auricle

Anterior trunk of triclad intestine

Pharynx, can be greatly lengthened and extended through mouth during feeding.

Posterior trunks and side branches of the intestine allow distribution of food to all parts of the body.

Ovary. Eggs are fertilized as discharged, then travel down oviduct.

Oviduct, conveys sperm anteriorly and fertilized eggs posteriorly

Yolk glands, produce yolk that is added to eggs

Lobules of testis

Vas deferens carries sperm posteriorly from testes to vesicles for storage

Pharyngeal sheath

Mouth

Seminal receptacle, receives sperm from a mating partner

Sperm duct and vesicles for sperm storage

Penis bulb containing penis transfers sperm to mating planarian

Common genital atrium leading to ventral genital pore. In the atrium, fertilized eggs and yolk are surrounded by a shell to form an egg capsule

A

B

Figure 7.1

Planaria, a freshwater turbellarian. **A,** Stained whole mount. **B,** Internal anatomy; half the anterior trunk of the intestine is removed to reveal the testes.

Written Report

In the lab report on p. 90, report your observations and the conclusions you drew from observing the live planarians.

Observation of Stained Whole Mounts

The stained whole mount of a planarian shows an animal that was fed food mixed with India ink, carmine, or some other suitable stain before being killed and fixed, resulting in a darkly stained gastrovascular tract.

Using a dissecting microscope or the low power of a compound microscope, study stained whole mounts of a freshwater planarian and the marine *Bdelloura* (de-lur′a), identifying the structures listed in the following sections.

Digestive System. As in cnidarians, the digestive tract of a turbellarian is a **gastrovascular cavity,** the branches of which fill most of the body. Because there is no anus, undigested food is ejected through the mouth.

The muscular **pharynx** is enclosed in a **pharyngeal sheath,** but its free end can be extended through the ventral **mouth** (Figure 7.1B). Ingestion occurs through the muscular sucking action of the pharynx. The pharynx opens into the intestine, which has one **anterior trunk** and two **posterior trunks,** one on each side of the pharynx. What might be an advantage to the digestive process of the branching diverticula that extend laterally from the intestinal trunks? ——————— Some digestion may occur within the lumen of the digestive cavity by means of enzymes secreted by intestinal gland cells **(extracellular digestion).** As in the cnidarians, digestion is completed within phagocytic cells of the gastrodermis **(intracellular digestion).**

Reproduction. Flatworms are monoecious, and their reproductive system is complex (Figure 7.1B). Because of the large digestive tract, most reproductive organs of turbellarians are obscured on stained whole mounts. However, the penis and genital pore may be seen on the *Bdelloura* slides. The functional anatomy of the planarian is illustrated in Figure 7.1.

Planarians and other freshwater turbellarians also reproduce asexually by transverse fission; in some species, chains of zooids are formed asexually.

Excretion and Osmoregulation. The excretory system, consisting of **excretory canals** and **protonephridia (flagellated flame cells),** cannot be seen in whole mounts; see your text for a discussion of this system. The instructor may prepare a demonstration of living flame cells from planaria (see Projects and Demonstrations later in this section). The main function of the protonephridial system may be regulation of the animal's internal fluid content (osmoregulation). The system is often absent in marine turbellarians.

Nervous System. *Bdelloura* is a marine turbellarian that lives as a commensal on the external surface of the horseshoe crab, *Limulus.* Properly stained slides show the **ladder type of nervous system,** as well as the triclad digestive tract.

On a stained whole mount of *Bdelloura,* find the **cerebral ganglia** at the anterior ends of the two lateral **nerve cords. Transverse nerves** connecting the cords and **lateral nerves** extending outward from the cords form the "rungs" of the ladder.

Sense Organs. A pair of **eyespots,** or **ocelli,** are light-sensitive pigment cups (Figure 7.1A). Chemoreceptive and tactile cells are abundant over the body surface, especially on the auricles.

Drawings

1. On separate paper, sketch the external morphology of a planarian and label fully.
2. The outline of *Bdelloura* on p. 91 contains an outline of the digestive system. Draw in and label the cerebal ganglia, nerve cords, and transverse and lateral nerves.

Further Study

Transverse Sections of a Planarian

The appearance of a cross section will depend on whether it is cut from the anterior, middle, or posterior part of the planarian (Figure 7.2).

The **epidermis** of cuboidal epithelial cells (derived from ectoderm) contains many dark, rodlike **rhabdites** (Gr. *rhabdos,* rod), which, when discharged in water, swell and form a protective mucous sheath around the body. Epithelial cells are ciliated only on the animal's ventral surface. What function do they serve? ——————————————————

Inside the epidermis is a layer of **circular muscles** and then a layer of **longitudinal muscles** (cut transversely and appearing as dark dots). **Dorsoventral muscle fibers** are also visible, particularly at the sides of the animal. Do these muscles explain the waves of contraction you saw in the living animal? ——————— Note that there is *no body cavity.* Flatworms are called **acoelomate** animals. **Parenchyma** (pair-en'ka-ma; Gr., anything poured in beside), largely of mesodermal origin, is the loose tissue filling up space between the organs.

Several hollow sections of the intestine and its diverticula (derived from endoderm) may be seen, depending on the location of the section. What kind of epithelial cells make up the intestinal walls? ——————— In a middle section, note the thick, circular **pharynx,** covered and lined with epithelium and containing layers of circular and longitudinal muscle similar to those in the body wall. The **pharyngeal chamber** is also lined with epithelium.

Nerve cords, reproductive and excretory ducts, testes, and ovaries are found in the parenchyma of adult animals, but they are difficult to identify.

Projects and Demonstrations

Observing Flame Cells

Flame cells of the excretory system are difficult to see in living planarians because they are usually obscured by pigment in the integument. To demonstrate flame cells, compress the body (or a sectioned portion of the body) of a planarian between a slide and coverslip to crush and partially disperse the tissues. Search the tissue debris with the high power of a compound microscope. Flame cells appear as a rapidly flickering movement of the ciliated tuft, resembling a wavering current of water. Once located (often several flame cells are found close together), switch to oil immersion for closer study.

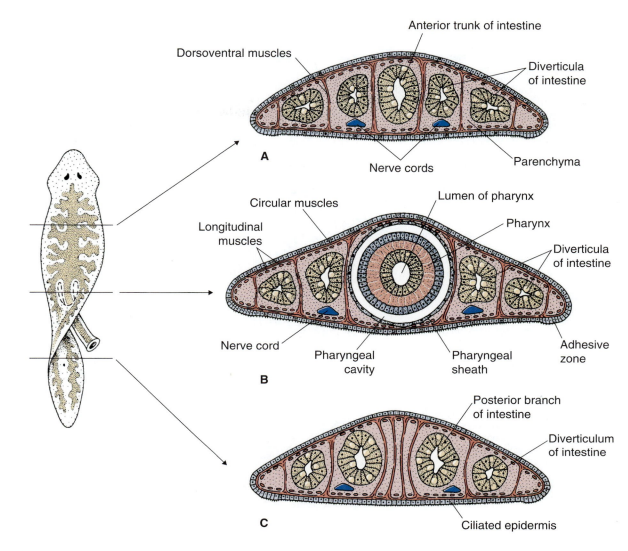

Figure 7.2
Cross sections of a planarian (diagrammatic). **A,** Anterior section. **B,** Section through the region of the pharynx. **C,** Posterior section.

EXERCISE 7B

Class Trematoda— Digenetic Flukes

Core Study

Clonorchis, the Liver Fluke of Humans

Phylum Platyhelminthes
    Class Trematoda
        Order Digenea
            Species: *Clonorchis sinensis* (human liver fluke)

Where Found

All trematodes are parasitic, harbored in or on a great variety of animals. Many of them have three different hosts in their life cycle.

The adult, or sexual, stage of *Clonorchis* lives in the human bile duct. It is widely distributed in Japan, Korea, China, Taiwan, and Vietnam, where it causes widespread suffering and economic loss. Prevalence ranges from 14% in cities such as Hong Kong to 80% in some rural areas. The asexual, or larval, stages are found in aquatic snails and fishes. Approximately 30 million people are infected with the fluke by recent (2009) estimates, making it the third most prevalent worm parasite of humans.

Study of a Stained Whole Mount

☞ Study a stained slide of an adult fluke, first with a hand lens, and then with the low power of the compound microscope.

How long is the specimen? _____ Compare it with the planarian in size and shape. Note the **oral**

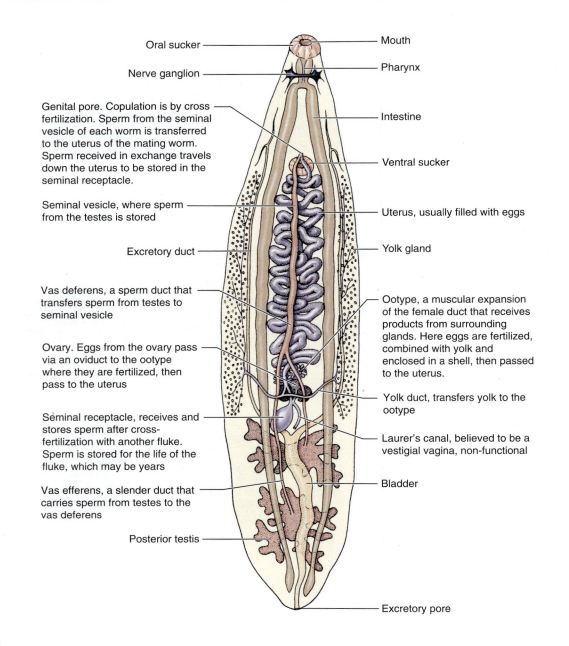

Oral sucker ———— **Mouth**

Nerve ganglion ———— **Pharynx**

Genital pore. Copulation is by cross fertilization. Sperm from the seminal vesicle of each worm is transferred to the uterus of the mating worm. Sperm received in exchange travels down the uterus to be stored in the seminal receptacle. ———— **Intestine**

———— **Ventral sucker**

Seminal vesicle, where sperm from the testes is stored ———— **Uterus,** usually filled with eggs

Excretory duct ———— **Yolk gland**

Vas deferens, a sperm duct that transfers sperm from testes to seminal vesicle ———— **Ootype,** a muscular expansion of the female duct that receives products from surrounding glands. Here eggs are fertilized, combined with yolk and enclosed in a shell, then passed to the uterus.

Ovary. Eggs from the ovary pass via an oviduct to the ootype where they are fertilized, then pass to the uterus

———— **Yolk duct,** transfers yolk to the ootype

Seminal receptacle, receives and stores sperm after cross-fertilization with another fluke. Sperm is stored for the life of the fluke, which may be years ———— **Laurer's canal,** believed to be a vestigial vagina, non-functional

Vas efferens, a slender duct that carries sperm from testes to the vas deferens ———— **Bladder**

Posterior testis ———— **Excretory pore**

Figure 7.3
Clonorchis, the liver fluke of humans.

sucker at the anterior end and the **ventral sucker (acetabulum)** on the ventral surface (Figure 7.3). What is their function?

Body Wall. The body covering of both flukes and tapeworms was formerly believed to be a nonliving cuticle, but the electron microscope reveals it to be a living syncytial **tegument** (L. *tegumentum,* to cover) consisting of cytoplasmic processes of cells that dip down into **parenchyma,** a meshwork of cells derived from mesoderm.

Circular and longitudinal **muscle layers** of the body wall are similar to those of turbellarians, and body spaces are filled with parenchyma.

Digestive System. The **mouth** lies anteriorly in the oral sucker. It leads to a muscular **pharynx** and a short **esophagus** that divides into two lateral branches of the digestive tract. The sucker serves for attachment and for the abrasion of the bile duct in which the fluke lives. Food—blood cells and lacerated cells from the inflamed bile duct—is aspirated by the muscular pharynx. There is no anus.

Protonephridial System. The **excretory pore** at the posterior end is the outlet of the **bladder.** Follow the bladder forward to see where it divides into two long tubules. The tubules connect to flame cells, which you will not be able to see.

Reproduction and Life Cycle. Flukes are monoecious, each animal having both male and female reproductive systems (Figure 7.3). Locate the testes and male reproductive organs and follow the passage of sperm from testes to genital pore as shown in Figure 7.3. Similarly, while referring to Figure 7.3, follow the movement of eggs from ovary to ootype, where the eggs are fertilized and surrounded with a shell, then passed to the uterus.

Eggs containing ciliated larvae are shed into the bile, carried to the host's intestine, and voided with the feces. If feces are passed into water, the larvae hatch and are eaten by snails, in which they multiply asexually through additional larval stages before leaving the snail as tadpole-like **cercariae.** If successful in finding a fish host, the cercariae bore into the muscle, where they encyst. Humans (and other mammals) are infected when they eat raw or undercooked fish bearing the cysts. Young flukes then emerge, travel to the bile ducts, and mature into adults. Consult your textbook for details of this animal's life cycle. Why do you think this parasite is prevalent in Southeast Asia?

Schistosoma, the Human Blood Fluke

Phylum Platyhelminthes
   Class Trematoda
      Subclass Digenea
         Order Strigeata
            Family Schistosomatidae
               Genus: *Schistosoma*
                  Species: *Schistosoma mansoni*
                  (or *Schistosoma haematobium*)

Where Found

Schistosomes of the genus *Schistosoma* are blood flukes of humans that affect an estimated 200 million people in Asia, Africa, the Caribbean (including Puerto Rico), and northeastern South America. As a major global health problem, **schistosomiasis** is exceeded only by malaria. In many areas, the disease is commonly known as "bilharzia" rather than the more correct schistosomiasis.

Three species of schistosomes are of enormous medical significance: *Schistosoma mansoni* lives primarily in the venules draining the large intestine; *S. haematobium* lives in the venules of the urinary bladder; *S. japonicum* inhabits the venules of the small intestine. All three species cling to the venule walls with their suckers and feed on blood.

The life cycle of *Schistosoma* is similar in all species (Figure 7.4). If discharged eggs from human feces or urine get in water, they hatch as ciliated **miracidia.** If they find the right snail, within a few hours they burrow in and transform into a saclike form, the **sporocyst.** These multiply asexually to produce another generation of sporocysts, within each of which develop numerous **cercariae.** Successive sexual generations of the sporocysts and the

cercariae they contain over a period of several months ensure an enormous increase in numbers: each miracidium may give rise to more than 200,000 cercariae.

Cercariae escape from the snail and swim about until they contact the bare skin of a human. They penetrate the skin, enter the circulatory system, and make their way to the hepatic portal system, where they develop before migrating to their characteristic sites. Here they mate with another worm, and the female begins producing eggs.[4]

Study of Prepared Slides

☞ Examine a slide of *Schistosoma mansoni* adults in copula.

Schistosomes are an exception to the rule that flatworms are monoecious. In fact, an odd feature of this genus is the strong sexual dimorphism (Figure 7.4, *inset*). Males are stouter than females and have a ventral longitudinal groove, the **gynecophoric canal** (gi'ne-ka-fore'ik; Gr. *gyne,* woman, + *pherein,* to carry). (The name *Schistosoma,* meaning "split body," refers to this canal.) Even more oddly, the thinner female normally resides there, permanently embraced by the male. Note the strong oral sucker and a secondary sucker near the anterior end called the **acetabulum.**

☞ Examine a slide of schistosome eggs.

Note the elliptical shape of the eggs, each of which bears a sharp spine (the spine is terminal in *S. haematobium* and lateral in *S. mansoni;* the eggs of *S. japonicum* lack a spine). The main ill effects of schistosomiasis result from the eggs, which, as they work their way out of the venules where the adults live, cause ulceration, abscesses, bloody diarrhea, and abdominal pain.

Calcified eggs of *S. haematobium* have been found in Egyptian mummies dating from 1200 B.C. There is a well-reasoned hypothesis that the curse Joshua placed on Jericho (after destroying the city and killing all of its inhabitants [Joshua 6:26]) was the introduction of *S. haematobium* into the communal well. The curse was removed after Jericho was abandoned and subsequent droughts killed the snail host. Today the people of Jericho (Ariha, in Jordan) are free of schistosomiasis.

Of Egypt's population, 30% to 40% suffered from the disease at the time of Napolean's invasion (1799–1801), and the disease became prevalent in his troops. It is still a problem today in Egypt. It is estimated that half the population of Egypt now has the infection.

---

[4]Although the life cycle is similar in all species of *Schistosoma,* it differs from that of other digeneans in that the cercariae penetrate the definitive host directly. There is no second intermediate host (unlike *Clonorchis,* for example, in which there are two intermediate hosts, a snail and a fish).

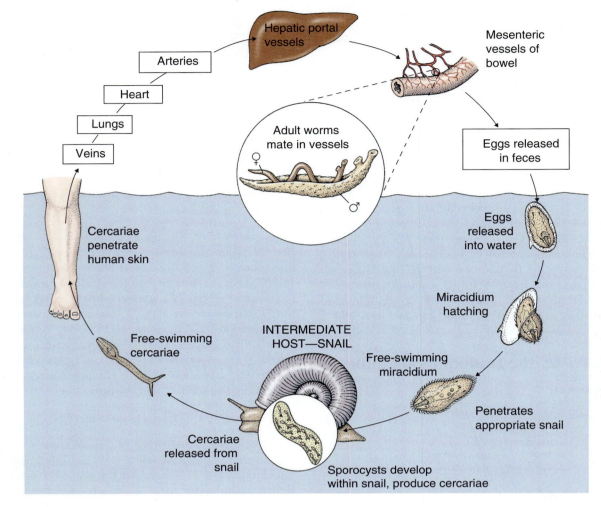

DEFINITIVE HOST—HUMAN

Hepatic portal vessels

Arteries

Heart

Lungs

Veins

Mesenteric vessels of bowel

Adult worms mate in vessels

Eggs released in feces

Cercariae penetrate human skin

Eggs released into water

Miracidium hatching

Free-swimming cercariae

INTERMEDIATE HOST—SNAIL

Free-swimming miracidium

Penetrates appropriate snail

Cercariae released from snail

Sporocysts develop within snail, produce cercariae

Figure 7.4
Life cycle of *Schistosoma mansoni*.

EXERCISE 7C
Class Cestoda—Tapeworms

Taenia or *Dipylidium*

Phylum Platyhelminthes
  Class Cestoda
    Order Cyclophyllidea
      Species *Taenia pisiformis* (dog tapeworm) or
      Dipylidium caninum (small dog tapeworm)

Tapeworms are all endoparasitic and show extreme adaptations for their parasitic existence. For example, they lack a digestive system and there is great emphasis on reproduction. Most require two hosts of different species, with the adult tapeworm characteristically living in the digestive tract of a vertebrate.

As their name implies, tapeworms are ribbon-like, and their long bodies (the **strobila**) are usually made up of segments called **proglottids** (Gr. *proglottis,* tongue tip). (Proglottids are not to be confused with metamerism in higher forms; they are formed by a continuous process of **budding** from the anterior end, or **scolex** [Gr. *skōlēx,* worm, grub].) As new proglottids are formed anteriorly, the older ones are pushed backward so that the oldest, or most mature, proglottids are always at the posterior end. The scolex, which serves as a holdfast, is usually equipped with **suckers, hooks,** or both. The body covering, or **tegument,** is similar to that of the flukes.

The forms described here are *Taenia pisiformis* (Gr. *tainia,* ribbon; L. *pisum,* pea, + *forma,* shape), a dog or cat tapeworm whose larval stage is found in the liver of rabbits (Figure 7.5), and *Dipylidium caninum* (Gr. *di,* two, + *pyle,* entrance, + *idion,* dim. suffix; L. *caninus,* belonging to a dog), a small tapeworm of dogs or cats (Figure 7.6), with fleas as the alternate host. Other forms may be substituted in the laboratory. Among those

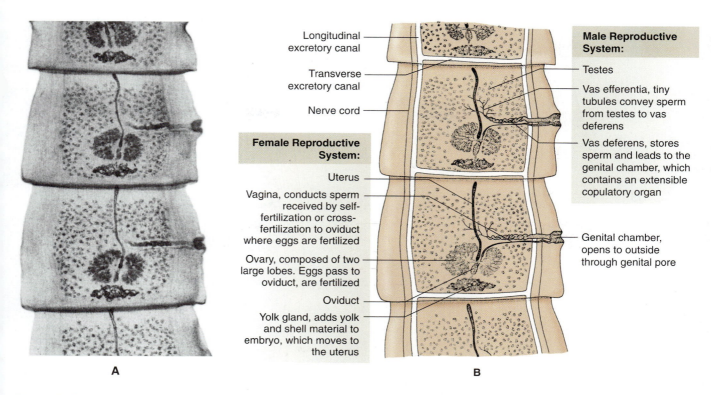

Longitudinal excretory canal

Transverse excretory canal

Nerve cord

Male Reproductive System:

Testes

Vas efferentia, tiny tubules convey sperm from testes to vas deferens

Vas deferens, stores sperm and leads to the genital chamber, which contains an extensible copulatory organ

Female Reproductive System:

Uterus

Vagina, conducts sperm received by self-fertilization or cross-fertilization to oviduct where eggs are fertilized

Ovary, composed of two large lobes. Eggs pass to oviduct, are fertilized

Oviduct

Yolk gland, adds yolk and shell material to embryo, which moves to the uterus

Genital chamber, opens to outside through genital pore

A   B

Figure 7.5

Photomicrograph and interpretive drawing of *Taenia pisiformis,* the dog tapeworm. The intermediate host is a rabbit. Genital pores may be oriented toward either right or left side.

that parasitize humans are *Taenia solium,* the larval stage of which encysts in the muscle of pigs and which, like *Taenia pisiformis,* possesses both hooks and suckers; *Taenia saginata,* which has cattle as the alternate host and has no hooks; and *Diphyllobothrium latum,* whose larval stage is found in crustaceans and fishes. *D. latum* is the largest of all human cestodes, some reaching a length of 20 m (65 ft)! Humans become infected by eating uncooked freshwater fish or uncooked salmon.

General Structure

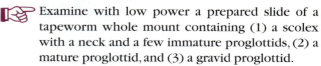 Study a preserved whole specimen or one embedded in plastic.

How long is the specimen? _____ Examine the **scolex** under a dissecting microscope and identify the **hooks** and **suckers.** Note the **neck,** from which new proglottids are budded off. As older proglottids are pushed back by the addition of new ones, they mature and become filled with reproductive organs. Can you distinguish maturing from very young proglottids by the presence of a **genital pore** at one side? As eggs develop and become fertilized, the proglottid becomes distended (gravid), with the uterus filled with embryos. Where do you find **gravid proglottids** (Figure 7.6)? _____ Such proglottids soon break off and are shed in the feces of the host. Ingested by a suitable alternate host, young

embryos of *T. pisiformis* migrate to the liver and encyst there as cysticerci (bladder worms). Eggs of *D. caninum* are eaten by larval fleas that, when mature are nipped or licked out of the fur of a dog or cat.

Microscopic Study

Examine with low power a prepared slide of a tapeworm whole mount containing (1) a scolex with a neck and a few immature proglottids, (2) a mature proglottid, and (3) a gravid proglottid.

Scolex. Note the **suckers** (how many? _____) and a **rostellum** bearing a circle of **hooks** (Figure 7.7). What is the function of hooks and suckers? _____ Is there a mouth? _____ Is the **neck** segmented? _____ Two lightly stained tubes, one on each side, are **excretory canals,** which extend the entire length of the animal. Some of the **immature proglottids** have developed **genital pores.**

Mature Proglottid

A mature proglottid contains complete hermaphroditic reproductive organs (both male and female). From your slide locate the genital pore (two pores in *D. caninum*), testes, vas deferens, vagina, ovaries, oviducts, yolk gland, and uterus. Study the functional anatomy of the

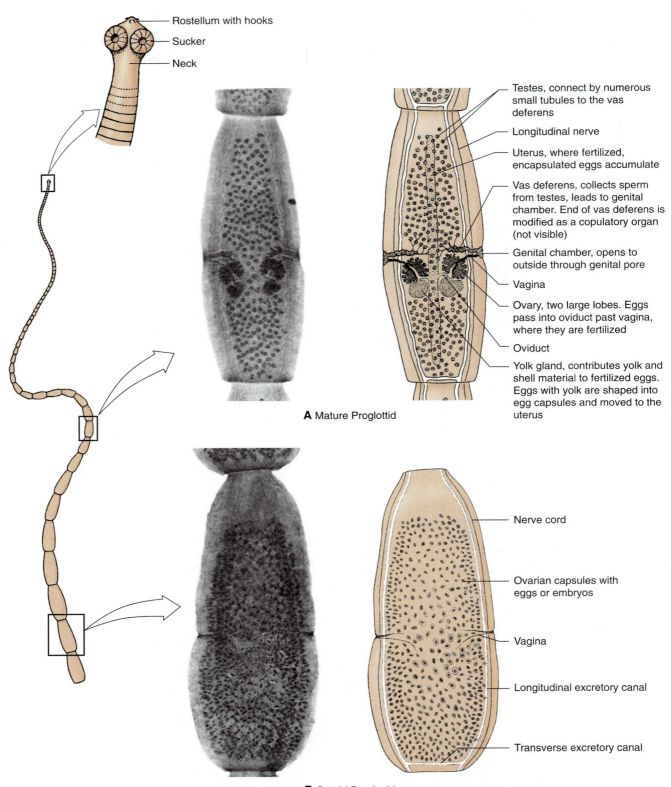

Rostellum with hooks

Sucker

Neck

Testes, connect by numerous small tubules to the vas deferens

Longitudinal nerve

Uterus, where fertilized, encapsulated eggs accumulate

Vas deferens, collects sperm from testes, leads to genital chamber. End of vas deferens is modified as a copulatory organ (not visible)

Genital chamber, opens to outside through genital pore

Vagina

Ovary, two large lobes. Eggs pass into oviduct past vagina, where they are fertilized

Oviduct

Yolk gland, contributes yolk and shell material to fertilized eggs. Eggs with yolk are shaped into egg capsules and moved to the uterus

A Mature Proglottid

Nerve cord

Ovarian capsules with eggs or embryos

Vagina

Longitudinal excretory canal

Transverse excretory canal

B Gravid Proglottid

Figure 7.6
Dipylidium caninum, the small tapeworm of dogs and cats. The alternate host is a flea.

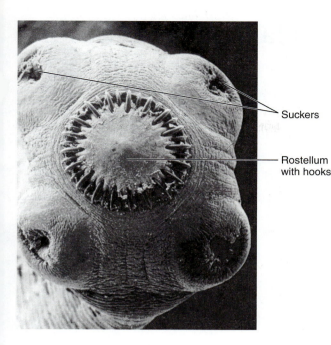

Figure 7.7

Scolex of *Taenia* sp., showing four suckers and a circlet of hooks on the rostellum. SEM × 125.

reproductive systems of *T. pisiformis* and *D. caninum* as explained in Figures 7.5 and 7.6.

Note **excretory canals** on each side of the proglottids of both tapeworm species. They are part of the protonephridial system and empty to the outside at the posterior end of the worm. They are connected by a **transverse canal** across the posterior margin of each proglottid.

Nerve cords run just lateral to the excretory ducts. The tapeworm nervous system resembles that of planarians and flukes, although it is less well developed.

Do you find any digestive organs in the tapeworm? _____ Why? _____ What is a tapeworm's food? _____ Where is it digested? _____ How does the worm obtain it? _____ How is the structure of tapeworms adaptive for their environment? _____ What control methods can be used against pork and beef tapeworms? _____

Gravid Proglottid

☞ Examine a **gravid proglottid.** What structures can you identify besides the distended **uterus?** If possible, study the **eggs** with high power. - (CAUTION! Be very careful when changing to high power because whole-mount slides are thicker than average.)

Can you distinguish in any eggs a six-hooked larval form called an **onchosphere** (Gr. *onkinos,* hook, + *sphaira,* globe)? _____

Drawings

✎ On separate paper, draw and label the scolex and neck, a mature proglottid, and a gravid proglottid. Include the scale of the drawings, and be sure to identify the species you are drawing.

Projects and Demonstrations

1. *Slides of cross sections of tapeworm.* Identify the tegument, parenchyma, excretory ducts, nerve cords, uterus, and other reproductive organs that appear in section.

2. *Preserved specimens or prepared slides.* Examine (a) preserved specimens or prepared slides of various species of tapeworms and (b) prepared slides of ova and cysts. Look for features described in this exercise.

Classification: Phylum Platyhelminthes

Class Turbellaria (tur′bel-lar′e-a; L. *turbellae* [pl.], stir, bustle, + *aria,* like or connected with). Turbellarians. Mostly free-living, with a ciliated epidermis. A paraphyletic grouping. Example: *Dugesia tigrina.*

Class Monogenea (mon′o-gen′e-a; Gr. *mono,* single, + *gene,* origin, birth). Monogenetic flukes. Adult body covered with syncytial tegument without cilia; leaflike to cylindrical in shape; posterior attachment organ with hooks, suckers, or clamps, usually in combination; all parasitic, mostly on skin or gills of fishes; single host; monoecious; usually free-swimming ciliated larva. Examples: *Polystoma, Gyrodactylus.*

Class Trematoda (trem′a-to′da; Gr. *trematodes,* with holes, + *eidos,* form). Digenetic flukes. Adult body covered with nonciliated syncytial tegument; leaflike or cylindrical in shape; usually with oral and ventral suckers, no hooks; development indirect, first host a mollusc, final host usually a vertebrate; parasitic in all classes of vertebrates. Examples: *Fasciola, Clonorchis, Schistosoma.*

Class Cestoda (ses-to′da; Gr. *kestos,* girdle, + *eidos,* form). Tapeworms. Adult body covered with nonciliated, syncytial tegument; scolex with suckers or hooks, sometimes both, for attachment; long, ribbon-like body, usually divided into a series of proglottids; no digestive organs; parasitic in digestive tract of all classes of vertebrates; first host may be invertebrate or vertebrate. Examples: *Taenia, Diphyllobothrium.*

Name_____

Date_____

Section_____

Planarian Behavior

Locomotion

Response to Touch (Thigmotaxis)

Response to Food (Chemotaxis)

Turbellarian Structure

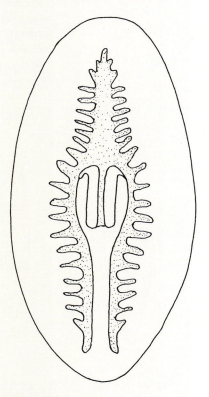

Nervous system of *Bdelloura* (digestive system is already shown).

EXPERIMENTING IN ZOOLOGY
Planaria Regeneration Experiment

Planarians are easy to work with and have remarkable powers of regeneration.

☞ Before starting, assemble your materials: sharp razor blade or scalpel blade, snap-cap vial or screw-cap specimen jar for *each* planarian, clean culture water (pond water or dechlorinated tap water), pipette, small artist's brush, lens paper, and ice cube. Decide how you want to cut the worms (Figure 7.8).

Fold a lens tissue over an ice cube and, with a camel's hair brush, transfer a planarian to the top surface of the ice cube; it will become quiescent almost immediately. Using a dissecting microscope, make the desired cut or cuts with the razor blade. If the worm is being partially split longitudinally, make certain the cut is clean and has completely separated the cut surfaces. Make a sketch in your notebook or on p. 93. With the pipette, rinse the pieces off the lens paper into the vial (previously half-filled with culture water). Label each vial with your name and the date, and prepare a sketch of the cut made. Put in a cool place.

Culture water *must* be changed every 3 or 4 days to keep the planaria healthy. *Do not* feed the animals. Remove any dead pieces, which will appear grayish or fuzzy. Additional tips for this experiment are found in Appendix A, p. 283.

Examine the worms about every 3 days for 2 weeks. To obtain two-headed or two-tailed planarians from the partial cuts shown in the second, third, and fifth sketches in Figure 7.8, you will need to recut them after 12 to 18 hours and probably again at 2 days to prevent the parts from fusing back together. Soon a blastema will form on the cut edge. Is the growth faster at the posterior end of a cephalic piece or the anterior end of a caudal piece? _____ On decapitated pieces, when do new eyes appear? _____ When can you distinguish new auricles? _____ Which is regenerated first, eyes or pharynx? _____ Do you find any evidence of anterior-posterior axis in the manner of growth of these pieces? _____

Drawings

✎ On the record sheet, p. 93, make a sketch of the shape of each piece. Examine the specimens twice a week if possible, each time recording the date and sketching each regenerating piece. Summarize the results on p. 94. In your report, describe (1) your technique, (2) any problems you encountered, (3) potential or real sources of error in the experimental approach, and (4) your results. If you were to repeat the experiment, what would you do differently?

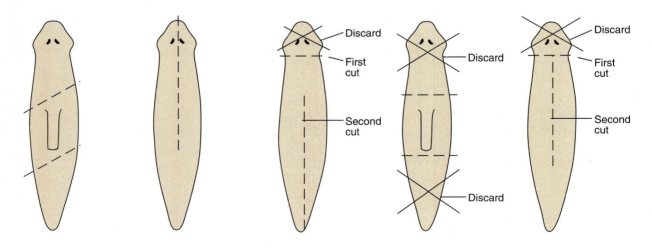

Figure 7.8
Some suggestions for regeneration experiments.

Regeneration Experiment with *Dugesia*

Name_____

Date_____

Section_____

Sketches of original cuts of specimens:

Dates	1	2	3	4	Control

Record of growth:

Dates	1	2	3	4	Control

LAB REPORT

Methods Used

Observations and Conclusions

NOTES

Nematodes and Four Small Protostome Phyla

Phylum Nematoda
Phylum Rotifera
Phylum Acanthocephala
Phylum Nematomorpha
Phylum Gastrotricha

The flatworms (Platyhelminthes) considered in Exercise 7 and the phyla considered in this exercise belong to the Protostomia, a clade that embraces the great majority of invertebrate phyla (see the cladogram on the inside front cover of the manual). The Protostomia are divided in turn into two smaller clades: Lophotrochozoa and Ecdysozoa. These two clades were originally separated using molecular genetic methods (each has a distinctive molecular signature), but each also has distinctive morphological characteristics. Members of the Lophotrochozoa share either an odd, horseshoe-shaped feeding structure, the lophophore, or a particular larval form, the trochophore. Members of the Ecdysozoa possess a cuticle that is shed at intervals as their bodies grow. Both lophotrochozoans and ecdysozoans are represented in the phyla considered in this exercise.

All bilateral animal phyla except the acoelomates possess a **body cavity** belonging to one of two types: (1) **true coelom,** in which a peritoneum (an epithelium of mesodermal origin) covers both the inner surface of the body wall and the outer surface of the visceral organs in the cavity, or (2) **pseudocoel,** a body cavity not entirely lined with peritoneum.

A body cavity of either type is an advantage because it provides room for organ development and storage and allows some freedom of movement within the body. Because the cavity is often fluid-filled, it also provides for a hydrostatic skeleton in those forms lacking a true skeleton.

Four of the five phyla considered in this exercise are pseudocoelomate (the exception, phylum Gastrotricha, is acoelomate). Of these four, phylum Nematoda is by far the largest and most important economically.

EXERCISE 8A
Phylum Nematoda—*Ascaris* and Others
Ascaris, Intestinal Roundworm
Some Free-living Nematodes
Some Parasitic Nematodes
Projects and Demonstrations

EXERCISE 8B
A Brief Look at Some Other Protostomes
Phylum Rotifera
Phylum Acanthocephala
Phylum Nematomorpha
Phylum Gastrotricha

EXERCISE 8A
Phylum Nematoda—*Ascaris* and Others

Core Study

Nematodes are an extensive group with worldwide distribution. They include terrestrial, freshwater, marine, and parasitic forms. They are elongated roundworms covered with a flexible, nonliving cuticle. Circular muscles are lacking in the body wall and, in *Ascaris,* longitudinal muscles are arranged in four groups separated by epidermal cords (some nematodes have six or eight groups of longitudinal muscles). Cilia are completely lacking. Are cilia present in any acoelomates? _____ In cnidarians? _____ Nematodes—both parasitic and free-living—are incredibly abundant. A handful of good garden soil contains thousands of

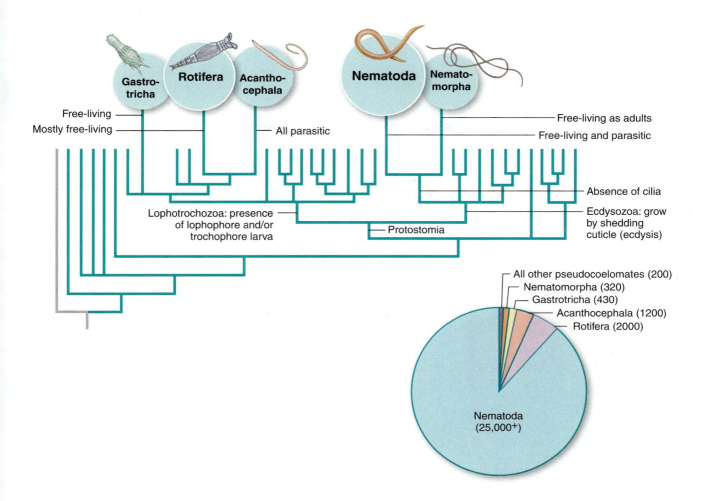

Free-living
Mostly free-living
All parasitic
Free-living as adults
Free-living and parasitic
Absence of cilia
Lophotrochozoa: presence of lophophore and/or trochophore larva
Protostomia
Ecdysozoa: grow by shedding cuticle (ecdysis)

Gastro-tricha
Rotifera
Acantho-cephala
Nematoda
Nemato-morpha

All other pseudocoelomates (200)
Nematomorpha (320)
Gastrotricha (430)
Acanthocephala (1200)
Rotifera (2000)

Nematoda (25,000+)

nematodes. Some 50 species of nematodes occur in humans, most of them nonpathogenic.

Ascaris, Intestinal Roundworm

Phylum Nematoda
    Class Rhabditea
        Order Ascaridida
            Genus *Ascaris*
                Species *Ascaris suum*

Where Found

Ascaris lumbricoides (Gr. *askaris,* intestinal worm) is a common intestinal parasite of humans. *A. suum,* which parasitizes pigs, is so similar to the human parasite that it was long considered merely a different strain of *Ascaris lumbricoides. Ascaris megalocephala* is common in horses.

    Ascaris infections are extremely common—nearly one-quarter of the human population is infected worldwide. These infections were once common in rural areas in the southeastern United States. The present number of infected persons in the United States is unknown but has been estimated to be as high as 4 million. Infections are most common among immigrants, travelers, and refugees. Careless defecation near habitation by individuals

harboring worms seeds the soil with eggs that may remain infective for years. The eggs are extraordinarily resistant to chemicals, remaining viable in 2% formalin and in 50% solutions of the common laboratory acids. Moderate to heavy infections cause malnutrition and underdevelopment in children, and the worms' metabolites may cause immune reactions. Heavy infections can lead to fatal intestinal blockage or cause overcrowded worms to wander into nasal passages; the psychological trauma caused by the latter can only be imagined.

General Features

☞ Place a preserved *Ascaris* in a dissecting pan and cover with water. A word of caution: although the chance of infection from eggs in formalin-preserved female *Ascaris* is remote, be sure to wash your hands after dissecting the worm.

    An *Ascaris* worm can be viewed as a bundle of adaptations that enable it to spend its life contentedly within the intestine of another animal. Females, which run 20 to 40 cm in length, are more numerous and larger than males, which average 15 to 31 cm in length. Males have a curved posterior end and two

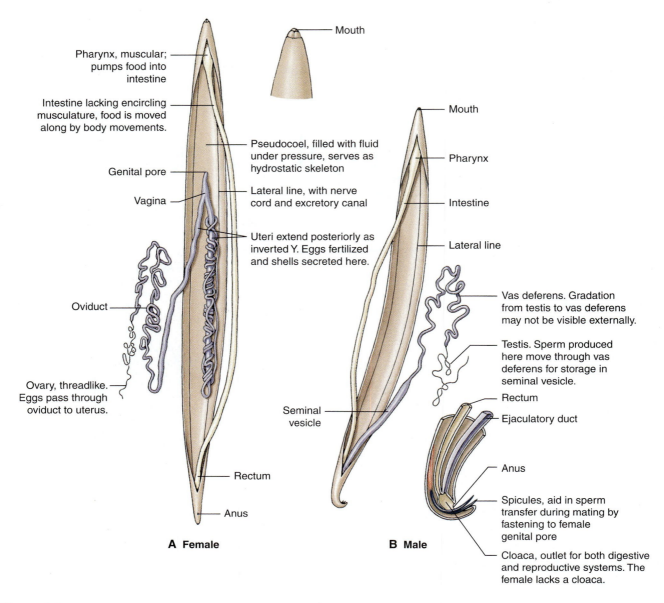

Figure 8.1

Ascaris. **A,** Internal structure of female, with insert showing detail of anterior end. **B,** Internal structure of male, with insert showing detail of cloacal region.

chitinous **spicules** projecting from the anal region. The spicules are used to hold the female's vulva open during copulation. How long is your *Ascaris* specimen? _____ Is the body segmented? _____ Compare your specimen with one of the other sex.

With a hand lens, find the **mouth,** with three **lips,** one dorsal and two ventral (Figure 8.1). Find the ventral **anus** at the posterior end. The anus in a male not only discharges feces from the rectum but also serves as a genital opening. The female genital opening **(vulva)** is located on the ventral side about one-third the length of the body from the anterior end. It may be

hard to distinguish from scars. Use a hand lens or dissecting microscope.

Note the shiny **cuticle** that covers the body wall. It is nonliving and consists primarily of **collagen,** which also is found in vertebrate connective tissue.

Four **longitudinal lines** run almost the entire length of the body—**dorsal** and **ventral median lines** and two **lateral lines.** Dorsal and ventral lines, which indicate the location of bundles of nerve fibers, are very difficult to see on preserved specimens. However, along the lateral lines, the body wall is thinner, and the lines usually appear somewhat transparent. Excretory canals are located inside the lateral lines.

Internal Structure

☞ Select a female specimen, place the worm in a dissecting pan, and cover it with water. Locate the lateral lines, where the body wall seems somewhat thinner. Now find the anus and vulva on the ventral side. This should help you identify the opposite, or middorsal line. Now, with a razor blade, slit open the body wall *along the middorsal line,* being careful to avoid injuring the internal structures. Pin back the body wall to expose the viscera, *slanting the pins outward* to allow room for dissection.

Body Wall and Pseudocoel. Note the body cavity. Why is it called a **pseudocoel?** _____ How does it differ from a true coelom? _____ _____ Note fluffy masses lining the body wall. These are large, nucleated cell bodies of the **longitudinal muscle cells,** whose fibers extend longitudinally in the body wall. With your teasing needle, tease out some of the fibers from the cut edge of the wall. Examine fibers and cells under the microscope. Absence of circular muscles accounts for the thrashing movements of these animals. Note the absence of muscle cells along the **lateral lines.**

Excretory System. Excretory canals located in the lateral lines unite just back of the mouth to empty ventrally through an **excretory pore.** The canals are largely osmoregulatory in function. Excretion also occurs through the cuticle. Flame cells are lacking in *Ascaris* and other nematodes, although they are found in some other pseudocoelomate phyla.

Digestive System. The mouth empties into a short muscular **pharynx,** which sucks food into the ribbon-like **intestine** (Figure 8.1A). The intestine is thin-walled for absorption of digested food products into the pseudocoel. Trace it to the **anus.** What is meant by "tube-within-a-tube" construction? _____ _____ Does *Ascaris* fit this description? _____ Does the planarian? _____
Digestion is begun extracellularly in the lumen of the intestine and is completed intracellularly in the cells of the intestinal wall.

There are no respiratory or circulatory organs. Oxygen is obtained mainly from the breakdown of glycogen within the body, and distribution is handled by the pseudocoelomic fluid.

Reproductive System. The female reproductive system fills most of the pseudocoel. The system is a Y-shaped set of long, convoluted tubes. Unravel them carefully with a probe. The short base of the Y, the **vagina,** opens to the outside at the **vulva.** The long arms of the inverted Y are the **uteri.** These extend posteriorly and then double back as slender, much-coiled **oviducts,** which connect the uteri with the threadlike terminal **ovaries.** Eggs pass from the ovaries through the oviducts to the uteri, where fertilization occurs and shells are secreted. Then they pass through the vagina and vulva to the outside. The uteri of an *Ascaris* may contain up to 27 million eggs at a time, with as many as 200,000 eggs being laid per day. Study the life history of *Ascaris* from your text. Is there an intermediate host in the life cycle? _____

The male reproductive system is essentially a single, long tube made up of a threadlike **testis,** which continues as a thicker **vas deferens.** Both are much coiled. The vas deferens connects with the wider **seminal vesicle,** which empties by a short, muscular **ejaculatory duct** into the anus. Thus the male anus serves as an outlet for both digestive system and reproductive system and is often called a **cloaca** (L., sewer). **Spicules** secreted by and contained in spicule pouches may be extended through the anus. In copulation, a male inserts the copulatory spicules into the vulva of a female and discharges spermatozoa through the ejaculatory duct into the vagina.

What is meant by "sexual dimorphism"? _____ _____ How does *Ascaris* illustrate this? _____ How is *Ascaris* transmitted from host to host? _____ How can infestation be prevented? _____ In what ways is *Ascaris* structurally and functionally adapted to life as a parasite in the intestine? _____ _____

Further Study

Transverse Sections of *Ascaris*

☞ Study a prepared stained slide, at first under low power. If both female and male cross sections are present, examine first the larger female cross section.

Note the thick, noncellular **cuticle** on the outside of the body wall (Figure 8.2). Below the cuticle is a thinner syncytial **epidermis,** which contains nuclei but few cell walls. The **longitudinal muscles** making up most of the body wall appear as fluffy, irregular masses dipping into the **pseudocoel,** with tips of the cells directed toward the nearest nerve cord. Muscle continuity is interrupted by the **longitudinal lines.** Look for **excretory canals** in the lateral lines and look for **dorsal** and **ventral nerve cords** in the dorsal and ventral lines. The lateral lines appear free of muscle cells. In the pseudocoel of the female, the large **uteri** (Figure 8.2) are filled with eggs enclosed in shells and in cleavage stages. The thin-walled **oviducts** also contain eggs, whereas the wheel-shaped **ovaries** are composed of tall epithelial cells and have small lumens. The **intestine** is

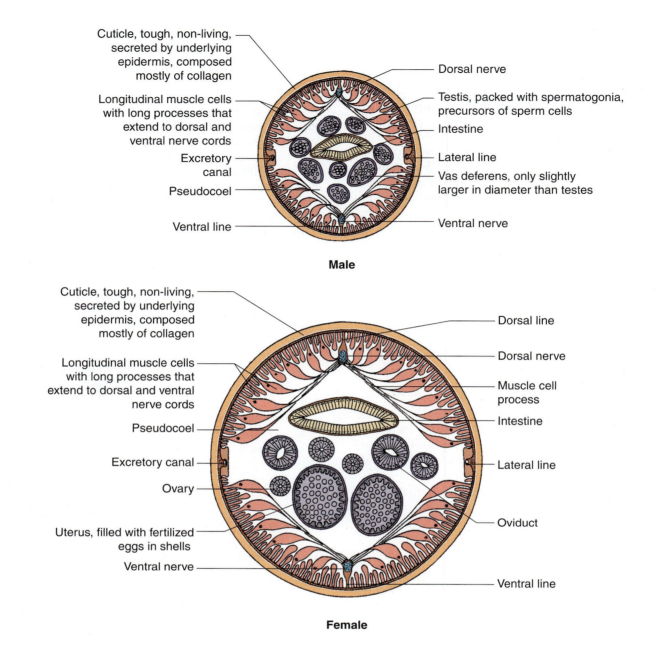

Figure 8.2

Transverse sections through male and female *Ascaris* worms.

composed of a single layer of tall columnar cells (endo-dermal). The pharynx and rectal region of the intestine are lined with cuticle. Why? _____ Why is *Ascaris* not digested in the human intestine?

_____

Examine the male cross section. It is similar to the female in all respects except for the reproductive system. You should see several rounded sections of **testes** packed with spermatogonia (precursors to male reproductive cells). There may also be several sections of **vas deferens** containing numerous spermatocytes, and possibly a section of a large **seminal vesicle** filled with

mature spermatozoa. Note that male reproductive structures visible in the cross section of the roundworm will depend on the body region from which the cross section is taken (refer to Figure 8.1B).

Lab Report

On pp. 106 and 107 is a table for comparing representatives of three of the metazoan phyla you have studied so far. Completing this table affords a survey of the development of these phyla and is an excellent form of review.

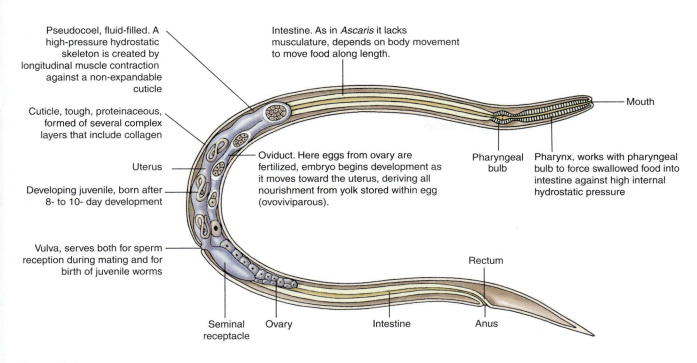

Figure 8.3
Turbatrix aceti, the vinegar eel (female).

The following labels appear in the figure:

Pseudocoel, fluid-filled. A high-pressure hydrostatic skeleton is created by longitudinal muscle contraction against a non-expandable cuticle

Cuticle, tough, proteinaceous, formed of several complex layers that include collagen

Uterus

Developing juvenile, born after 8- to 10- day development

Vulva, serves both for sperm reception during mating and for birth of juvenile worms

Intestine. As in *Ascaris* it lacks musculature, depends on body movement to move food along length.

Mouth

Oviduct. Here eggs from ovary are fertilized, embryo begins development as it moves toward the uterus, deriving all nourishment from yolk stored within egg (ovoviviparous).

Pharyngeal bulb

Pharynx, works with pharyngeal bulb to force swallowed food into intestine against high internal hydrostatic pressure

Rectum

Seminal receptacle

Ovary

Intestine

Anus

Alternative Core Study

Some Free-Living Nematodes

Vinegar Eels

Turbatrix aceti (tur-ba′tricks; L., female that disturbs), the vinegar eel (Figure 8.3), is often found in fermented fruit juices, particularly the sediment of nonpasteurized vinegar, where it feeds on the yeasts and bacteria found there.[1] As in *Ascaris,* females are larger than males.

☞ Place a drop of the culture containing worms on a clean glass slide and examine with a compound microscope.

Notice the violent thrashing movements, the body bending in **S** and **C** shapes. Because they have only longitudinal muscles in the body wall, vinegar eels, like other nematodes, can flex their body only from side to side. Locomotion is especially inefficient in free water, in which they can make little directional progress.

☞ Add a pinch of sand to the slide and observe again. The worms are now able to use the sand grains as fulcra to make forward progress. In this

manner, nematodes can readily move through the interstitial fluids of soils or between the cells of tissues. Now place another drop of culture on a clean slide and cover with a coverslip. Quiet the worms for study by *gently* warming the slide over a small flame or an incandescent lamp. Worms also may be quieted by adding a drop of 1N HCl to the culture.

Note the blunt anterior end, the **mouth,** the **pharynx,** and the **pharyngeal bulb.** These lead into a long, straight **intestine,** which ends at the ventral **anus,** located a short distance from the pointed tail end.

Select a female specimen. Large females will contain two to five developing juveniles in the **uterus.** *Turbatrix* is **ovoviviparous** (L. *ovum,* egg, + *vivere,* to live, + *parere,* to bring forth), giving birth to living young. During its average life span of 10 months, a female vinegar eel typically produces about 45 living young. Note the **vulva** on the ventral side, which receives sperm in copulation and through which the young worms are born. The **seminal receptacle** lies just behind the uterus. The **ovary** extends anteriorly from the uterus and doubles back dorsally.

Males can usually be identified by their **copulatory spicules,** which may protrude from the anal opening. They are used for holding the female during copulation. It may be possible to see the filamentous **testis** extending forward from about the middle of the body, and then bending back on itself. It leads, by way of a sperm duct, to the cloaca.

---

[1]Although vinegar eels once were present in all vinegar containers, commercial vinegars are now filtered and pasteurized. Vinegar eels, harmless and nonparasitic, still are raised for fish food for the fry of tropical fish and for classroom use.

Further Study

Some Parasitic Nematodes

Hookworm

Hookworms, *Necator americanus* ("American killer"), live in the intestines of their vertebrate hosts. They attach themselves to the mucosa and suck up the blood and tissue fluids from it. The species most important to humans are *N. americanus* and *Ancylostoma duodenale* (an-ke-los′ta-muh; Gr. *ankylos,* crooked, + *stoma,* mouth). Hookworms infect about 4% of the population in the southern United States, where 95% of the cases are *Necator* infections. Caucasians are 10 times more susceptible to hookworm than are African Americans. *A. caninum* is the common hookworm of domestic dogs and cats (Figure 8.4).

Hookworms mature and mate in the small intestine of the host. Developing eggs are passed out in the feces. On the ground they require warmth (preferably 20° to 30°C), shade, and moisture for continued development. They hatch in 24 to 48 hours into young juveniles, which feed on fecal matter, molt their cuticles twice, and in a week or so are ready to infect a new host.

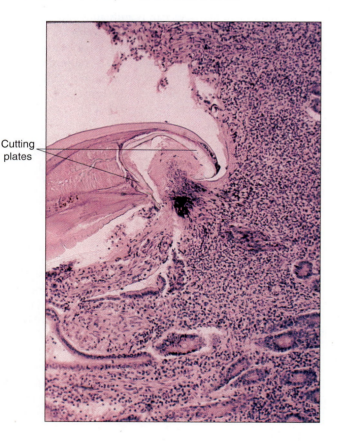

Cutting plates

Figure 8.4

Section through anterior end of *Ancylostoma caninum,* the dog hookworm, in the intestine. Note the lacerating cutting plates, used for attachment to the intestinal mucosa.

If the ground surface is dry, they migrate into the soil, but after a rain or morning dew they move to the surface, extend their bodies in a snakelike fashion, and wave back and forth. Thousands may group together, waving rhythmically in unison. Under ideal conditions, they may live for several weeks.

Infection occurs when the juveniles contact the skin of the host and burrow into it. Those that reach blood vessels travel to the heart and then to the lungs. Here they are carried by ciliary action up the respiratory passages to the glottis and swallowed. In the small intestine, they grow, molt, mature, and mate. By 5 weeks after entry, they are producing eggs.

Whether infection results in hookworm disease depends on the number of worms present and the nutritional condition of the infected person. Massive infections in the lungs may cause coughing, sore throat, and lung infection. In the intestinal phase, moderate infections cause iron-deficiency anemia. Severe infections may result in severe protein deficiency. When accompanied by chronic malnutrition, as occurs in many tropical countries, damage may be irreversible, resulting in stunted growth and below-average intelligence.

☞ **Examine prepared slides of hookworms.**

Adult males of *Necator* are typically 7 to 9 mm long, and females 9 to 11 mm long. Specimens of *Ancylostoma* are slightly longer. The anterior end curves dorsally, giving the worm a hooklike appearance. Note the large buccal capsule, which bears a pair of dorsal and a pair of ventral cutting plates surrounding its margin (Figure 8.4). A stout muscular esophagus serves as a powerful pump.

Note on the male a conspicuous copulatory bursa consisting of two lateral lobes and a smaller dorsal lobe, all supported by fleshy rays. Needle-like spicules are present, which in *Necator* are fused at the distal ends to form a characteristic hook. In the female, the vulva is located in about the middle of the body.

Trichina Worm

The trichina worm, *Trichinella spiralis* (Gr. *trichinos,* of hair, + *ella,* dim. suffix), is a nematode parasite in humans, hogs, rats, and other omnivorous or carnivorous mammals.

☞ **Study a slide of larvae encysted in pork muscle (Figure 8.5).**

How many cysts does your slide show? _____ Would you be able to see the cysts in meat with the naked eye? _____ The cyst wall is made of fibrous tissue that gradually becomes calcified due to the host's immune reactions. How many worms are coiled in a cyst? _____ Study the life cycle of *Trichinella* in your text. Are they oviparous or ovoviviparous? _____ How can you prevent trichinosis?

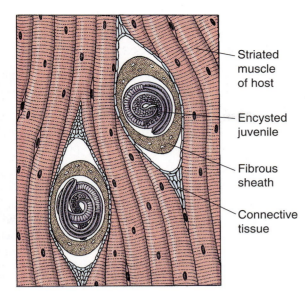

Figure 8.5
Trichinella spiralis, the trichina worm. Juveniles are shown encysted in skeletal muscle.

Labels on figure:
Striated muscle of host
Encysted juvenile
Fibrous sheath
Connective tissue

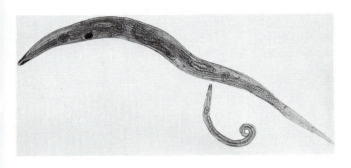

Figure 8.6
Pinworms, *Enterobius vermicularis.* The male is smaller and has a curled posterior end.

_____ Study a slide showing male and female adults. Where would the adults live in the human host? _____

Pinworm

Pinworms, *Enterobius vermicularis* (en-te-robe′ee-us; Gr. *enteron,* intestine, + *bios,* life) (Figure 8.6), are the most common nematode parasites of humans in the United States. They live in the large intestine and cecum. Females, which may reach up to 12 mm in length, lay their eggs at night around the anal region of their host. A single female may lay 4600 to 16,000 eggs. Scratching contaminates the hands and bedding of the host. The eggs, when swallowed, hatch in the duodenum and mature in the intestine.

Unlike hookworms, pinworms are often found at high socioeconomic levels. Although approximately one-third of the human population in the United States is infected with pinworms, their presence is mainly an irritation and an embarrassment because pinworms cause no obvious debilitating effects. But unmeasurable is the mental stress suffered by families in their efforts to rid their households of the worms.

Projects and Demonstrations

1. *Wuchereria bancrofti.* Examine prepared slides of the filarial worm, *Wuchereria bancrofti* (wu-ka-rir′ee-a, after Otto Wucherer, nineteenth-century German physician). What disease is caused by this nematode? _____ In what climate is infestation common? _____ What is the alternate host? _____ What control methods might be used? _____

2. *Dracunculus medinensis.* Examine prepared slides of the guinea worm, *Dracunculus medinensis* (dra-kunk′u-les, L. dim. of *draco,* dragon). In what part of the world is this nematode common? _____ Where do the larvae develop? _____ How is this parasite acquired? _____ What control methods would you use to prevent infestation? _____

3. *Other nematode parasites.* Examine prepared slides of eggs, larvae, or cysts of any of the nematode parasites.

4. *Dirofilaria immitis (dog heartworm).* Examine a prepared slide of microfilariae in a smear of dog blood. Heartworm is especially prevalent along the Atlantic and Gulf Coast states and northward along the Mississippi River drainage, where dogs are infected by mosquitos that ingest and transmit the microfilariae with their blood meals. The worms mature in the right side of the heart and pulmonary artery. Heavy infestations cause cardiopulmonary failure.

EXERCISE 8B
A Brief Look at Some Other Protostomes

Core Study

Phylum Rotifera

Philodina or Others

☞ Place a drop of rotifer culture on a depression slide, cover, and examine with subdued light under low power.

How does the animal attach itself? _____ Is it free-swimming? _____ Does it have a definite head end? _____ Note the anterior discs of cilia **(corona)** that give the impression of wheels turning. Are they retractile? _____ The cilia function both in swimming and in feeding.

The tail end (or **foot**) bears slender toes. How many? _____ The foot contains a pedal gland that secretes a cement used for clinging to objects.

Locate the pharynx **(mastax),** which is fitted with jaws for grinding up food particles. The mastax is conspicuous in living rotifers because of its rhythmic contractions (Figure 8.7). Rotifers feed on small plankton organisms swept in by cilia. Can you identify the digestive tract?

In *Philodina* (fill-uh-dine'uh; Gr. *philos,* fond of, loving, + *dinos,* whirling) (Figure 8.7), the **cuticle** is ringed (annulated) so that it appears segmented. From watching its movements, would you say it had circular muscles? Longitudinal? Oblique? _____ Estimate the size of the rotifers. Is there more than one variety in the culture? _____

In many rotifers, the cuticle is thickened and rigid and is called a **lorica.** *Monostyla* and *Platyias* are examples. Some rotifers, such as *Floscularia* (Figure 8.7), live in a secreted tube. Most rotifers live in fresh water.

Phylum Gastrotricha
Chaetonotus or Others

Gastrotrichs include both freshwater and marine organisms. They are similar in size and general habits to the rotifers and are often found in the same cultures.

☞ Place a drop of culture on a slide, cover, and study first with low and then with high power.

Observe the manner of locomotion of the gastrotrichs. They glide along on a substrate by means of ventral cilia. *Chaetonotus* (NL. *chaeta,* bristle, + Gr. *nōtos,* back), a common genus, is covered with short, curved dorsal spines (Figure 8.7). The rounded head bears cilia and little tufts of sensory bristles. The tail end is forked and contains cement glands similar to those of rotifers. Do gastrotrichs use the forked tail in the same manner as rotifers use the toes? _____

Gastrotrichs have a syncytial epidermis covered with a cuticle. In feeding, they use the head cilia to sweep algae, detritus, and protozoans into the mouth. How long are the specimens? _____

Most marine gastrotrichs are hermaphroditic, but in freshwater species only parthenogenetic females are known.

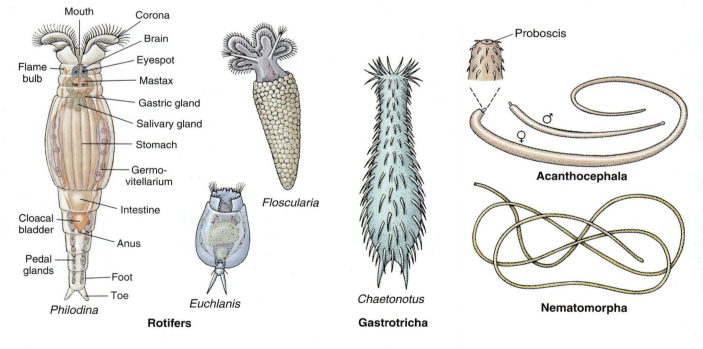

Figure 8.7
Some representative pseudocoelomates.

Phylum Nematomorpha

The Threadworms, or "Horsehair" Worms

Threadworms, or "horsehair" worms of phylum Nematomorpha (Figure 8.7), such as *Paragordius* and *Gordius,*[2] are long, cylindrical, hair-like worms often found wriggling in watering troughs, puddles, ponds, and quiet streams. Most of them range between 0.5 and 3 mm in diameter and from 10 to 300 mm in length, but some may reach a length of 1 m.

How long are your specimens? _____ Is the diameter uniform throughout? _____ If you have live specimens, what would you conclude about their muscular makeup, judging from their movements? _____

Nematomorphs have no lateral lines or excretory system, and in the adults the digestive system is degenerate. They differ from nematodes in having a cloaca in both sexes. Females lay long, gelatinous strings of eggs on water plants. Larvae, which are encysted on plants, are sometimes eaten by arthropods, in whom the larvae are parasitic for a while.

Phylum Acanthocephala

The Spiny-Headed Worms, *Macracanthorhynchus*

The adult *Macracanthorhynchus hirudinaceus* (mak'ruh-kan-thuh-rink'us; Gr. *makros,* long, + *akantha,* thorn, + *rhynchos,* snout) is parasitic in the small intestine of pigs, where it attaches to the intestinal lining and absorbs digested food of its host. Like a tapeworm, it has no digestive system at all.

The body is cylindrical and widest near the anterior end. A small, spiny proboscis on the anterior end bears six rows of recurved hooks for attachment to the intestinal wall (Figure 8.7, *top right*). The proboscis is hollow and can be partially retracted into a proboscis sheath.

The worms are dioecious. The male is much smaller than the female and has a genital bursa at the posterior end that may be partly evaginated through the genital pore and is used in copulation. Eggs discharged by the female into the host feces may be eaten by white grubs (larvae of the beetle family *Scarabeidae*) in whom they develop. Pigs are infected by eating the grubs or the adult beetles.

---

[2]L. *Gordius,* after Gordius, king of Phrygia, who tied an intricate knot that, according to legend, could be untied only by one destined to rule Asia. Many generic names have been taken from mythology, and in many cases their application is quite obscure. However, because horsehair worms have a habit of becoming inextricably entangled in knots comprising several worms, the mythological application in this case is apt.

Name_____

Date_____

Section_____

Comparing Representatives of Three Phyla

Features	Cnidaria (Hydra)	Platyhelminthes (Planaria)	Nematoda (*Ascaris*)
Symmetry			
Shape			
Germ layers			
Body covering			
Cephalization (present or absent)			
Coelomic cavity (If present, state what type)			

Features	Cnidaria (Hydra)	Platyhelminthes (Planaria)	Nematoda (*Ascaris*)
Musculature (layers present and how arranged)			
Digestive tract and digestion			
Excretion			
Nervous system			
Sense organs			
Reproduction, sexual			
Reproduction, asexual			

LAB REPORT

The Molluscs
Phylum Mollusca

The molluscs, with over 90,000 species, rank next to the arthropods in number of named species. They include chitons, snails, slugs, clams, oysters, squids, octopuses, cuttlefish, and some others. They have retained the basic features introduced by the preceding phyla, such as triploblastic structure, bilateral symmetry, cephalization, and a body cavity. The body cavity, though small, is now a **true coelom,** a characteristic shared by all remaining phyla. All organ systems are present. Molluscs, with spiral cleavage and an ancestral trochophore larva, belong to the Lophotrochozoa clade of the Animal Kingdom.

The name Mollusca means "soft-bodied" and because of their soft, plump bodies, molluscs such as clams, oysters, snails, and octopuses have been widely

EXERCISE 9A
Class Bivalvia
(= Pelecypoda)—Freshwater Clams
Freshwater Clam
Projects and Demonstrations

EXERCISE 9B
Class Gastropoda—Pulmonate Land Snails
Land Snail
Projects and Demonstrations

EXERCISE 9C
Class Cephalopoda—*Loligo,* the Squid
Loligo
Demonstrations

Classification: Phylum Mollusca

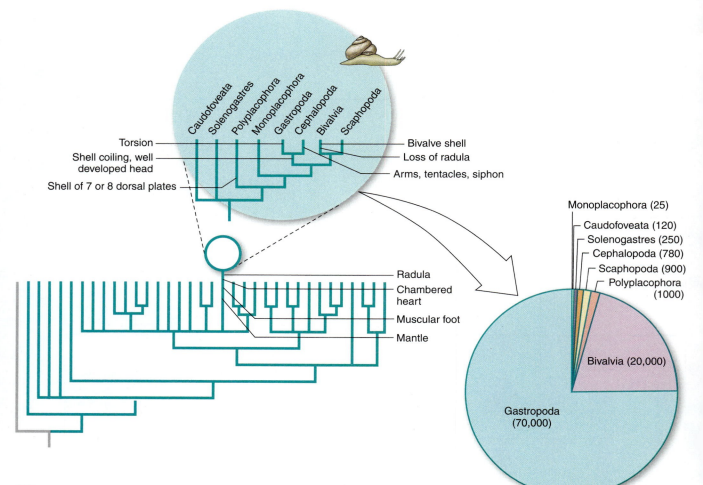

used for human food. Although the ancestral molluscan body has undergone enormous radiation, all have a muscular **foot,** generally used in locomotion. Two folds of skin from the dorsal body wall form a protective **mantle** that encloses a space between the mantle and body wall, the **mantle cavity.** The mantle cavity typically houses **gills** (a lung in some) and the mantle itself may secrete a protective **shell.**

Molluscs have left an extensive fossil record, indicating that their evolution has been a long one. They occupy numerous ecological niches and are found in the sea, in fresh water, and on land. They range from sedentary herbivores to fast-swimming predators.

EXERCISE 9A
Class Bivalvia (= Pelecypoda)— Freshwater Clams
Core Study

Freshwater Clam

Phylum Mollusca
    Class Bivalvia
        Subclass Palaeoheterodonta
            Order Unionoida
                Genus *Anodonta,* others

Where Found

Bivalves are found in both fresh water and salt water. Many of them spend most of their existence partly or wholly buried in mud or sand. Freshwater clams (also called mussels) are found in rivers, lakes, and streams and were once particularly abundant in the Mississippi River watershed before stream pollution and increasing acidity from acid rain greatly depleted their populations.[1] Native bivalves are also the biggest losers in the recent invasion of zebra mussels, unintentionally introduced from eastern Europe into the Great Lakes in 1988 and now running amok throughout the central United States and Canada. Freshwater clams are overwhelmed by zebra mussels, which attach to their shells in enormous numbers. Some formerly common freshwater genera are *Anodonta, Lampsilis, Elliptio,* and *Quadrula.*

Traditionally, freshwater clams are used in general laboratories because of their availability, but the sea clam, *Spisula,* is quite similar to the freshwater clams and makes a good substitute. Marine mussels, *Mytilus,* or quahogs, *Mercenaria,* can also be substituted.

---

[1]Clams are especially vulnerable to aquatic pollution. Approximately 10% of all freshwater clams have become extinct in the last century, and most of the rest are endangered.

Behavior and General Features

 Observe living bivalves in an aquarium if they are available.

Freshwater clams lie half buried in the sand. They are sluggish, and their reactions are slow.

What is the natural position of the clam at rest? _____ When moving? _____ Note that it leaves a furrow in the sand when it moves. The soft body is protected by a hard exoskeleton composed of a pair of **valves,** or shells, hinged on the dorsal side. When the animal is at rest, the valves are slightly agape ventrally, and you can see at the posterior end the fringed edges of the **mantle** that lines the valves. The posterior edges of the mantle are shaped so as to form two openings **(apertures)** to the inside of the mantle cavity (Figure 9.1).

With a pipette or hypodermic syringe and needle, carefully introduce a small amount of carmine dye into the water near the apertures. Watch what happens to it.

Which of the apertures has an incurrent flow, and which has an excurrent flow? _____ A steady flow of water through these apertures is necessary to bring oxygen and food to the animal and to carry away wastes. Most bivalves are filter feeders that filter minute food particles from the water, trap them in mucus, and carry them by ciliary action to the mouth.

Some marine clams have the mantle drawn out into long muscular **siphons.** When the animal burrows deeply into the mud or sand, the siphons extend to the surface to bring clear water into the mantle cavity.

Gently touch the mantle edge with a glass rod.

What happens? The mantle around the apertures is highly sensitive, not only to touch but also to

Figure 9.1
Freshwater clam, showing apertures in mantle. The dorsal, or upper, aperture is excurrent; the larger, lower aperture is incurrent.

chemical stimuli, allowing the animal to close its valves to exclude water containing unpleasant or harmful substances.

 Lift a clam out of the sand to observe the hasty withdrawal of the foot. Then lay it on its side on the sand to see if it will right itself.

The foot is as soft, flexible, and sensitive as the human tongue. Mucous glands keep the foot well protected with mucus.

 If your lab has a marine tank containing scallops, compare the scallops' method of locomotion with that of the clams.

Written Report

 On separate paper, report your observations on clam behavior.

External Structure

The bivalve shell protects the animal from predators, serves as a skeleton for muscle attachment, and, in burrowing forms, helps keep mud and sand out of the mantle cavity.

The two valves of the clam are attached by a **hinge ligament** on the dorsal side; the ventral side is free for the protrusion of the foot. A swollen hump, the **umbo** (pl. **umbones**), near the anterior end of the hinge, is the oldest and thickest part of the shell and the part most resistant to boring gastropod predators. Concentric **lines of growth** around the umbo indicate growth periods. Where is the youngest part of the shell? _____

The outer, horny layer of the shell is the **periostracum,** which protects the underlying calcium carbonate from being dissolved by acid in the water. The periostracum is thin and tends to be worn away from older parts of the shell.

Now determine the correct orientation of the clam. Note that the umbones are dorsal and located toward the anterior end. Identify the right and left valves.

Internal Structure

The directions that follow apply to living clams (see Appendix A, p. 284) but may be easily adapted to preserved specimens.

 Obtain a living clam that has been heated to about 40°C (104°F), causing the valves to gap slightly. *Caution! Do not overheat.* Your instructor or teaching assistant will open the clam by inserting a strong, short-bladed knife (*not* a scalpel) between the *left* valve and the *left* mantle at the posterior end and cutting the posterior adductor muscle as close as possible to the left valve (Figure 9.2A). Cut

the anterior adductor muscle in the same manner. Hold the clam against a firm surface when cutting, and keep your hand clear of the knife blade. The valve will gap open because of the action of the hinge ligament. Separate the mantle completely from the left valve and lift the loosened left valve. Place the clam in a dissecting pan and flood the body with pond or dechlorinated water, allowing the right valve to serve as a container.

Examine the inner surface of the left valve.

The inner, iridescent mother-of-pearl surface is the **nacreous layer,** which lies next to the mantle and is secreted continuously by the mantle surface. Between the inner nacreous layer and outer periostracum is a **prismatic layer,** made up of crystalline calcium carbonate. It is secreted by glands in the edge of the mantle.

With the aid of Figure 9.3, locate the **hinge ligament,** which acts as a spring to force the shells apart, and scars of the **anterior** and **posterior adductor muscles,** which pull the shells closed. Near the adductors are smaller muscles that retract and extend the foot. The **pallial line** marks the location of the pallial muscle that retracts the edge of the mantle (see Figure 9.4). Along the dorsal margin of the shell is a ridge of lateral **hinge teeth** that interlock the valves to prevent them from slipping apart.

The Mantle

 With the left valve removed, examine the thin **mantle** covering all the soft tissues of the clam.

Posteriorly, the edges of the two mantles are thickened, darkly pigmented, and fused together dorsally to form the ventral **incurrent aperture** and dorsal **excurrent aperture.** The apertures permit a continuous flow of water through the mantle cavity. In many burrowing clams, the apertures are extended into siphons.

 Examine the edge of the mantle, which forms three parallel folds: an outer, a middle, and an inner fold (Figure 9.4).

The outer fold (closest to the shell) secretes the hornlike periostracum as well as the prismatic layer of the shell. The middle lobe is sensory in function; in some bivalves (scallops, for example), it is drawn out into specialized sensory structures, such as tentacles and eyes. The ciliated surface of the inner fold assists in water circulation within the mantle cavity and sweeps out debris; the inner fold also seals the mantle cavity when the clam closes its shell. It is this inner fold that, on the posterior side, forms the clam's incurrent and excurrent apertures.

Sometimes a foreign object, such as a sand grain or parasite, becomes lodged between the mantle and shell.

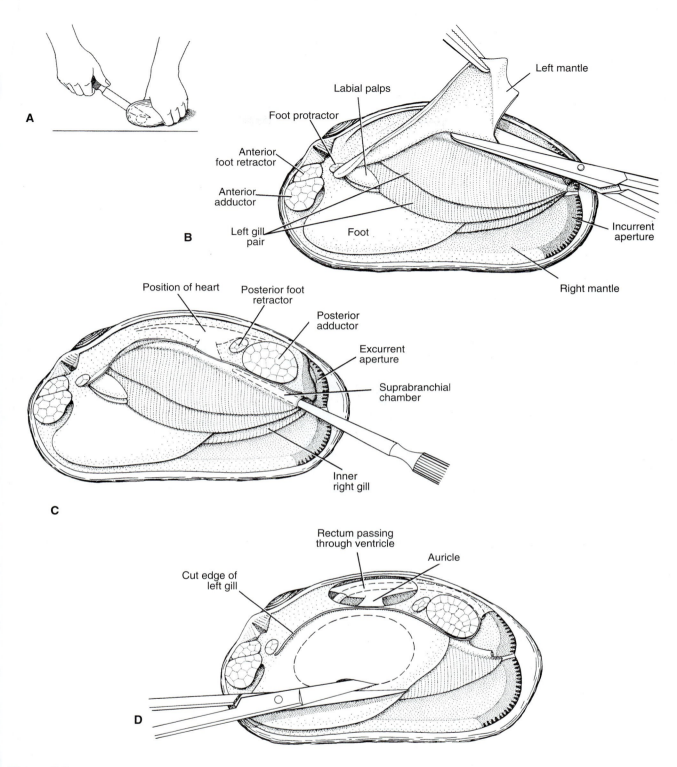

Figure 9.2

Dissection of a freshwater clam. **A,** Cutting the posterior adductor. **B,** Trimming off the left mantle. **C,** Probing the suprabranchial chamber. **D,** Cutting to expose the visceral mass.

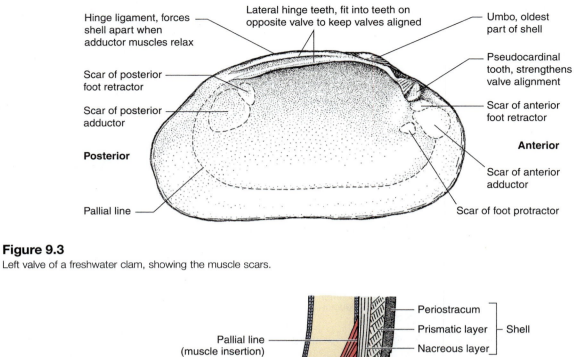

Figure 9.3
Left valve of a freshwater clam, showing the muscle scars.

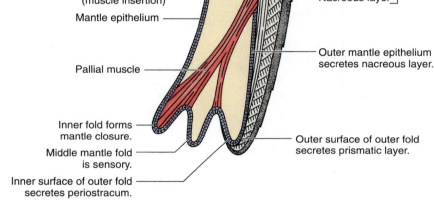

Figure 9.4
Section through margin of freshwater clam shell and mantle.

The outer epithelium of the mantle then secretes nacre around the object, forming a pearl. Often, however, the formative pearl becomes fused with the nacreous layer of the shell. Do you find any pearls or other evidence of such irritation in the valve or mantle?

Muscles. Locate the large **adductor muscles,** which close the valves (Figure 9.5). Slightly dorsal to them are the **foot retractor muscles.** The **foot protractor muscle** is small and will be found lying in the visceral mass just posterior to the anterior adductor.

Pericardium. On the dorsal side of the animal, locate the delicate, almost transparent **pericardial membrane** surrounding the heart. In living clams, the beating ventricle is visible through the pericardium. Do not open the pericardium until instructed to do so later.

Mantle Cavity

Lift up the mantle to expose the outer pair of **gills** and the body mass beneath. The entire space between the right and left lobes of the mantle is the **mantle cavity.** Cilia on both mantle and gills keep water flowing through the mantle cavity. If the outer gill is much thicker than the inner gill, the animal is probably a female in which the gill is serving as a **brood chamber** for developing embryos. Are clams monoecious or dioecious? _____

The soft portion of the body is the **visceral mass.** The muscular **foot** lies ventral to the visceral mass; it will be retracted but will contract even farther if you touch it with a probe. The foot, much like the mammalian tongue, is a **muscular hydrostat.** It operates by a combination of muscular contraction and hydraulic mechanisms. A clam can extend or enlarge its foot

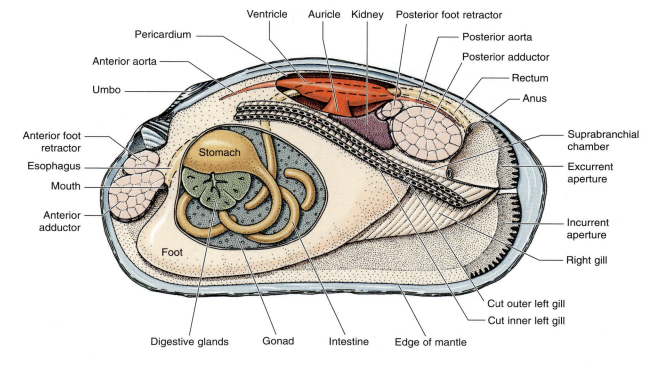

Figure 9.5
Anatomy of a unionid clam.

hydraulically by engorgement with blood and uses the extended foot for anchoring itself or drawing its body forward.

While most molluscs have a well-developed head, the clam has none—unless the mouth and labial palps can be considered a head. The two pairs of **labial palps,** a pair on each side of the body, are ciliated and serve to guide food particles trapped in mucus into the mouth. Bivalves also lack a radula, which would be of no use to a filter feeder.

Respiration and the Gills

☞ Carefully trim off the mantle where it is attached dorsally to the gills (see Figure 9.2B).

With the mantle removed, the gills are conspicuous. The freshwater clam, like most bivalve molluscs, is a **filter feeder** that depends on highly modified gills to gather food materials, as well as for respiration. There are two gills on each side of the visceral mass. Note that only the dorsal margins of the gills are attached. The ventral edges of the gills hang free in the mantle cavity.

Strong currents produced by cilia on the **gill filaments** draw water in the incurrent siphon, then into the **water tubes** through numerous tiny pores (**ostia**) between the gill filaments (Figure 9.6). Food particles, mostly phytoplankton and organic debris, are carried by ciliary currents and mucus down the gill surface toward the food groove on the lower edge of the gill. Here, mucus

and food particles are rolled into a food string that passes into the mouth by way of the labial palps. Large particles are dropped to the edge of the mantle and discarded.

To observe the ciliary action that maintains and controls water flow, sprinkle a few carmine granules on the gills, on the labial palps, and on the inside of the mantle.

To see a video showing the action of cilia of the clam gill, visit this link: http://www.linkpublishing.com/video-cell.htm#Cilia

☞ Cut a transverse section about 1 mm wide from the lower part of the gill. Lay the section on a slide and examine with a dissecting microscope. Identify lamellae, filaments, and water tubes, as well as the partitions (**interlamellar junctions**) that hold the two lamellae apart (Figure 9.6).

Water tubes connect dorsally with the **suprabranchial chamber,** which in turn empties to the outside through the excurrent aperture.

☞ From the excurrent aperture, run a probe into the suprabranchial chamber (see Figure 9.2C). With scissors, slit the suprabranchial chamber to see the tops of the water tubes of the outer gill.

A sedentary animal, such as a clam, which lives half-buried in mud or sand, must have some means of clearing

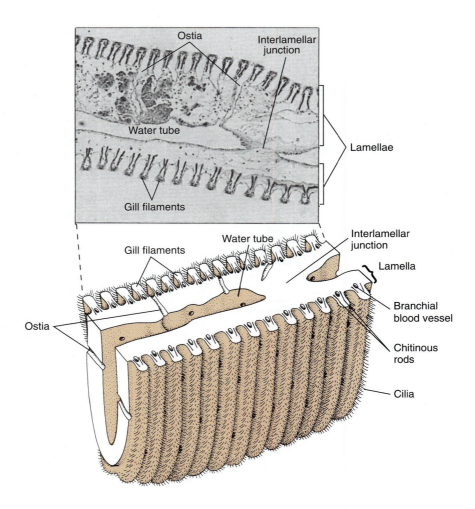

Figure 9.6
Cross section through a portion of the gill of a freshwater clam.

the water passing through the gills of sediment, detritus, and fecal matter. This is accomplished partially by the small size of the ostia and partially by mucus secreted by glands in the roof of the mantle cavity, which traps particles too large for the ostia. Larger debris drops off the gills, while smaller food particles are carried toward the mouth.

The water tubes in the female serve as brood pouches for eggs or larvae during breeding season.

Further Study

Circulatory System
Near the dorsal midline, just below the hinge, is the thin-walled **pericardial sac,** within which lies the **heart.**

☞ Carefully slit open the pericardium, using fine-tipped surgical scissors. Be cautious not to injure the delicate heart inside.

The three-chambered heart is composed of a single **ventricle** and a pair of **auricles** (see Figure 9.5).

☞ If the ventricle is still beating, count the rate (beats/minute).

Note that the ventricle surrounds the intestine. Two aortae leave the ventricle; the **anterior aorta** passes to the visceral mass and intestine, and the **posterior aorta** runs along the ventral side of the rectum to the mantle.

The paired auricles are fan-shaped and very thin-walled. Pass a probe under the uppermost (left) auricle and lift carefully to see its connection to the ventricle.

At the instructor's option, you can inject a small amount of carmine solution into the ventricle using a tuberculin syringe and a small (26 or 27 gauge) hypodermic needle. The carmine solution will flow into and reveal the two aortae.

The pericardial space around the heart is a part of the **coelomic cavity,** greatly reduced in molluscs.

Almost all molluscs (most celphalopod molluscs are an exception) have an **open circulatory system** (Figure 9.7). In a clam, the two aortae carry blood into

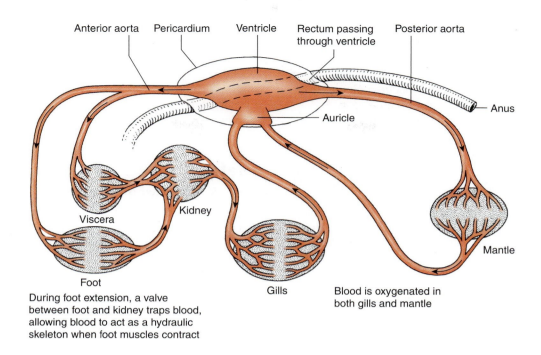

Anterior aorta Pericardium Ventricle Rectum passing through ventricle Posterior aorta

Anus

Auricle

Kidney

Viscera

Mantle

Foot

Gills

Blood is oxygenated in both gills and mantle

During foot extension, a valve between foot and kidney traps blood, allowing blood to act as a hydraulic skeleton when foot muscles contract

Figure 9.7
Scheme of circulation of a freshwater clam.

smaller and smaller vessels and finally into sinus spaces that bathe the tissues directly. There are no capillaries as present in a closed circulatory system. However, the sinuses are not spacious cavities but rather are narrow clefts and channels in connective tissue surrounding the muscular and nervous tissues. Although the system is "open," it functions much like a system of capillaries.

Coelom

Although molluscs have a true coelom, it is small. The pericardial cavity is part of the coelom, as is the small space around the gonads. A true coelom, you recall, is distinguished from other cavities by being lined with epithelium that arises from the mesoderm. What is the name of the lining of the coelom? _____

Excretory System

A pair of dark kidneys lie under the floor of the pericardial sinus. They are roughly U-shaped tubes. The kidneys pick up waste from blood vessels, with which they are richly supplied, and from the pericardial sinus, with which they connect. Waste is discharged into the suprabranchial chamber and carried away with the exhalant current.

Digestive System

Locate again the **labial palps** and the **mouth.**

 To reveal the alimentary canal, cut through the surface tissue on one side of the visceral mass and foot and strip away the epithelium (see Figure 9.2D).

The mouth leads into a short **esophagus** that widens into the **stomach,** surrounded by greenish brown **digestive glands.** The stomach narrows into a tubular **intestine,** which can be seen looping back and forth through the visceral mass. The intestine connects to the rectum, seen earlier passing through the ventricle. Trace the rectum as it passes dorsal to the posterior adductor muscle to its end, the **anus,** which empties feces into the exhalant current. Surrounding the intestine is light brown tissue of the gonad (see Figure 9.5).

In freshly collected clams, you may find a solid, gelatinous rod, the **crystalline style,** projecting into the stomach. It is composed of mucoproteins and digestive enzymes (chiefly amylase), which are released into the food. The crystalline style disappears within a few days after clams are collected and usually is absent from specimens purchased from biological supply houses. Digestion is mostly intracellular.

Reproductive System

Sexes are separate but are difficult to distinguish, except by the swollen gills of the pregnant female. The **gonads** (**ovaries** or **testes**) are a brownish mass of minute tubes filling the space between the coils of the intestine.

 Make a wet mount of gonadal tissue and determine whether there are eggs or sperm in it.

The gonads discharge their products into the suprabranchial chamber. Spermatozoa pass into the surrounding water. They enter a female with the inhalant current and fertilize eggs in the suprabranchial chamber. The **zygotes** settle into the water tubes of the outer gill (brood pouch), where each zygote develops into a tiny bivalved larval form known as a **glochidium** (Gr. *glochis,* point, + *idion,* dim.) (found only in freshwater clams) (Figure 9.8). Glochidia, about the size of dust particles, escape through the excurrent siphon. Glochidia have valves bearing hooks, by which they fasten themselves to gills, fins, or skin of a passing fish. Here they encyst and live as parasites for several weeks. Carried by their fish host, the larvae are dispersed upstream into lakes and rivers, enabling bivalve mollusc populations to invade and flourish in virtually all freshwater environments. After a growth period, the young clams break loose and sink to the bottom sand, where they develop as free-living adults.

☞ If a female with a swollen brood pouch is available, make a wet mount of some gill contents and examine with a microscope to determine whether eggs or glochidia are present.

Nervous System

The nervous system of a clam is not highly centralized. Dissection of the nervous system is difficult and often impractical. Three pairs of **ganglia** (small groups of nerve cells) are connected to each other by nerves. **Cerebropleural ganglia** are found one on each side of the esophagus on the posterior surface of the anterior adductor muscle. **Pedal ganglia** are fused and are found in the anterior part of the foot. **Visceral ganglia** are fused into a star-shaped body just ventral to the posterior adductor muscle. They are covered by a yellowish membrane and are connected to the cerebropleural ganglia by nerves.

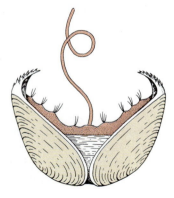

Figure 9.8
Glochidium, the larval form of freshwater bivalve molluscs. Each glochidium is about 0.3 mm in diameter.

Sense organs are poorly developed in clams. They are involved with touch, chemical sensitivity, balance, and light sensitivity. They are most numerous on the edge of the mantle, particularly around the incurrent aperture, but you will not be able to see them. However, in scallops, such as *Pecten,* the ocelli are large and numerous, forming a distinctive row of steel-blue "eyes" along the edges of each mantle.

Oral Report

Be prepared to demonstrate your dissection and explain the clam's structures and their functions.

Projects and Demonstrations

1. *Cross section of entire clam after removal from the shell.* This study is made from prepared slides. What you will see in it depends to some extent on the region through which the body was cut. Identify **mantle, mantle cavity, gills, lamellae, water tubes, intestine, foot, suprabranchial space, gonad,** and other structures revealed in the cross section.

2. *Oysters, scallops, and sea clams.* Examine the shells of oysters, scallops, and sea clams.

3. *Shipworm, Teredo.*[2] Examine a piece of wood into which the wormlike molluscs, *Teredo,* have bored. Examine a preserved specimen of the shipworm. Note the small valves, the small body, and the prolonged siphon that make up the bulk of the shipworm.

EXERCISE 9B
Class Gastropoda— Pulmonate Land Snail

Core Study

Land Snail
Phylum Mollusca
    Class Gastropoda
        Subclass Pulmonata
            Superorder Stylommatophora
                Genus *Helix, Polygyra,* others

[2]See Lane, C. E. 1961. The teredo. Sci. Amer. (Feb.); see also, interesting Wikipedia entries to shipworms on the web.

Where Found

Most land snails (Figure 9.9) prefer fairly moist habitats. They are common in wooded areas, where they spend their days in the damp leaf mold on the ground, coming out to feed on the vegetation at night. *Helix* (Gr. *helix,* twisted) (Figure 9.9, top) is a large snail from southern Europe, and introduced all around the world. The most common species, *H. aspera,* has become a garden pest, especially in California. *Polygyra* (Gr. *poly,* many, + *gyros,* round) is an American land snail with approximately 30 recognized species.

The subclass Pulmonata constitutes one of three major groups of gastropods. Two other subclasses—Prosobranchia and Opisthobranchia—contain mainly marine snails and marine nudibranchs and tectibranchs.

Behavior

☞ Place a land snail on a moistened glass plate and invert the plate over a dish, so that you can watch the ventral side of the foot under a dissecting microscope.

Do you see waves of motion? Can you see evidence of mucus? Of ciliary action? _____ If you are fortunate, you may see the action of the **radula** (L., scraper) in the mouth. The radula is a series of tiny teeth attached to a ribbonlike organ that moves rapidly back and forth with an action like that of a rasp or file (Figure 9.10).

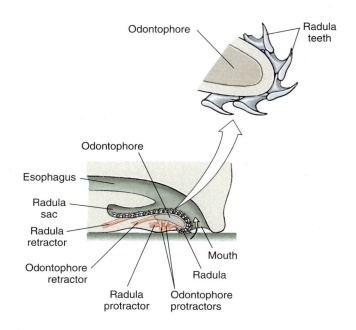

Figure 9.10
Diagrammatic longitudinal section of a gastropod head, showing the radula and the radula sac. The radula moves back and forth over the odontophore cartilage. As the animal grazes, the mouth opens, the odontophore is thrust forward against the substratum, the teeth scrape food into the pharynx, the odontophore retracts, and the mouth opens. The sequence is repeated rhythmically.

☞ Prop up the plate in a vertical position. In which direction does the snail move? Now rotate the plate 90°, so that the snail is at right angles to its former position. When it resumes its travels, in which direction does it move? Now rotate the plate again and observe.

Do you think the snail is influenced in its movements by the forces of gravity (geotaxis; refer to p. 48 for an explanation of taxes and the difference between a positive and a negative taxis)? _____ Try several snails. Do they respond in a similar manner?

☞ Place a snail in a finger bowl and, using the following suggestions, observe its behavior.

Place a piece of lettuce leaf near a snail and see whether it will eat. How long does it take the snail to find the food? _____ Note the **head** with its **tentacles.** How do the two pairs of tentacles differ? _____ Where are the **eyes** located? _____ Touch an anterior tentacle gently and describe what happens. _____ Touch some other parts of the body and note results. _____ Pick up a snail very carefully from the glass on which it is moving and observe exactly what it does.

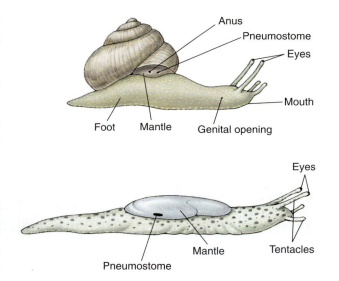

Figure 9.9
Common pulmonates. **Top,** *Helix,* a common land snail. **Bottom,** *Limax,* a common garden slug.

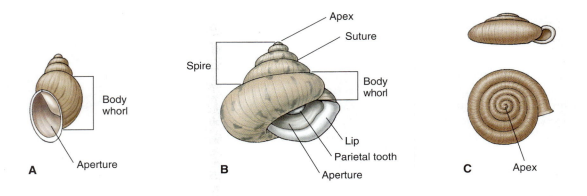

Figure 9.11
Structure of a snail shell. **A,** *Physa,* a sinistral, or left-handed, freshwater snail with lymnaeiform shell (height exceeds width). **B,** *Mesodon* (= *Polygyra*), a dextral (right-handed) snail with heliciform shell (width exceeds height). **C,** *Helicodiscus,* a land snail with planospiral, or flattened, spire (two views).

The Shell

Examine the shell of a preserved specimen. Note the nature of the spiral shell. Is it symmetrical? _____ To what part of the clam shell does the **apex** correspond? _____ The body of the snail extends through the **aperture.** A **whorl** is one complete spiral turn of the shell. On the whorls are fine **lines of growth** running parallel to the edge of the aperture. Holding the apex of the shell upward and the aperture toward you, note whether the aperture is at the right or left. **Dextral,** or right-handed, shells will have the aperture toward the right, and **sinistral,** or left-handed, shells toward the left (Figure 9.11). Is your snail dextral or sinistral? _____ Examine a piece of broken shell for the characteristic three layers—the outer **periostracum,** the middle **prismatic layer,** and the shiny inner **nacre.**

Further Study

Surface Anatomy

☞ If the snail has not contracted enough during killing and preservation[3] to allow you to slip it carefully from the shell, you may use scissors to cut carefully around the spiral between the whorls, removing pieces of shell as you go and

_____

[3]Specimens from biological supply houses are usually narcotized before killing, which leaves the animal relaxed and expanded. This is best done by sealing snails in a jar of water, capped so as to exclude all air. Boiling (and then cooling) the water beforehand to drive out oxygen will shorten the asphyxiation time. Animals thus treated will be fully relaxed with antennae and foot extended.

leaving only the central parts, or **columella** (L., pillar). Try not to damage the coiled part of the visceral mass.

The body is made up of **head,** muscular **foot,** and coiled **visceral hump** (Figure 9.12). Identify the **tentacles, eyes,** ventral **mouth** with three lips, and **genital pore** just above and behind the right side of the mouth. The foot bears a **mucous gland** just below the mouth. Mucous secretions aid in locomotion.

Note the thin **mantle** that covers the visceral hump and forms the roof of the **mantle cavity.** It is thickened anteriorly to form the **collar** that secretes the shell.

Find a small opening, the **pneumostome** (Gr. *pneuma,* air, + *stoma,* mouth), under the edge of the collar (see Figure 9.9). It opens into a highly vascular portion of the mantle cavity, located in the first half-turn of the spiral, that serves as a respiratory chamber **(lung)** in pulmonates. Here, diffusion of gases occurs between air and blood. Oxygen is carried by the pigment hemocyanin. Most aquatic gastropods possess gills. The mantle cavity in the second half-turn contains the heart and a large kidney.

The rest of the coiled visceral mass contains the dark lobes of the digestive gland, the intestine, the lighter-colored albumin gland (part of the reproductive system), and the ovotestis (Figure 9.12).

Projects and Demonstrations

1. *Example of a pond snail.* Watch a pond snail attached to the glass side of an aquarium. Note the broad foot by which it clings to the glass. Can you see the motion of the **radula** as the animal eats algae that has settled on the glass?

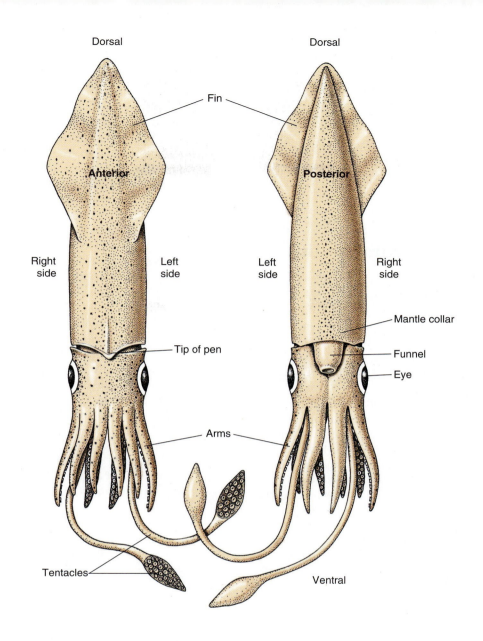

Dorsal

Dorsal

Fin

Anterior

Posterior

Right
side

Left
side

Left
side

Right
side

Mantle collar

Tip of pen

Funnel

Eye

Arms

Tentacles

Ventral

Figure 9.13
External structure of a squid. **Left,** anterior view. **Right,** posterior view.

This valve allows hydrostatic pressure to build up in the mantle cavity before a jet stream of water is ejected through the funnel.

Locate a large pair of **funnel retractor muscles,** and beneath them, the even larger **head retractor muscles.** Locate the free end of the **rectum** with its **anus** near the inner opening of the funnel. Between it and the visceral mass is the **ink sac.** Do not puncture it. When a squid is endangered, it can send out a cloud of black ink through the funnel as it darts off in another direction.

A pair of long **gills** is attached at one end to the visceral mass and at the other to the mantle. The gills are located so that water entering the mantle cavity passes directly over them. The gills are not ciliated as they are in the bivalves. The mantle cavity is ventilated by action of the mantle itself. Contraction of the radial muscles in the body wall causes the wall to become thinner and the capacity of the mantle cavity to become greater, so that water flows in around the collar. Water is then expelled through the funnel when mantle muscles contract. This movement, also used in locomotion, permits very efficient ventilation of the gills.

In most introductory laboratories, observation of the squid ends with the study of the mantle cavity.

Demonstrations

1. Microslides showing spermatophores of *Loligo*.
2. Preserved octopuses and cuttlefish *(Sepia)*.
3. Shells of *Nautilus*.
4. Dried cuttlebone of *Sepia*.
5. Dissection of an injected cephalopod to show circulatory system.
6. Dissection of a cephalopod brain.
7. Living cephalopod, if available.

Classification: Phylum Mollusca

Class Monoplacophora (mon′o-pla-kof′o-ra; Gr. *monos,* one, + *plax,* plate, + *phora,* bearing). Body bilaterally symmetrical, with broad flat foot; a single dome-shaped shell; five or six pairs of gills in shallow mantle cavity; radula present; separate sexes. Example: *Neopilina*.

Class Polyplacophora (pol′y-pla-kof′o-ra; Gr. *polys,* many, + *plax,* plate, + *phora,* bearing). Chitons. Elongated, dorsally flattened body with reduced head; bilaterally symmetrical; radula present; shell of eight dorsal plates; foot broad and flat; gills multiple, along sides of body between foot and mantle edge; sexes usually separate. Examples: *Katharina, Mopalia, Chaetopleura*.

Class Caudofoveata (kaw′do-fo-ve-at′a; L. *cauda,* tail, + *fovea,* small pit). Wormlike; shell, head, and excretory organs absent; radula usually present; mantle with chitinous cuticle and calcareous scales; oral pedal shield near anterior mouth; mantle cavity at posterior end with pair of gills; sexes separate; formerly united with solenogasters in class Aplacophora. Examples: *Chaetoderma, Limifossor*.

Class Solenogastres (so-len′o-gas′trez; Gr. *solen,* pipe, + *gaster,* stomach). Solenogasters. Wormlike; shell, head, and excretory organs absent; radula usually absent; mantle usually covered with scales or spicules; mantle cavity posterior, without true gills but sometimes with secondary respiratory structures; foot represented by long, narrow, ventral pedal groove; hermaphroditic. Example: *Neomenia*.

Class Scaphopoda (ska-fop′o-da; Gr. *skaphe,* boat, + *pous, podos,* foot). Tusk shells. Body enclosed in a one-piece, tubular shell open at both ends; conical foot; mouth with radula and tentacles; head absent; mantle for respiration; sexes separate. Example: *Dentalium*.

Class Gastropoda (gas-trop′o-da; Gr. *gaster,* belly, + *pous, podos,* foot). Snails, slugs, conchs, whelks, and others. Body asymmetrical, usually in a coiled shell (shell uncoiled or absent in some); head well developed, with radula; foot large and flat; one or two gills, or with mantle modified into secondary gills or lung; dioecious or monoecious. Examples: *Busycon, Physa, Helix, Aplysia*.

Class Bivalvia (bi-val′vi-a; L. *bi,* two, + *valva,* valve). Bivalves. Body enclosed in a two-lobed mantle; shell composed of two lateral valves of variable size and form, with dorsal hinge; cephalization much reduced; no radula; foot usually wedge-shaped; gills platelike; sexes usually separate. Examples: *Anodonta, Venus, Tagelus, Teredo*.

Class Cephalopoda (sef′a-lop′a-da; Gr. *kephalē,* head, + *pous, podos,* foot). Squids, nautiloids, and octopuses. Shell often reduced or absent; head well developed with eyes and radula; foot modified into arms or tentacles; siphon present; sexes separate. Examples: *Loligo, Octopus, Sepia*.

The Annelids
Phylum Annelida

The annelids include a variety of earthworms, leeches, and marine polychaetes. Their various adaptations fit them for freshwater, marine, terrestrial, and parasitic living. They are typically elongate, wormlike animals, circular in cross section, with muscular body walls. The most distinguishing characteristic that sets them apart from other wormlike creatures is their **segmentation;** in fact, they are often referred to collectively as the "segmented worms." This repetition of body parts, also called **metamerism** (me-ta′me-ri′sum; Gr. *meta,* between, + *meros,* part), is not only external but is also seen internally in the serial repetition of body organs. Development of segmentation is significant, because along with segmentation comes the opportunity for segments to become specialized for certain functions. Such specialization is not as advanced in annelids as in arthropods, but the introduction of metamerism presaged the rapid evolution seen in arthropods and chordates, the only other phyla emphasizing segmentation.

Division of the coelomic cavity into fluid-filled compartments also has increased the usefulness of hydrostatic pressure in the locomotion of annelids. Coordination between their well-developed neuromuscular system and their more efficient hydrostatic skeleton makes the annelids proficient in swimming, creeping, and burrowing.

Annelids have a complete mouth-to-anus digestive tract with muscular walls so that digestive tract movements are independent of body movements. They also have a well-developed closed circulatory system with pumping vessels, a high degree of cephalization, and an excretory system of nephridia. Some annelids have respiratory organs.

EXERCISE 10A
Class Polychaeta—Clamworms

Core Study

Nereis

Phylum Annelida
    Class Polychaeta
        Order Phyllodocida
            Family Nereididae
                Genus *Nereis*

EXERCISE 10A
Class Polychaeta—Clamworms
Nereis
Other Polychaetes

EXERCISE 10B
Class Oligochaeta—Earthworms
Lumbricus, the Common Earthworm

EXERCISE 10C
Class Hirudinida—Leeches
Hirudo, the Medicinal Leech

Classification: Phylum Annelida

EXPERIMENTING IN ZOOLOGY
Behavior of Medicinal Leeches, *Hirudo medicinalis*

Where Found

Clamworms (also called sandworms or ragworms) are strictly marine. They live in the mud and debris of shallow coastal waters, often in burrows lined with mucus. Largely nocturnal in habit, they are usually concealed by day under stones, in coral crevices, or in their burrows. The common clamworm *Nereis virens* (Gr. *Nēreis,* a sea nymph) may reach a length of one-half meter.

Behavior

☞ If living nereid worms are available, study their patterns of locomotion.

When the animal is quiescent or moving slowly, note that the lateral appendages, called **parapodia** (Gr. *para,* beside, + *pous, podos,* foot), undergo a circular motion that involves an effective stroke and a recovery stroke, each parapodium describing an ellipse during each two-stroke cycle. In the effective stroke, the parapodium makes contact with the substrate, lifting the body slightly off the ground. The two parapodia of each segment act alternately, and successive waves of parapodial activity pass along the worm.

For more rapid locomotion, the worm uses undulatory movements of the body produced by muscular

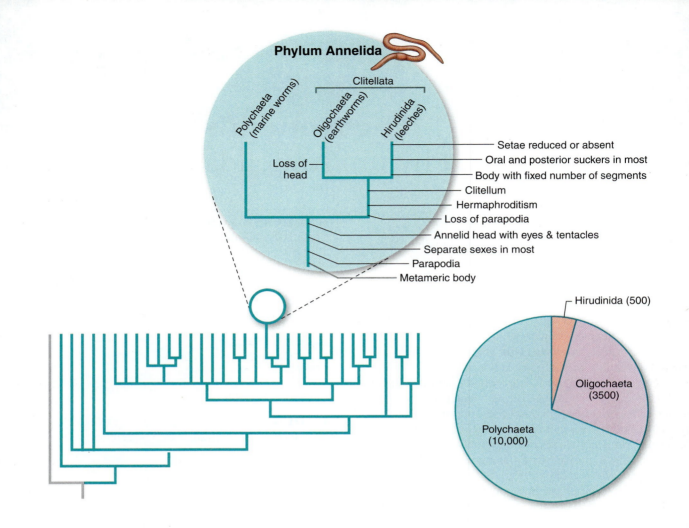

Phylum Annelida

Clitellata

Polychaeta (marine worms)

Oligochaeta (earthworms)

Hirudinida (leeches)

Setae reduced or absent
Oral and posterior suckers in most
Body with fixed number of segments
Clitellum
Hermaphroditism
Loss of parapodia
Annelid head with eyes & tentacles
Separate sexes in most
Parapodia
Metameric body

Loss of head

Hirudinida (500)

Oligochaeta (3500)

Polychaeta (10,000)

contraction and relaxation in addition to parapodial action. As the parapodia on one side move forward in the recovery stroke, the longitudinal muscles on that side contract; as the parapodia sweep backward in their effective stroke, the muscles relax. Watch a worm in action and note these waves of undulatory movements. Can the worm move swiftly? Does it seek cover? Does it maintain contact with the substrate, or does it swim freely?

Place near the worm a glass tube with an opening a little wider than the worm. Does the worm enter it? Why? Place a bit of fresh mollusc or fish meat near the entrance of the tube. You may be able to observe feeding reactions.

Written Report (Optional)

Record your observations on separate paper.

External Features

☞ Place a preserved clamworm in a dissecting pan and cover it with water.

Note the body with its specialized **head,** variable number of segments bearing **parapodia,** and the caudal

segment bearing the **anus** and a pair of feelers, or **cirri** (sing., **cirrus;** sir′us; L., curl). Compare the length and number of segments of your specimen with those of other specimens at your table. Do they vary? The posterior segments are the smallest because they are the youngest. As the animal grows, new segments are added just anterior to the caudal segment.

Along with segmentation, one of the great advancements of the annelids is a well-developed head. The two-part head consists of the **prostomium** (Gr. *pro,* before, + *stoma,* mouth) and the **peristomium** (Gr. *peri,* around, + *stoma,* mouth). The prostomium is a small protuberance bearing the head's sensory structures of touch, taste, and photoreception: two small, median **tentacles,** a pair of flesy **palps,** and four small, dark **eyes.** The peristomium bears sensory tentacles and an eversible pharynx.

If the pharynx is everted on your specimen, do not confuse it with head structures. The pharynx is large and muscular, bearing a number of small, horny teeth and a pair of dark, pincerlike, chitinous **jaws.** If the pharynx is fully everted, the jaws will be exposed. If not, you may be able to probe into the pharynx to find them.

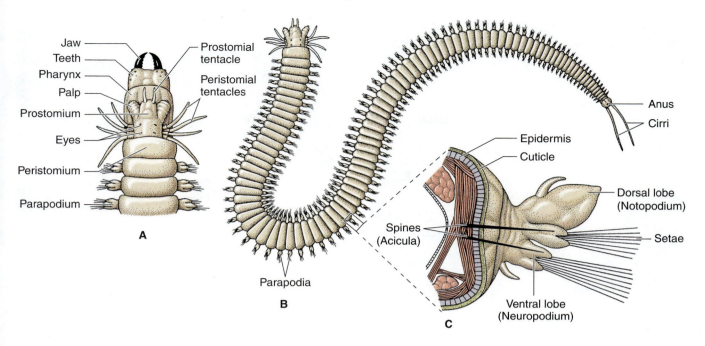

Figure 10.1
Structure of the clamworm *Nereis*. **A,** Anterior view with pharynx extended. **B,** Dorsal view of the entire worm with pharynx withdrawn. **C,** Structure of parapodium.

👉 Cut off, close to the body, a parapodium from the posterior third of the body. Mount it in water on a slide, cover with a coverslip, and examine it with a hand lens or a dissecting microscope.

Parapodia are used for respiration as well as for locomotion. Are they all identical? _____ Each parapodium has a dorsal lobe and a ventral lobe (Figure 10.1C). Each lobe bears a bundle of bristles called **setae** (see'tae; sing., **seta;** L. bristle) and a long, chitinous, deeply embedded spine. The spines are the supporting structures of the parapodia (they are more conspicuous in the posterior parapodia). Each spine is attached by muscles that can protrude it as the parapodium goes into its effective stroke and retract it during the recovery stroke. How might the spines help in locomotion? _____

👉 Peel off a piece of the thin cuticle that covers the animal and study it in a wet mount under the microscope.

The cuticle is fibrous, and its iridescence is caused by its cross striations. It is full of small pores through which the gland cells of the underlying epidermis discharge their products.

Internal Structure

Because the internal structure of polychaetes is similar to that of oligochaetes, our study of their internal anatomy will be limited to that of the earthworm.

Other Polychaetes

There are over 10,000 species of polychaetes, most of them marine. They include some unusual and fascinating animals. Besides the errant, or free-moving, worms such as *Nereis* and its relatives, there are many sedentary species, including the burrowing and tube-building forms (Figure 10.2).

👉 Examine the marine aquarium for living polychaetes and also examine the preserved material on the demonstration table.

EXERCISE 10B
Class Oligochaeta—Earthworms

Core Study

Lumbricus, the Common Earthworm

Phylum Annelida
   Class Oligochaeta
      Order Haplotaxina
         Family Lumbricidae
            Genus *Lumbricus*
               Species *Lumbricus terrestris*

Where Found

Earthworms prefer moist, rich soil that is not too dry or sandy. They are found all over the earth. They are chiefly

A

B

Figure 10.2
Annelid tubeworms secrete the tubes in which they live. They are sessile suspension feeders that have feathered crowns well adapted for capture of food. **A,** Social featherduster worm, *Bispira brunnea.* **B,** Christmas tree worms, *Spirobranchus giganteus.*

nocturnal and come out of their burrows at night to forage. A good way to find them is to search with a flashlight around the rich soil of lawn shrubbery. The large "night crawler" is easily found this way, especially during warm, moist nights of spring and early summer. *Lumbricus terrestris* (L. *lumbricum,* earthworm), named by Linnaeus, is one of the most common earthworms in Europe, Asia, and North America and has been introduced all over the world.

Behavior

☞ **Wet the center of a paper towel with pond or dechlorinated water, leaving the rest of the paper dry. Place a live earthworm on the moist area. Using the following criteria, observe its behavior.**

Is the skin of the worm dry or moist? _____
Do you find any obvious respiratory organs? _____ Where do you think exchange of gases occurs? _____ Would this necessitate a dry or a damp environment? _____ _____ Does the worm respond positively or negatively to moisture? _____ _____

Notice the mechanics of crawling. The earthworm's body wall contains well-developed layers of circular and longitudinal muscles. As it crawls, notice the progressing peristaltic waves of contraction. These are produced by alternate contraction and relaxation of longitudinal and circular muscles in the body wall acting against noncompressible coelomic fluid. An earthworm's body is divided internally into segments by septa. When the longitudinal muscles of a segment contract, the segment becomes shorter and thicker because the volume within each body segment remains constant. Conversely, when the circular muscles contract, the segment elongates. Do the waves of circular muscle contraction move anteriorly or posteriorly when the worm as a whole is moving forward? _____ How far apart (what proportion of total body length) are the waves of contraction? _____ Watch the anterior end as the worm advances. Do the short and thick regions or the long and thin regions advance the head end forward? _____

Run a finger along the side of the worm. Do you detect the presence of small setae (bristles)? How might these setae be used? _____

How does the animal respond when you gently touch its anterior end? _____ Its posterior end? _____ Draw the towel to the edge of the desk and see what happens when the worm's head projects over the edge of the table. Is it positively or negatively thigmotactic (responsive to touch)? _____ Turn the worm over and see if and how it can right itself.

Can you devise a means of determining whether the earthworm is positively or negatively geotactic (responsive to gravity)?

☞ **Place the earthworm on a large plate of wet glass. Does this difference in substrate affect its locomotion? Is friction important in earthworm locomotion?**

Does the earthworm have eyes or other obvious sensory organs? _____ Can you devise a means of determining whether it responds positively or negatively to light?

Written Report

In the lab report on p. 133, record the responses of the earthworm. Comment on hydrotaxis, locomotion, thigmotaxis, phototaxis, and the importance of friction.

External Structure

☞ Anesthetize an earthworm by immersing it for 30 to 40 minutes in 7% ethanol.[1] Once the worm is completely limp when picked up, transfer it to a dissecting tray that has been dampened with water. Examine the worm with a dissecting microscope or hand lens as necessary.

What are the most obvious differences between the earthworm and the clamworm? List two or three here: _____

_____

The first four segments make up the head region. The first segment is the **peristomium.** It bears the **mouth,** which is overhung by a lobe, the **prostomium.** The head of the earthworm, lacking specialized sense organs, is considered degenerate and is not a typical annelid head.

Find the **anus** in the last segment. Observe the saddlelike **clitellum** (L. *clitellae,* packsaddle), which in mature worms secretes the egg capsules into which the eggs are laid. In what segments does it occur?

Because leeches share with earthworms the presence of a clitellum, both are included in the clade Clitellata. Unlike the earthworm's clitellum, which is always present, that of the leech is visible only during the reproductive season.

How many pairs of setae are on each segment, and where are they located? _____ Use the hand lens or dissecting microscope to determine this. What does the name "Oligochaeta" mean? _____ "Polychaeta"? _____ Are these names well chosen?

Earthworms are monoecious. Related to that characteristic, they have many external openings besides the mouth and anus. Note the **male pores** on the ventral surface of somite 15. These are conspicuous openings of the sperm ducts from which spermatozoa are discharged. Also note the two long **seminal grooves** extending between the male pores and the clitellum. These guide the flow of spermatozoa during copulation. Viewing the small **female pores** on the ventral side of segment 14 will require a

hand lens. Here the oviducts discharge eggs. You may not be able to see the openings of 2 pairs of **seminal receptacles** in the grooves between segments 9 and 10 and 10 and 11, or the paired excretory openings, **nephridiopores,** located on the lateroventral surface of each segment (except the first 3 and the last one).

A **dorsal pore** from the coelomic cavity is located at the anterior edge of the middorsal line on each segment, from 8 or 9 to the last one. Many earthworms eject a malodorous coelomic fluid through the dorsal pores in response to mechanical or chemical irritation or when subjected to extremes of heat or cold.

Drawings

✐ Complete the external ventral view of the earthworm in the lab report on p. 132. Draw in and label prostomium, peristomium, mouth, setae, male pores, female pores, seminal grooves, clitellum, and anus.

Internal Structure and Function

☞ Reanesthetize the earthworm in 7% ethanol if necessary. Place the anesthetized worm dorsal side up in a dissecting pan and straighten it by passing one pin through the fourth or fifth segment (just behind the peristomium) and another pin through any segment near the posterior end of the worm. With a razor blade or new scalpel blade, and beginning at about the fortieth segment (just behind the clitellum), cut through the body wall at a point just to one side of the dark middorsal line (the **dorsal blood vessel**). Use fine-tipped scissors to complete the middorsal cut all the way to the head, pulling up on the scissors as you proceed to avoid damaging internal organs. Keep your incision slightly to one side of the dorsal blood vessel. With a pipette, squirt some isotonic salt solution on the internal organs to keep them moist. Now, starting at the posterior end, of the incision, pin the animal open. You will need to break the septa (partitions between the metameres) with a needle as you proceed anteriorly. When you have finished, remove all the pins except those anchoring the worm at the anterior end. Stretch out the worm by pulling gently on the posterior end, and repin, placing the pins at an oblique angle. If you are using a dissecting microscope rather than a hand lens, you may need to position the worm to one side of the dissecting tray for viewing. Now flood the tray with enough isotonic saline (0.6% NaCl) to completely cover the earthworm.

Note the peristaltic movements of the **digestive tract,** which propel food posteriorly. Find three pairs of cream-colored **seminal vesicles** in somites 9 to 12 (Figure 10.3), two pairs of glistening white **seminal**

---

[1]Prepared by diluting 74 ml of 95% ethanol with 1 liter of water. See Appendix A, p. 285, for a note on the use of live versus preserved earthworms for this exercise.

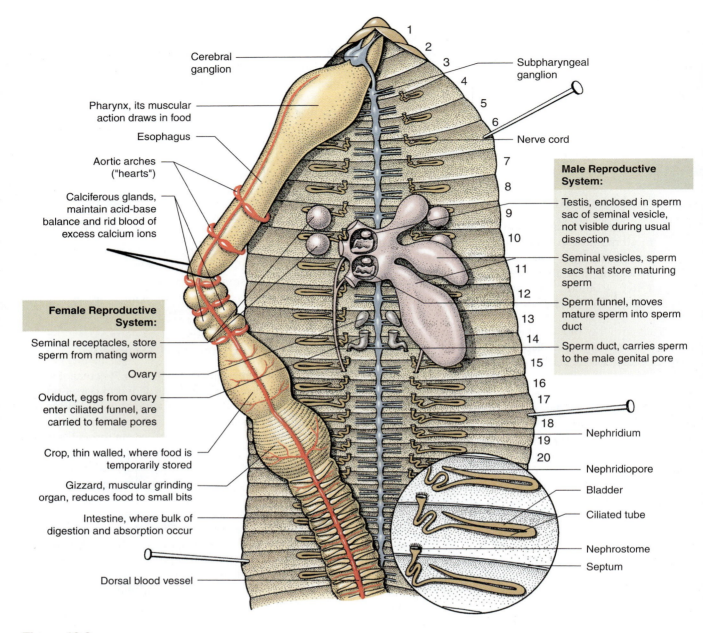

Figure 10.3
Internal structure of *Lumbricus,* dorsal view.

The following labels appear in the figure:

Cerebral ganglion

Pharynx, its muscular action draws in food

Esophagus

Aortic arches ("hearts")

Calciferous glands, maintain acid-base balance and rid blood of excess calcium ions

Female Reproductive System:

Seminal receptacles, store sperm from mating worm

Ovary

Oviduct, eggs from ovary enter ciliated funnel, are carried to female pores

Crop, thin walled, where food is temporarily stored

Gizzard, muscular grinding organ, reduces food to small bits

Intestine, where bulk of digestion and absorption occur

Dorsal blood vessel

Subpharyngeal ganglion

Nerve cord

Male Reproductive System:

Testis, enclosed in sperm sac of seminal vesicle, not visible during usual dissection

Seminal vesicles, sperm sacs that store maturing sperm

Sperm funnel, moves mature sperm into sperm duct

Sperm duct, carries sperm to the male genital pore

Nephridium

Nephridiopore

Bladder

Ciliated tube

Nephrostome

Septum

receptacles in somites 9 and 10, and a pair of delicate, almost transparent tubular **nephridia** in the coelomic cavity of each segment. Note the **dorsal blood vessel** riding on the digestive tract. In which direction is the blood flowing in this vessel? A total of five pairs of pulsating **aortic arches,** sometimes called "hearts," surround the **esophagus** in somites 7 to 11 (some of these arches are covered by the seminal vesicles).

Digestive System. Identify the **mouth;** the muscular **pharynx;** the slender **esophagus** in segments 6 to 13; the thin-walled **crop;** and the muscular **gizzard, intestine,**

and **anus.** Note the functions of these organs as summarized with the labels in Figure 10.3. Locate the **calciferous glands** that lie on both sides of the esophagus (usually partly concealed by the seminal vesicles). Bright yellow or green **chloragogue cells** often cover the intestine and much of the dorsal vessel. They are known to store glycogen and lipids but probably have other functions as well, similar to those of the vertebrate liver. Make an off-center longitudinal cut into the intestine in the region of the clitellum to expose the **typhlosole** (Gr. *typlos,* blind, + *sōlēn,* channel), a ridgelike structure projecting into the lumen of the intestine. The typhlosole increases the surface available for digestive enzyme production and absorption.

Circulatory System. The earthworm has a **closed circulatory system.** Note that both the **dorsal blood vessel** and the **aortic arches** (identified earlier) are contractile, with the dorsal vessel being the chief pumping organ and the arches maintaining a steady flow of blood into the **ventral vessel** beneath the digestive tract.

☞ Retract the digestive tract. Lift up the white nerve cord in the ventral wall. Note the **subneural vessel** clinging to its lower surface and a pair of **lateroneural vessels,** with one located on each side of the nerve cord. Be able to trace blood flow from the dorsal vessel to the intestinal wall and back, to the epidermis and back, and to the nerve cord and back.

Reproductive System. The earthworm is monoecious; it has both male and female organs in the same individual, but cross-fertilization occurs during copulation. First, consider the **male organs** (Figure 10.3). The 3 pairs of **seminal vesicles** (sperm sacs in which spermatozoa mature and are stored before copulation) are attached in somites 9, 11, and 12; they lie close to the esophagus. The 2 pairs of small, branched **testes** are housed in special reservoirs in the seminal vesicles, and the two small sperm ducts connect the testes with the **male pores** in somite 15; however, both testes and ducts are too small to be found easily. The **female organs** are also small. The two pairs of small, round **seminal receptacles,** easily seen in somites 9 and 10, store spermatozoa after copulation. You should be able to find the paired **ovaries** that lie ventral to the third pair of seminal vesicles. You probably will not be able to see the paired **oviducts** with ciliated funnels that carry eggs to the female pores in the next segment.

Earthworm Copulation. When mating, two earthworms, attracted to each other by glandular secretions, extend their anterior ends from their burrows and, with their heads pointing in opposite directions, join their ventral surfaces in such a way that the seminal receptacle openings of one worm lie in opposition to the clitellum of the other (Figure 10.4). Both worms secrete quantities of mucus so that each is enveloped in a slime tube extending from segment 9 to the posterior end of the clitellum. Seminal fluid discharged from the sperm ducts of each worm is carried along the seminal grooves by contraction of longitudinal muscles and enters the seminal receptacles of the mate. After copulation, the worms separate, and each clitellum produces a secretion that finally hardens over its outer surface. The worm moves backward, drawing the hardened tube over its head (Figure 10.4C). As it is moved forward, the tube receives eggs from the oviducts, sperm from the seminal receptacles, and a nutritive albuminous fluid from skin glands. Fertilization occurs in the cocoon. As the worm withdraws, the cocoon closes and is deposited on the ground. Young worms hatch in 2 to 3 weeks.

Excretory System. A pair of tubular **nephridia** lies in each somite except the first three and the last one. Each nephridium begins with a ciliated, funnel-shaped **nephrostome,** which projects through the anterior septum of the segment and opens into the next anterior segment.

☞ Use a dissecting microscope to examine a nephridium. They are largest in the region just posterior to the clitellum. With fine-tipped scissors, carefully remove a nephridium along with a small portion of the septum through which the nephrostome projects. Mount it on a slide with a drop or two of saline solution, cover with a coverslip, and examine with a compound microscope.

Note the slender tubule passing from the nephrostome to the looped nephridium, and observe the ciliary activity in one narrow portion of the tubule. You may also see parasitic nematodes in the large bladder segment of the tubule. Coelomic fluid is drawn by ciliary activity into the nephrostome and then flows through the narrow tubule where ions, especially sodium and chloride, are reabsorbed. Urine, containing wastes, collects in the bladder, which empties to the outside through a **nephridiopore.**

Nervous System

☞ If you have not already done so, extend the dorsal incision to the first segment.

Find a small pair of white **cerebral ganglia** (the brain), lying on the anterior end of the pharynx and partially hidden by dilator muscles; small white **nerves** from the ganglia to the prostomium; a pair of **circumpharyngeal connectives,** extending from the ganglia and encircling the pharynx to reach the **subpharyngeal ganglia** under the pharynx; and a **ventral nerve cord,** extending posteriorly from the subpharyngeal ganglia for the entire length of the animal. Remove or lay aside the digestive tract, and examine the nerve cord with a hand lens to see in each body segment a slightly enlarged **ganglion** and **lateral nerves.**

Oral Report

Be prepared to (1) demonstrate your dissection to the instructor, (2) point out both the external and internal structures you have studied, and (3) explain their functions.

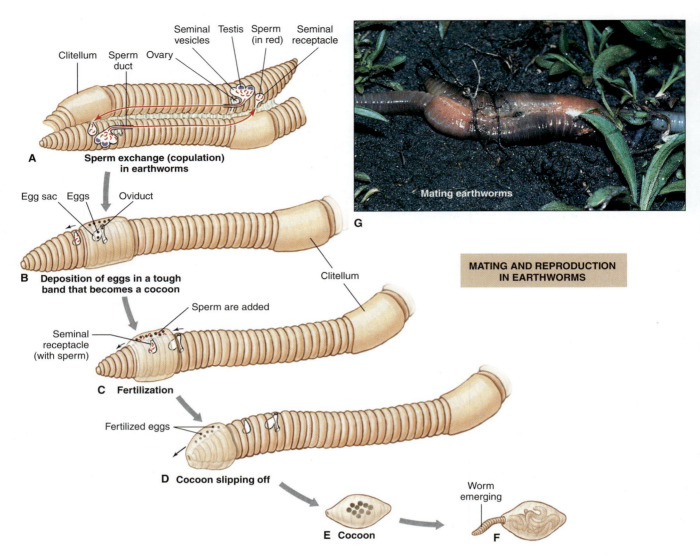

Figure 10.4
Earthworm copulation and formation of egg cocoons. **A,** Mutual insemination; sperm from genital pore (segment 15) pass along seminal grooves to seminal receptacles (segments 9 and 10) of each mate. **B,** and **C,** After the worms separate, the clitellum secretes first a mucous tube and then a tough band that forms a cocoon. The developing cocoon passes forward to receive eggs from oviducts and sperm from seminal receptacles. **D,** As the cocoon slips off over the anterior end, its ends close and seal. **E,** The cocoon is deposited near a burrow entrance. **F,** Young worms emerge in 2 to 3 weeks. **G,** Two earthworms in copulation. Their anterior ends point in opposite directions as their ventral surfaces are held together by mucous bands secreted by the clitella.

Further Study

Histology of Cross Section

☞ Examine a stained slide with low power. Note the tube-within-a-tube arrangement of intestine and body wall (Figure 10.5). Identify the following:

Cuticle. Thin, noncellular, and secreted by the epidermis.

Epidermis (ectodermal). Columnar epithelium containing mucous gland cells. Mucus prevents the skin from drying out.

Circular muscle layer. Smooth muscle fibers running around the circumference of the body. How does their contraction affect the body shape?

Longitudinal muscle layer. Thick layer of featherlike fibers that run longitudinally. The muscle layers may be interrupted by the setae and the dorsal pore.

Peritoneum (mesodermal). The peritoneum (Gr. *peritonaios,* stretched around), a thin epithelial layer lining the body wall and covering the visceral organs (Figure 10.5B). Peritoneum lining the body

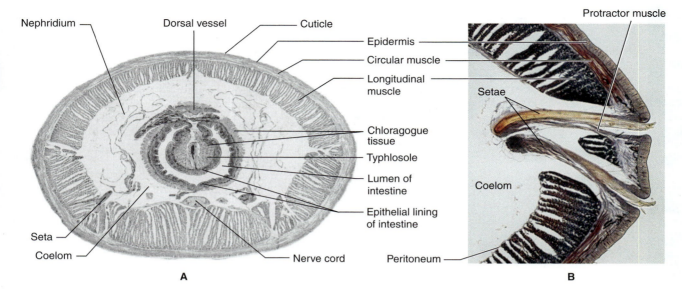

Figure 10.5
A, Cross section of an earthworm through the intestinal region. **B,** Portion of a cross section of an earthworm, showing one set of setae with their protractor muscles.

wall is called parietal (L. *paries,* wall) peritoneum. Peritoneum covering the digestive tract and other visceral organs is called visceral (L. *viscera,* bowels) peritoneum.

Setae. If present, they are brownish spines in a sheath secreted by epidermis. They are moved by tiny muscles (Figure 10.5B).

Coelom. Space between the **parietal peritoneum,** which lines the body wall, and the **visceral peritoneum,** which covers the intestine and other organs.

Alimentary canal. The intestine, surrounded by chloragogue tissue, which plays a role in intermediary metabolism similar to that of the liver in vertebrates. Inside the chloragogue layer is a layer of **longitudinal muscle.** Why does it appear as a circle of dots? Next is a **circular muscle layer,** followed by a layer of ciliated columnar epithelium (endodermal), which lines the intestine. Intestinal contents are moved along by peristaltic movement.

Is such movement possible without longitudinal and circular muscles? _____ Is peristalsis possible in the intestine of the flatworm or *Ascaris?* _____

Ventral nerve cord. Use high power to identify the three **giant fibers** in the dorsal side of the nerve cord, and the nerve cells and fibers in the rest of the cord.

Blood vessels. Identify the **dorsal vessel** above the typhlosole, the **ventral vessel** below the intestine, the **subneural vessel** below the nerve cord, and the **lateral neurals** beside the nerve cord.

Some slides may also reveal parts of **nephridia, septa, mesenteries,** and other structures.

Drawing

In the lab report on page 132, sketch and label a cross section of the earthworm as it appears on your slide.

Name_____

Date_____

Section_____

Phylum_____

Genus_____

Earthworm

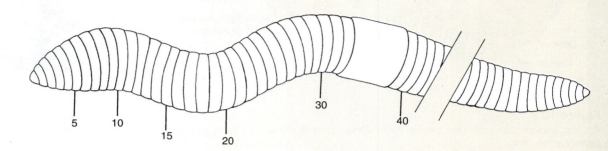

External structure of the earthworm, ventral view

Cross section through the body of an earthworm

Observations on the Behavioral Responses of the Earthworm

Hydrotaxis _____
_____
_____
_____
_____
_____
_____

Locomotion _____
_____
_____
_____
_____
_____
_____

Thigmotaxis (response to touch) _____
_____
_____
_____
_____
_____
_____
_____

Phototaxis _____
_____
_____
_____
_____
_____
_____

Importance of friction _____
_____
_____
_____
_____
_____
_____
_____

EXERCISE 10C

Class Hirudinida—Leeches

Hirudo, the Medicinal Leech

Phylum Annelida
    Class Hirudinida
        Order Gnathobdellida
            Genus *Hirudo*
                Species *Hirudo medicinalis*

Where Found

Leeches are predaceous and mostly fluid feeders. Some are true bloodsuckers, attaching themselves to a host during feeding periods.

Hirudo is one of the freshwater leeches found in lakes, ponds, streams, and marshes. Most marine leeches feed on turtles, fishes, dolphins, and other species. *H. medicinalis* (Figure 10.6) is often referred to as the "medicinal leech" because it was formerly used in blood-letting and is still occasionally used in plastic surgery and in the treatment of hematomas. This leech feeds on the blood of vertebrates, to which it attaches itself periodically. Living leeches are readily available from biological supply houses.

Behavior

Note the use of the ventral **suckers**—the smaller oral sucker at the anterior end and the larger caudal sucker at the posterior end.

☞ Attempt to pull an attached leech free from its substrate.

Are the suckers powerful? To creep, the leech combines the use of its suckers with muscular body contractions.

☞ Place a leech on a glass plate and watch it move.

The oral sucker attaches and the body contracts; then the caudal sucker attaches and the body extends forward before the oral sucker attaches again. How does the movement of a leech on the glass plate compare to the movement of the earthworm you previously observed? _____

_____

☞ Drop a leech into the water to observe its undulating, free-swimming motions.

Medicinal leeches are attracted to heat. Why would this be? _____
To investigate this further, see the Experimenting in Zoology feature at the end of the chapter.

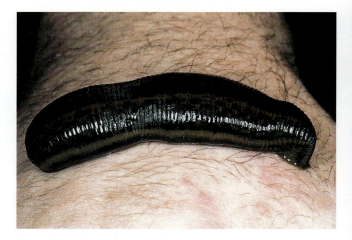

Figure 10.6
Medicinal leech, *Hirudo medicinalis,* feeding on blood from an obliging student.

External Features

☞ Study a preserved specimen of *Hirudo*.

How does its shape compare with that of the earthworm? _____ Can you distinguish between the dorsal and ventral surfaces? _____ How? _____ _____ The leech is segmented inside and outside, but externally the segments are also marked off into 1 to 5 **annuli** each, the larger number found in segments 9 to 23 inclusive (Figure 10.7). Each true segment bears a pair of **nephridiopores** and, on one of its annuli, a row of **sensillae** (used mainly to detect water movements) or (at the anterior end) eyespots.

The oral sucker contains the **mouth.** The **anus** is located in the middorsal line at the junction with the caudal sucker. The **male genital pore** is located ventrally on segment 11, and the **female pore** is on segment 12. Segments 10, 11, and 12 serve as the functional **clitellum,** secreting the capsule in which the eggs develop. Leeches are monoecious.

Classification

The three groups of annelids are classified chiefly on the basis of presence or absence of a clitellum, parapodia, setae, annuli, and other features.

Phylum Annelida

Class Polychaeta (pol′e-ke′ta) (Gr. *polys,* many, + *chaitē,* long hair). Segmented inside and out; parapodia with many setae; distinct head with eyes, palps, and tentacles; no clitellum; separate sexes; trochophore larva usually present; mostly marine. Examples: *Nereis, Chaetopterus.*

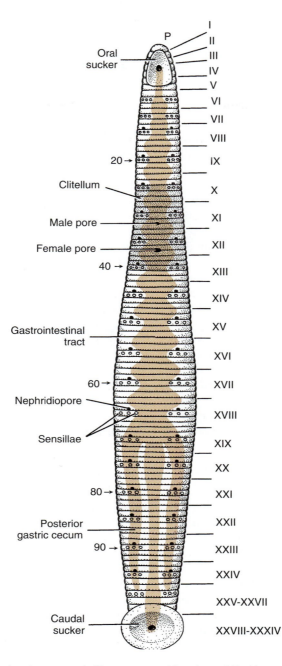

Figure 10.7

External anatomy of *Hirudo medicinalis,* showing segments (Roman numerals) and annuli (Arabic numerals). The position of the gut, not visible externally, is indicated in color.

Class Oligochaeta (ol'i-go-ke'ta) (Gr. *oligos,* few, + *chaitē,* long hair). Body segmented inside and out; number of segments variable; clitellum present; few setae; no parapodia, head poorly developed; coelom spacious and usually divided by intersegmental septa; direct development; chiefly terrestrial and freshwater. Examples: *Lumbricus, Tubifex.*

Class Hirudinida (hir'u-din'i-da) (L. *hirudo,* leech, + *ea,* characterized by). Segments 33 or 34 in number, with many annuli; clitellum present; anterior and posterior suckers; setae absent (except *Acanthobdella*); parapodia absent; coelom closely packed with connective tissue and muscle; terrestrial, freshwater, and marine. Examples: *Hirudo, Placobdella.*

EXPERIMENTING IN ZOOLOGY
Behavior of Medicinal Leeches, *Hirudo medicinalis*

Medicinal leeches, *Hirudo medicinalis,* are best known because of their use in medical procedures. Historically, these leeches were used to draw blood from victims of bites or stings from insects or venomous animals. In addition, during the nineteenth century, leeches were used to draw out "corrupt blood" from people suffering from disease. Today these annelids are frequently used to restore venous circulation after reconstructive surgical procedures, particularly when a finger, a toe, or an ear is surgically reattached after an accident. In 2004 the Food and Drug Administration (FDA), for the first time, cleared the commercial marketing of leeches for medicinal purposes. Aside from their use in modern medicine, medicinal leeches are wild animals that have evolved specific adaptations for locating their mammalian prey. Because mammals are endotherms, their body temperatures typically are above ambient water temperatures. Medicinal leeches are able to locate mammals that have entered their ponds or lakes because they have evolved the ability to detect the slightly elevated water temperatures generated by the presence of a mammal.

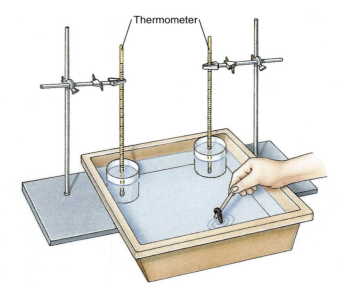

Figure 10.8
Setup for observing the behavior of leeches.

Getting Ready

Work with a partner for this exercise. Obtain a plastic tub approximately 40 cm wide by 60 cm long. Wash the tub and fill it to a depth of 8 to 10 cm using pond water, bottled water, or dechlorinated tap water. After taking the temperature of the water in the tub, fill two 500 ml beakers with water of the same temperature. Place the two filled beakers onto opposite sides of the tub about 30 cm from one end (Figure 10.8). Use ring stands and clamps to suspend thermometers for monitoring the temperatures throughout the experiment. Have a stopwatch available for timing leech behavior.

How to Proceed

Use a short wooden dowel to choose an active medicinal leech. The leech will attach itself to the dowel or balance momentarily on it while you transfer it to your experimental tub. If the leech will not stay on the dowel, gently place it on a paper towel to transfer it to the tub. Release the leech at the end of the tub. Try to release the leech so as not to bias its behavior toward either of the beakers. How does the leech behave when you release it? Does it go toward the beakers? When leeches find potential prey items or simply want to

crawl on a substrate, they first attach to an object using their anterior sucker. Does the leech attach itself to the side of the tub? Does it attach itself to one of the beakers? If so, how much time expires between when you first release the leech and when it attaches itself to a beaker? Once the leech attaches itself to a beaker or the tub, gently remove the leech with the wooden dowel and repeat the trial by replacing the leech at the end of the tub opposite the beakers.

After three trials with the beakers at room temperature, replace the water in one beaker with water warmed to 40°C. Again monitor the temperatures in the beakers. Watch that the temperature in the warm beaker remains at approximately 40°C. Release the leech at the end of the tub opposite the beakers. Does the leech attach itself to one of the beakers? To which beaker does it attach? How much time elapses between release and attachment? Remove the leech from the beaker and conduct at least two more trials.

Does the leech appear to be attracted to the beaker filled with warm water? Compare your results with those of your classmates. Your instructor may want you to combine your data with your classmates' and compare control treatments with experimental treatments, using a statistical test. How many times out of the total control trials does a leech attach to a beaker? How many times in the warm beaker trials? Do the leeches attach to the warm water beakers faster than when there are only beakers filled with room temperature water?

Questions for Independent Investigation

Medicinal leeches are hearty animals that are ideal specimens for these types of experiments. While repeated trials using the same animal violates the important statistical assumption of independence (see Exercise 1), you may be able to borrow five or six of the other leeches ordered for use by your classmates to conduct an independent investigation on one of the following questions:

1. How small may the temperature difference be between "prey" and the environment to be detected by the leech? How do leeches respond when presented with two beakers, each filled with warm water differing by only a degree or two?

2. Leeches are known to be sensitive to chemical stimuli in the water. Medicinal leeches will feed on beef liver. Can they detect the odor of beef liver? Explore this question by using a setup similar to that described for examining their sensitivities to temperature differences. Place beef liver inside a plastic beaker that has been pierced repeatedly with a heated pin or needle.

3. How do leeches respond when presented with a choice of attaching either to a beaker filled with warm water (thermal cues) or to one that is releasing food odor (chemical cues)? Which appears to be the more attractive stimulus?

References

Dickinson, M. H., and C. M. Lent. 1984. Feeding behavior of the medicinal leech, *Hirudo medicinalis*. Jour. Comp. Phys. 154:449–455.

Lent, C. M., and M. H. Dickinson. 1987. On the termination of ingestive behavior by the medicinal leech. Jour. Exp. Bio. **131:**1–15.

Lent, C. M., K. H. Fliegner, E. Freedman, and M. H. Dickinson. 1988. Ingestive behavior and physiology of the medicinal leech. Jour. Exp. Bio. **137:**513–527.

The Chelicerate Arthropods
Phylum Arthropoda

There are approximately a million named species of arthropods, a huge and diverse phylum that embraces the chelicerates (spiders, scorpions, and their allies), the crustaceans, and the uniramous arthropods (millipedes, centipedes, insects, and their kin).

Arthropods have **jointed appendages,** a characteristic that has given them their name (Gr. *arthron,* joint, + *pous, podos,* foot). Like the annelids, the arthropod body is constructed of an extended series of repeated segments. This design principle is called **metamerism,** or serial segmentation. However, unlike the annelids, the

---

EXERCISE 11

The Chelicerate Arthropods—Horseshoe Crab and Garden Spider
Horseshoe Crab
Garden Spider

Classification: Phylum Arthropoda

---

initial array of repeated segments has, through fusion, reduction, and specialization, evolved into the many divergent anatomies of advanced arthropods. A major

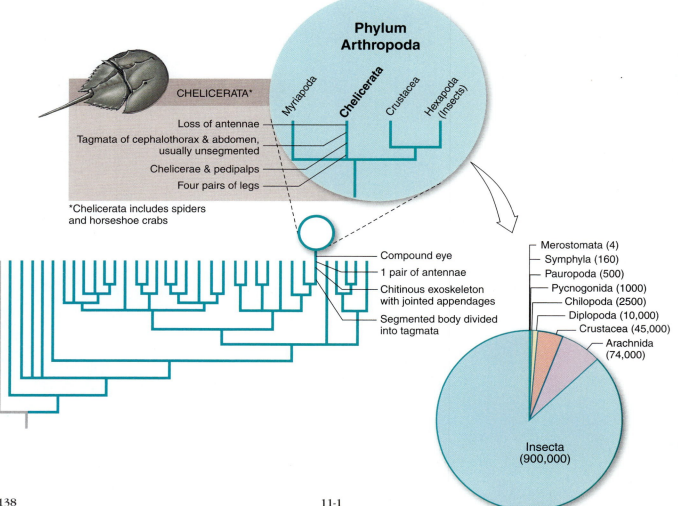

CHELICERATA*

Loss of antennae
Tagmata of cephalothorax & abdomen, usually unsegmented
Chelicerae & pedipalps
Four pairs of legs

*Chelicerata includes spiders and horseshoe crabs

Phylum Arthropoda

Myriapoda   Chelicerata   Crustacea   Hexapoda (Insects)

Compound eye
1 pair of antennae
Chitinous exoskeleton with jointed appendages
Segmented body divided into tagmata

Merostomata (4)
Symphyla (160)
Pauropoda (500)
Pycnogonida (1000)
Chilopoda (2500)
Diplopoda (10,000)
Crustacea (45,000)
Arachnida (74,000)
Insecta (900,000)

evolutionary shift was the fusion of segments into discrete, functional units called **tagmata** (sing., **tagma**) (Gr. *tagma,* arrangement, order, row)—for example, the head, thorax, and abdomen of insects, or the cephalothorax and abdomen of spiders and ticks. Another important evolutionary trend is the specialization of appendages. You will see several examples of appendage specialization among the three arthropod subphyla to be studied in this and the following two exercises.

In addition to features shared by other phyla, such as **triploblastic development,** a **true coelom, bilateral symmetry, cephalization,** and **all organ systems,** arthropods have developed **striated muscle** for rapid movement; an **exoskeleton,** or cuticle, containing the tough nitrogenous polysaccharide **chitin** for support and protection; **gills** and a very efficient **tracheal system** for gas exchange; and **greater specialization** of body organs, especially very specialized form and function in the appendages. The coelom is much reduced. Instead, the major body space is a **hemocoel** derived from the embryonic blastocoel. It is filled with **hemolymph,** which circulates through the arthropod's "open" circulatory system.

EXERCISE 11
The Chelicerate Arthropods— Horseshoe Crab and Garden Spider

The chelicerates include horseshoe crabs, sea spiders, spiders, scorpions, mites, ticks, and some others. They are arthropods that do not possess mandibles (jaws) for chewing. Instead, the first pair of appendages, called chelicerae (ke-lis'er-e), are feeding appendages adapted for seizing and tearing. Most chelicerates have a two-part body composed of a cephalothorax (prosoma) and an abdomen (opisthosoma). There are no antennae.

Core Study

Horseshoe Crab

Phylum Arthropoda
    Subphylum Chelicerata
        Class Merostomata
            Subclass Xiphosurida
                Genus *Limulus*
                    Species *Limulus polyphemus*

Where Found

Horseshoe crabs are not really crabs at all but members of an ancient group of chelicerates, the Merostomata (mer'o-sto'ma-ta; Gr. *mēros,* thigh, + *stoma,* mouth), most of which are long extinct.

Horseshoe crabs are called "living fossils" because they are survivors of a group that was abundant during the Silurian period, some 425 million years ago. The genus *Limulus,* to which the living horseshoe crab *Limulus polyphemus*[1] belongs, arose 250 million years ago and has changed little during this vast span of time. Horseshoe crabs are marine bottom-dwellers that feed on molluscs, worms, and dead fish. They live along the Atlantic coast from Nova Scotia to Mexico's Yucatan peninsula, but are especially abundant along the coasts of Virginia, Delaware, and New Jersey where they come ashore in great numbers during spring high tides to spawn. Their eggs are important food for millions of migratory shorebirds, who feast on the eggs and may more than double their weight before continuing their northward migration. Until recently, farmers used horseshoe crabs as fertilizer and animal feed.[2] Today, the blood of horseshoe crabs is important for biomedical research and for screening vaccines and IV fluids for pathogens.

External Features

The entire body of the horseshoe crab is covered with a tough, leathery **exoskeleton** that contains chitin, a tough polysaccharide. Among the products made from the chitin of a horseshoe crab is a surgical suture thread that actually promotes healing as it slowly dissolves. As the animal grows, it must shed (molt) the exoskeleton, a process called **ecdysis.**

    Cephalothorax (Prosoma). Covering the cephalothorax dorsally and laterally is a hard, horseshoe-shaped **carapace** (F. from Sp. *carapacho,* shell), concave below and convex above. A pair of lateral compound eyes and a pair of median simple eyes are on the dorsal side (Figure 11.1A).

    On the ventral side are six pairs of appendages, located around the mouth. Using Figure 11.1 as a guide, and noting function, find the first pair of appendages, the **chelicerae** (ke-liss'uh-ree; sing. **chelicera;** Gr. *chēlē,* claw, + *keras,* horn); the second pair, the **pedipalps** (L. *pes, pedis,* foot, + *palpus,* stroking); and the **walking legs.** All appendages except the last pair of walking legs and pedipalps of the male are chelate—that is, they bear pincers, or **chelae** (ke'lee; sing. **chela;** Gr. *chēlē,* claw). Lacking specialized appendages for chewing food, such as the mandibles of other arthropods, the walking legs bear spiny processes called **gnathobases** (Gr. *gnathos,* jaw, + *basis,* base) on their basal segments that serve a similar function. Move the appendages and note how these processes would tear up food and move it toward the mouth. The chelae of the appendages pick up food and pass it to the

---

[1]The curious generic name of the horseshoe crab, *Limulus,* derives from the Latin meaning "a little askew" or "odd." The specific epithet, *polyphemus,* derives from the one-eyed giant of Greek mythology and was based on the mistaken notion that the animal had a single eye.

[2]Although farmers no longer harvest significant numbers of horseshoe crabs for fertilizer, eel and conch fishermen collect horseshoe crabs for bait. Responding to concern from scientists, environmentalists, and the biochemical industry that populations were declining rapidly, the Atlantic States Marine Fisheries Commission adopted an interstate management plan in 1998 that has helped to reduce capture of horseshoe crabs for bait.

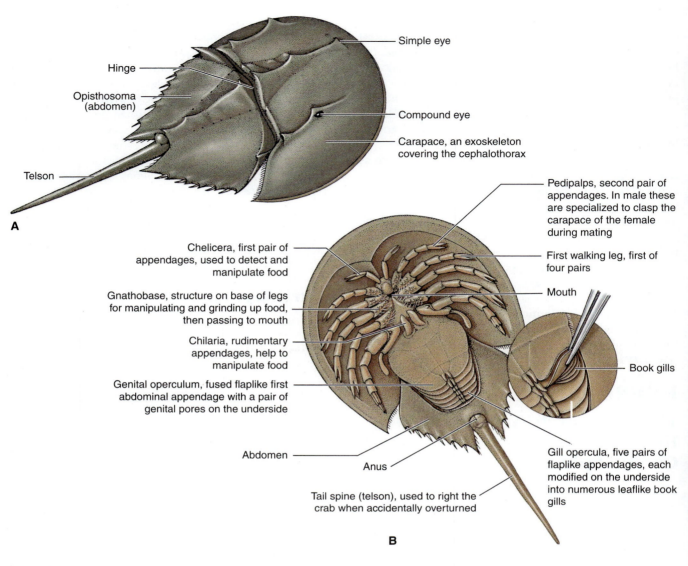

Figure 11.1
A, Dorsal view of *Limulus,* the horseshoe crab. **B,** Ventral view of a female. The inset shows an operculum lifted to reveal the location of book gills.

gnathobases. Note that the last pair of walking legs has no chelae; instead, each has four movable, bladelike processes, one of which is tipped with a pair of spines. These legs are used to push against the sand to help in forward movement and in burrowing. Between the last pair of walking legs is a small, rudimentary pair of appendages called **chilaria.**

Abdomen (Opisthosoma). The abdomen bears six pairs of spines along the sides and, on its ventral side, six pairs of flat, platelike appendages. The first of these forms the **genital operculum,** on the underside of which are two **genital pores.**

The other five abdominal appendages are modified as **gills.** Lift up one of these flaps to see the many (100 to 150) leaflike folds called **lamellae** (la-mel′ee; sing. **lamella;** L. dim. of *lamina,* plate). Because of this leaflike arrangement, such gills are called **book gills** (Figure 11.1).

Exchange of gases between blood and surrounding water takes place in the lamellae. Movement of the gills not only circulates water over them but also pumps blood in and out of the lamellae. The blood contains **hemocyanin,** a respiratory pigment used in oxygen transport, as well as an amebocyte that kills invading bacteria. A chemical extracted from the amebocytes is now used to assure absence of harmful bacteria in intravenous drugs, vaccines, and implantable medical devices. Each summer, blood is collected from thousands of horseshoe crabs for extraction of amebocytes; the crabs are returned to the sea after bleeding.

Beating of the abdominal flaps can also be used in swimming and may aid the animal in burrowing by creating a water current that washes out mud or sand posteriorly.

Telson. The long, slender **telson** (Gr., extremity), or tail spine, is used for anchoring when the animal is burrowing or plowing through the sand or in righting itself when turned over. The **anus** is located under the proximal end of the telson.

Reproduction

During mating and egg-laying, horseshoe crabs aggregate in shallow water during high tides of full and new moons in spring and summer. A male clasps a female's carapace with his modified pedipalps and is carried around by the larger female. The female lays eggs in a depression in the sand near the high-tide mark while the male sprays sperm on the eggs as they emerge. After several weeks, the eggs hatch as free-swimming **trilobite larvae** (Figure 11.2), so named because of their superficial resemblance to a trilobite. A larva looks much like an adult except that it lacks the tail spine and has only two of the five pairs of book gills. As it develops through a series of molts, segments and appendages are added until the young animal reaches adult form.

Behavior

☞ If live horseshoe crabs are available in a marine aquarium, you may make some observations.

Where do you find the resting animals—swimming about, resting on the sand, or covered with sand? Before disturbing them, drop some bits of fresh shrimp, oyster, or fish meat near them. Do they respond? How? (Horseshoe crabs normally do most of their feeding at night.)

To observe their respiratory gill movements, lift up a crab in the water so that you can watch the beating of the gills. The gill movement is probably faster than when the animal is resting. Why? Each time the flaps move forward, blood flows into the gill lamellae; as the flaps move backward, the blood flows out.

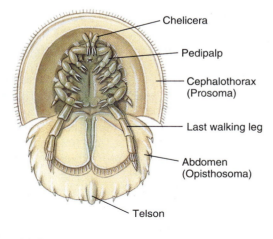

Figure 11.2
Trilobite larva of the horseshoe crab.

Free the animal near the surface of the water and see how it swims. Does it turn over? Which appendages does it use in swimming? As it settles to the sandy bottom, watch its reactions. Does it try to burrow? How? Can you see the use of the last pair of legs? Does the animal arch its back? Does it use the telson?

Written Report

✍ Record your observations on separate paper.

Further Study

Garden Spider
Phylum Arthropoda
  Subphylum Chelicerata
    Class Arachnida
      Order Araneae
        Genus *Argiope*

Where Found

Spiders are distributed in all kinds of habitats, including forests, deserts, mountains, swamps, land, and water. The garden spider (*Argiope,* ar-jī′uh-pee; Gr., nymph of mythology), which builds its orb webs in sunny places in gardens and tall grass, is found throughout the United States and southern Canada. Two common species are *A. aurantia,* with a mottled black and yellow abdomen (Figure 11.3), and *A. trifasciata,* with black and yellow bands on its abdomen. The males are about one-fourth the size of the females (4 to 8 mm) and are rarely seen. They spin smaller webs near those of the female.

Behavior

The garden spider will spin its orb web in captivity, and if kept in a large glass container provided with tall grass or twigs and covered with screen or cheesecloth, the process can be watched. The spider spins a symmetrical orb web and prefers to hang head downward in the center, holding its forelegs and hindlegs close together (Figure 11.3). If possible, examine bits of the silk under a microscope. Is there more than one kind of silk? _____

Adding insects to the terrarium may allow you to watch the spider capture an insect in the net, bite the prey to paralyze it, and then secure the prey with the silken thread.

Spiders secrete enzymes to begin the digestive process outside the body. *Argiope* can apparently crush and tear the prey as well as suck out the liquid parts. How long does it take the spider to complete a meal? _____

Figure 11.3
The black and yellow garden spider, *Argiope aurantia*, builds its orb web in gardens or tall grasses and then hangs there, head down, awaiting an unwary insect. The dense vertical band of silk makes the web conspicuous, reducing the chance a bird will accidentally fly through the web.

External Features

☞ Study a preserved specimen, handling it gently because there is always danger of breaking off appendages or the abdomen from the rest of the body. Use the hand lens or a dissecting microscope to examine the parts. Keeping the specimen moist will help prevent its becoming brittle.

The chitinous **exoskeleton** is hard, thin, and somewhat flexible. **Sensory hairs** project on all parts of the body. The tagmata of the arachnid include the anterior **cephalothorax** and the posterior **abdomen** joined by a slender waist, or **pedicel** (Figure 11.4A).

Cephalothorax. Most spiders have six to eight **eyes** on the anterior dorsal surface, but some have fewer. Spiders do not have compound eyes; all are simple ocelli. How many eyes does your specimen have? _____ Use the dissecting microscope to find them.

Can you see the eyes? _____ Most spiders have poor vision but are covered with sensory hairs and are very sensitive to touch and vibration.

Identify the paired chelicerae, which are vertically oriented on the front of the face. The terminal segment of a chelicera is a **fang** by which the spider ejects poison from its poison gland (Figure 11.4B). Does the spider have true jaws (mandibles)? _____ The pedipalps are six-jointed and used for gripping the prey. In the male, the pedipalp is modified as an intromittent organ to transfer sperm to the female. The basal parts (coxal endites) of the pedipalps are used to squeeze and chew the food. Find the mouth between the pedipalps. Can the spider ingest only liquid food?

How many pairs of **walking legs** are there? _____ Each leg is made up of seven segments, as follows (from base to distal end): **coxa, trochanter, femur, patella, tibia, metatarsus,** and **tarsus.** The tarsus has claws and a tuft of hair at its terminal end.

Abdomen. On the ventral surface at the anterior end, find two lateral, slitlike openings that mark the location of the **book lungs.** Book lungs are fashioned very much like the book gills of the horseshoe crab, except that they are enclosed internally in pockets. The inner walls of these pockets are folded into long, thin plates (the leaves of the "book") held apart by bars so that there are always air spaces between them. Gas exchange occurs between the blood circulating inside the lamellae and the air flowing in the spaces between the lamellae. These air spaces connect with a small air chamber in each lung that opens to the outside through the slitlike openings, or spiracles, already observed (Figure 11.4A). Some spiders—tarantula, for example—have two pairs of book lungs. By peeling forward the cover of the lung and examining with the dissecting microscope, you should be able to see the many thin respiratory leaves.

Between the spiracles, locate the **epigynum** (e-pij'in-um; Gr. *epi,* on, + *gynē,* woman), which conceals the female genital pore. Preserved specimens are probably all females.

Posteriorly on the abdomen, just in front of the spinnerets, is a small **tracheal spiracle.** This is an opening into a small chamber from which tracheal tubes extend into the body. What is the function of the tracheal system? _____ Arachnid tracheal systems are similar to those of insects but less extensive. The garden spider has both book lungs and tracheae, but not all spiders do; some have only one type of respiratory organ.

There are three pairs of **spinnerets** on a raised surface of the posterior abdomen. The middle pair is

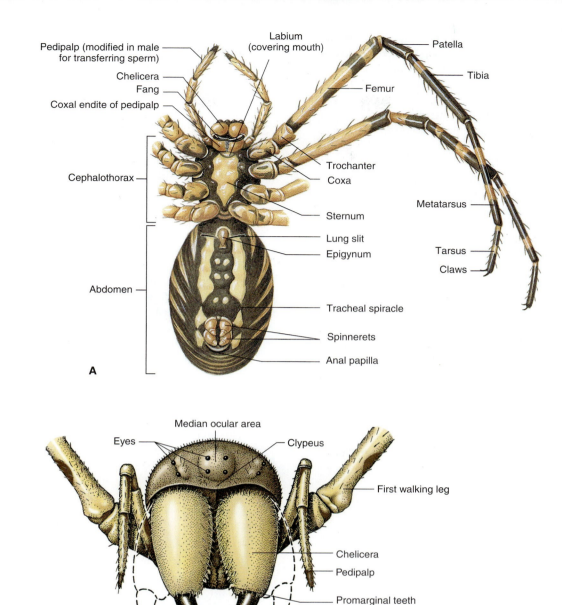

Figure 11.4

A, Ventral view of a female garden spider, *Argiope aurantia*. **B,** Anterior, or "face," view.

quite small, but the other two pairs are rather large and conical and in life are readily movable. The ends of the spinnerets have a variety of tiny silk spouts, each producing a particular type of silk. The silk is secreted as a fluid by silk glands located in the posterior abdomen. The silk hardens on exposure to the air. Examine the spinnerets with a dissecting microscope.

A small, fleshy papilla just posterior to the spinnerets bears the **anus.**

Classification

Phylum Arthropoda

Subphylum Trilobita (tri′lo-bi′ta) (Gr. *tri,* three, + *lobos,* lobe). Trilobites. All extinct forms; Cambrian to Carboniferous; body divided by two longitudinal furrows into three lobes; distinct head, thorax, and abdomen; biramous (two-branched) appendages.

Subphylum Chelicerata (ke-liss'uh-ra'ta) (Gr. *chēlē,* claw, + *keras,* horn, + *ata,* group suffix). Eurypterids, horseshoe crabs, spiders, and ticks. First pair of appendages modified to form chelicerae; pair of pedipalps and four pairs of legs; no antennae, no mandibles; cephalothorax and abdomen usually unsegmented.

Class Merostomata (mer'o-sto'ma-ta) (Gr. *mēros,* thigh, + *stoma,* mouth, + *ata,* group suffix). Aquatic chelicerates that include horseshoe crabs (*Limulus*) and extinct Eurypterida.

Class Pycnogonida (pik'no-gon'i-da) (Gr. *pyknos,* compact, + *gony,* knee, angle). Sea spiders.

Class Arachnida (ar-ack'ni-da) (Gr. *arachnē,* spider). Spiders, scorpions, and their allies. Segments fused into cephalothorax; head with paired chelicerae and pedipalps; four pairs of legs; abdomen segmented or unsegmented, with or without appendages; respiration by gills, tracheae, or book lungs. Example: *Argiope.*

Subphylum Crustacea (crus-ta'she-a) (L. *crusta,* shell, + *acea,* group suffix). Crustaceans. With gills; body covered with carapace; exoskeleton with limy salts; appendages currently considered biramous and variously modified for different functions; head with two pairs of antennae. Examples: *Cambarus, Homarus.*

Subphylum Myriapoda (mir-ee-ap'o-da) (Gr. *myrias,* a myriad, + *podus,* foot). Centipedes, millipedes, pauropods, and symphylans. All appendages uniramous; head appendages consisting of one pair of antennae, one pair of mandibles, and one or two pairs of maxillae.

Class Diplopoda (di-plop'o-da) (Gr. *diploos,* double, + *pous, podos,* foot). Millipedes. Subcylindric body elongated and wormlike; variable number of segments; usually two pairs of legs to a segment. Example: *Spirobolus.*

Class Chilopoda (ki-lop'o-da) (Gr. *cheilos,* lip, + *pous, podos,* foot). Centipedes. Elongated with dorsoventrally flattened body; variable number of segments, each with pair of legs; tracheae present. Example: *Lithobius.*

Class Pauropoda (pau-rop'o-da) (Gr. *pauros,* small, + *pous, podos,* foot). Pauropods. Minute, soft-bodied forms with 12 segments and 9 or 10 pairs of legs. Example: *Pauropus.*

Class Symphyla (sym'fy-la) (Gr. *syn,* together, + *phylon,* tribe). Garden centipedes. Centipede-like bodies of 15 to 22 segments and usually 12 pairs of legs. Example: garden centipede, *Scutigerella.*

Subphylum Hexapoda (hek-sap'oda) (Gr. *hex,* six + *pous, podus,* foot). Insects. Body with distinct head, thorax, and abdomen; thorax usually with two pairs of wings; three pairs of jointed legs. Example: lubber grasshopper, *Romalea.*

The Crustacean Arthropods

EXERCISE 12
Subphylum Crustacea—Crayfish, Lobsters, and Other Crustaceans

U ntil recently, the crustaceans, the insects, and the myriopods (millipedes and centipedes) were all placed in the single subphylum Mandibulata of the phylum Arthropoda because all possess mandibles. However, a growing number of zoologists contend that the mandibles of the crustaceans and the mandibles of insects and myriopods are not homologous, but are the result of convergent evolution from separate origins. Consequently, the crustaceans are now placed in a separate subphylum, Crustacea.

Crustaceans are **gill-breathing arthropods,** with **two pairs of antennae** and **two pairs of maxillae** on the head, and usually a pair of appendages on each body segment. Some of the appendages of present-day adult crustaceans are biramous (two-branched).

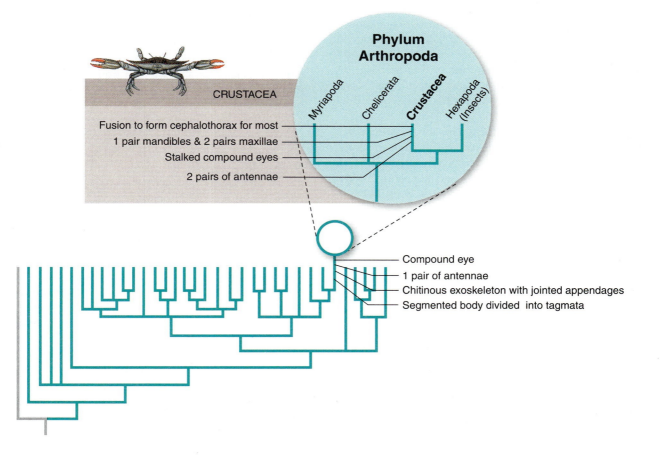

CRUSTACEA

Fusion to form cephalothorax for most
1 pair mandibles & 2 pairs maxillae
Stalked compound eyes
2 pairs of antennae

Phylum Arthropoda

Myriapoda   Chelicerata   Crustacea   Hexapoda (Insects)

Compound eye
1 pair of antennae
Chitinous exoskeleton with jointed appendages
Segmented body divided into tagmata

Core Study

Crayfish or Lobster

Phylum Arthropoda
  Subphylum Crustacea
    Class Malacostraca
      Order Decapoda
        Genus *Cambarus* (and others)

This description will apply to either the crayfishes (*Cambarus, Procambarus, Pacifastacus,* and *Orconectes*) or the lobster (*Homarus* and others) because crayfishes and lobsters are very similar except in size.

Where Found

Crayfish, called locally and variously crawfish, crawdads, or mudbugs, are found in freshwater streams and ponds all over the world. There are about 350 species of crayfish in the United States, more than in all the rest of the world. The southern United States, especially the Gulf Coast, has the largest number of crayfish species. This area also produces most of the edible crayfish, and millions are raised each year on watery crayfish "farms" in Louisiana.

Crayfish are omnivorous, feeding on all kinds of succulent aquatic vegetation and on animal foods such as snails, worms, and small vertebrates. Crayfish are hardy animals, able to survive in almost any type of freshwater habitat. They are frequently sold as bait to fishermen and may either escape from bait buckets or be intentionally released. When crayfish are introduced into new habitats, they quickly become established and can harm the biological community by feeding on native freshwater invertebrates and amphibians. Crayfish should not be introduced into habitats where they are not naturally found.

The lobster is a marine form. *Homarus americanus* (F. *homard,* lobster) is found along the eastern North American coast, from Labrador to North Carolina. Both lobsters and crayfish are widely used for food in various parts of the world. The spiny, or rock lobster, *Panulirus* (anagram of *Palinurus,* Gr. *palin,* backwards, + *oura,* tail), from the West Coast and southern Atlantic coast, has no pincers and differs in several other respects.

Behavior

☞ Place a live crayfish or lobster in a bowl or small aquarium partly filled with water. Provide some shells or stones for shelter.

How are the antennae used? _____
What is the response when you stroke the antennae?
_____ Notice the compound eyes on the ends of stalks. Are the eyestalks movable?

_____ Of what advantage would such eyes be to the animal? _____ What evidence of segmentation do you see on the dorsal side of the animal? _____

Note the five pairs of legs. The first pair are called chelipeds because they bear the large claws (chelae). How are the chelipeds used? _____ When the animal is startled, what does it do? _____ Is the escape movement forward or backward? _____ How is it accomplished? _____ Is the tail fan involved? _____

Lift up a crayfish and notice the swimmerets on the abdomen. Release it near the surface of the water. Can it swim? _____ How? _____ The females use the swimmerets to carry and aerate their eggs during the breeding season.

Drop a bit of meat or fish near an undisturbed crayfish in an aquarium. What appendages does the animal use to pick up and handle food? _____ Notice the activity of the small mouth appendages as the crayfish feeds.

External Features

☞ Place a preserved crayfish (or lobster) in a dissecting pan and add water to the pan.

The **exoskeleton** is a cuticle secreted by the epidermis and hardened with a nitrogenous polysaccharide called **chitin,** with the addition of mineral salts such as calcium carbonate. The cuticle must be shed or **molted** (ecdysis) several times while the crayfish is growing, each time being replaced by a new soft exoskeleton that soon hardens. During and immediately after molting, crayfish spend a great deal of time immobile and hiding. Why? _____

The **cephalothorax** is covered by a hard **carapace.** A transverse **cervical groove** marks the head-thorax fusion line (Figure 12.1A). Posterior to this groove are two grooves that separate the median middorsal cardiac area of the carapace, which covers the heart, from the broad lateral extensions of the carapace, which cover the gills. These lateral extensions are also called **branchiostegites** ("gill-covers"). Lift up the edge of a branchiostegite to disclose the **gill chamber** and the feathery **gills.**

Extending anteriorly in the head region is the pointed **rostrum** (L., bill, snout), and coming from under the rostrum are the stalked **eyes** and two pairs of **antennae** (the second pair of antennae are also called antennules).

On the ventral side, the five fused segments of the head bear the two pairs of antennae and three pairs of small mouthparts. The thorax, made up of eight segments, bears three pairs of **maxillipeds** and five pairs of walking legs.

The **abdomen** is made up of six segments, the first five of which bear the **swimmerets** (Figure 12.1B). The

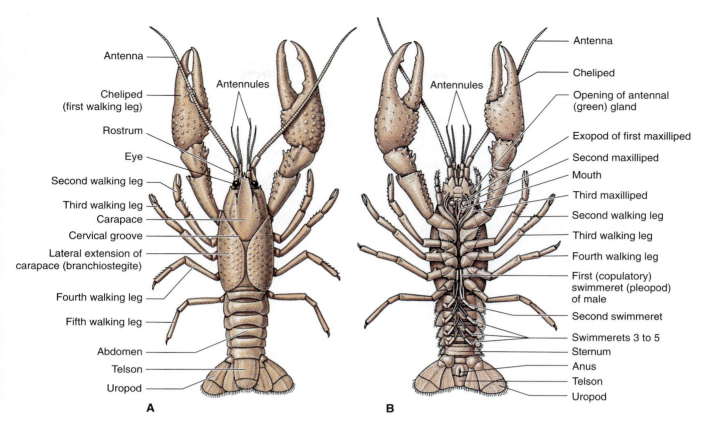

Figure 12.1

External structure of the crayfish. **A,** Dorsal view. **B,** Ventral view.

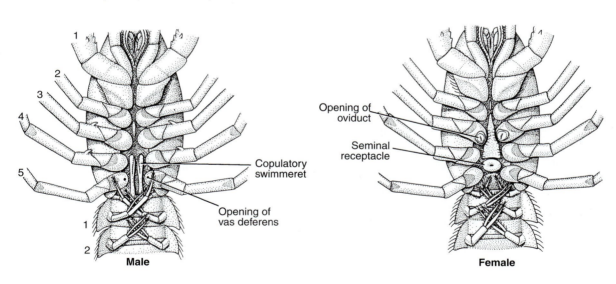

Figure 12.2

Ventral views of crayfish, male and female.

sixth is a flat process, the **telson** (Gr., extremity), which bears the **anus** on the ventral side and a pair of fan-shaped **uropods.**

   Genital Openings. In the male, the genital openings of the sperm ducts are located medially at the base of each of the fifth walking legs (Figure 12.2). In the female, the genital openings of the oviducts are located at the base of each third walking leg. Where is the opening of the seminal receptacle? _____ During copulation, the male turns the female over and presses the sperm along grooves of the specialized first (copulatory) swimmerets into the female's seminal receptacle.

Dissection of the Appendages

As you dissect crayfish appendages, you will see how they illustrate the **principle of serial homology.** In crayfish and their relatives, the common biramous appendage became modified into mouthparts, walking legs, chelipeds, and swimmerets with different functions. The basic plan of the biramous appendage in its least modified form is most clearly seen in the swimmerets of the crayfish. During evolution, some of the biramous branches became lost or modified, and new parts were added. This condition is called **serial homology,** the evolution of a series of structures all homologous to each other but modified for different functions. Crayfish and their relatives possess the best examples of serial homology in the animal kingdom.

☞ From the animal's left side, remove and study each of the appendages. However, do not remove any appendages until you have read the instructions.

Although appendages are numbered consecutively beginning at the anterior end of the animal, it is easier to remove them by beginning at the **posterior** end and proceeding forward. To remove an appendage, grasp it at the base with forceps as near the body as possible and work it loose gradually by gently manipulating the forceps back and forth. Some of the appendages are quite small and feathery, and some have attached gills. *Be very careful to remove all of the parts of each appendage together* (Figure 12.3). As you remove each one, identify its **medial** and **lateral** sides. Pin the appendages in order on a sheet of paper in the bottom of a dissecting pan. Keep covered with water. Alternatively, the appendages may be glued (e.g., with Elmer's Glue-All) in order on a piece of white cardboard and labeled. Each biramous appendage, at least in its early growth stages, consists of a basal part, the **protopod,** which bears a medial (inner) branch called the **endopod,** and a lateral (outer) branch, the **exopod.**

Uropods. The broad uropods (Gr. *oura,* tail, + *pous,* foot) on the most posterior segment are biramous; together with the medial telson, they make up the strong **tail fan** that is used in the rapid backward movements so important in escape. The tail fan also helps protect eggs and young on the female's swimmerets.

Swimmerets (Pleopods). Swimmerets illustrate the unmodified biramous plan. They aid in forward locomotion and, in females, serve as a place of attachment for the eggs.

The first two pairs in males are modified as **copulatory organs.** They are large and grooved and used to direct sperm onto females (see Figure 12.2). The first pair in the female is usually reduced in size.

Walking Legs (Pereiopods). The first (most anterior) pair of walking legs are **chelipeds,** with enlarged claws **(chelae).** Note that the two chelae (keé-lee) are not identical. The heavier chela with rounded teeth is used for crushing; the other, more slender, chela with sharp teeth is used for tearing and seizing prey. This difference is more distinctive in lobsters than in crayfish. The other four pairs are used for walking and food handling. There is no exopod, so they are called **uniramous** (single-branched). The first four pairs bear attached gills. Which of the four pairs of walking legs bear chelae? _____

Maxillipeds. The three paired maxillipeds (L. *maxilla,* jaw, + *pes,* foot) are feeding appendages with bristly edges and toothed processes that tear the food as it is passed to the mouth. The second and third maxillipeds have an attached gill.

Maxillae. The maxillae anterior to the maxillipeds are foliaceous (thin and leaflike) and direct food toward the mouth. On the second maxilla, the exopod forms a long blade called the **gill bailer,** which beats to draw currents of water from the gill chambers.

Mandibles. The toothed mandibles work from side to side (unlike vertebrate jaws that move up and down) to direct food into the mouth and hold the food while the maxillae tear it up. A little palp is folded above each tooth margin. Pry the mandibles apart and remove the left mandible carefully. A strong mandibular muscle attached to the base may tear off with the mandible.

Antennae. The endopod of each antenna is a very long, many-jointed filament. The exopod is a broad, sharp, movable projection near the base. On its broad protopod on the ventral side of the head is the **renal opening** from the excretory gland. Antennae and antennules are sensitive to touch, vibrations, and the chemistry of the water (taste). When you fed the crayfish, how were the antennae used? _____

Antennules. Each antennule has a three-jointed protopod as well as two long, many-jointed filaments. The antennules are also concerned with equilibrium.

Branchial (Respiratory) System

Remove part of the carapace on the animal's right side and note the feathery **gills** lying in the branchial chamber. You have already seen that some of the gills are attached to certain appendages. These outer gills are called the foot gills. How many appendages have gills attached? _____ Move the appendages to determine this. Separate the gills carefully, laying aside the foot gills. Another row of gills underneath is attached to membranes that hold the appendages to the body. These gills are called the joint gills. Some genera, but not *Cambarus,* have a third row of gills (side gills) attached to the body wall.

☞ Remove a gill, place it in a watch glass, cover with water, and examine with a hand lens. Now cut the gill in two and look at one of the cut ends.

The **central axis** bears **gill filaments** that give it a feathery appearance. Notice that the central axis and also the little filaments contain canals. These represent

Appendage		Function

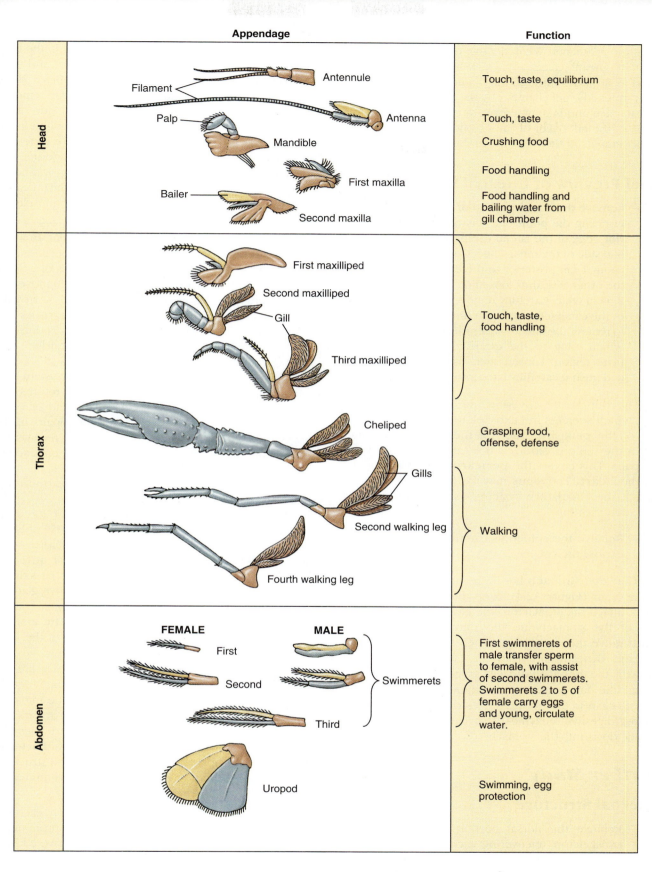

Head

Antennule — Touch, taste, equilibrium

Filament

Antenna — Touch, taste

Palp

Mandible — Crushing food

First maxilla — Food handling

Bailer

Second maxilla — Food handling and bailing water from gill chamber

Thorax

First maxilliped

Second maxilliped

Gill

Third maxilliped — Touch, taste, food handling

Cheliped — Grasping food, offense, defense

Gills

Second walking leg — Walking

Fourth walking leg

Abdomen

FEMALE   MALE

First

Second

Third — Swimmerets — First swimmerets of male transfer sperm to female, with assist of second swimmerets. Swimmerets 2 to 5 of female carry eggs and young, circulate water.

Uropod — Swimming, egg protection

Figure 12.3
Appendages of the crayfish. Protopod in pink; endopod in blue; exopod in yellow.

blood vessels (afferent and efferent) that enter and leave the filaments. What do the terms "afferent" and "efferent" mean? _____ What happens to the blood as it passes through the filaments? _____ Water enters the gill chamber by the free ventral edge of the carapace and is drawn forward over the gills by the action of the gill bailer of the second maxilla, facilitated by movements of the other appendages.

Brief Preview of Internal Structure

 Remove the dorsal portion of the exoskeleton by inserting the point of a scissors under the posterior edge of the lateral carapace about 1.3 cm to one side of the medial line. Cut forward to a point about 1 cm posterior to the eye. Do the same on the other side, thus loosening a dorsal strip about 2.5 cm wide. Carefully remove this center portion of the carapace, a little at a time, being careful not to remove the underlying **epidermis** and **muscles,** which cling to the carapace, especially in the head region. Loosen such tissue with a scalpel and push it carefully back into place.

The thin tissue covering the viscera is **epidermis,** which secretes the exoskeleton. Notice the position of the pinkish portion of the epidermis, which lies in the same position as the cardiac region of the carapace. This covers the **pericardial sinus** containing the **heart.** The sinus may be filled with colored latex if the circulatory system has been injected for easy identification.

 Remove the epidermis carefully with forceps to expose the viscera.

The large **stomach** lies in the head region, anterior to the heart (Figure 12.4). Note the **gastric muscles,** which attach the stomach to the carapace. Lying to the side of the stomach are the **mandibular muscles,** which move the mandibles. On each side of the stomach and beneath the heart are large, cream-colored lobes of the **hepatopancreas** (digestive gland). They extend the full length of the thorax. This gland, the largest organ in the body, secretes digestive juices, containing digestive enzymes, that are poured through hepatic ducts into the stomach.

Further Study

Internal Structure

 Remove the dorsal portion of the abdominal exoskeleton, uncovering each segment carefully so as not to destroy the long **extensor muscle** lying underneath.

Muscular system. As you removed the carapace you noticed the tendency of certain muscles in the head region to cling to it. These are gastric muscles and mandibular muscles mentioned in the preliminary study. One of the mandibular muscles may have been removed when the mandible was removed. On each side of the thorax, a narrow band of muscles runs longitudinally to the distal end of the abdomen. These are **extensor muscles,** which are used to straighten the abdomen. In the abdomen, lying ventrally to the extensors and nearly filling the abdomen, are **flexor muscles,** which flex the abdomen (bend it ventrally). Why are these muscles so large? _____

Circulatory System. The small, angular **heart** is just posterior to the stomach. In your specimen, the heart may be filled and covered with a mass of colored injection fluid or latex, filling the sinus. Remove the latex carefully bit by bit, being careful not to destroy the tiny blood vessels leaving the heart. The heart lies in a cavity called the **pericardial sinus,** which is enclosed in a membrane, the **pericardium.** The pericardium may have been removed with the carapace.

This is an "open" type of circulatory system. The hemolymph leaves the heart in arteries but returns to it by way of venous sinuses, or spaces, instead of veins. Hemolymph enters the heart through three pairs of slit-like openings, the **ostia,** which open to receive the hemolymph and then close when the heart contracts to force out the hemolymph through the arteries.

Your instructor may demonstrate the heartbeat of a living crayfish (instructions given on p. 155).

Five arteries leave the anterior end of the heart (Figure 12.4): a median **ophthalmic artery** (Gr. *ophthalmos,* eye) extends forward to supply the cardiac stomach, the esophagus, and the head; a pair of **antennal arteries,** one on each side of the ophthalmic, pass diagonally forward and downward over the digestive gland to supply the stomach, the antennae, the antennal glands, and parts of the head; and a pair of **hepatic arteries** (Gr. *hepatos,* liver) from the ventral surface of the heart supply the hepatopancreas.

Leaving the posterior end of the heart are the **dorsal abdominal artery,** which extends the length of the abdomen, lying on the dorsal side of the intestine, and the **sternal artery,** which runs straight down (ventrally) to beneath the nerve cord, where it divides into the **ventral thoracic** and **ventral abdominal arteries** that supply the appendages and other ventral structures (Figure 12.4). Do not attempt to find these ventral arteries now. They will be referred to later.

Reproductive System. The **gonads** in each sex lie just under the heart. Their size and prominence depend on the season in which the animals were killed. To find them, lay aside the heart and abdominal extensor muscle bands. Sometimes the gonads are difficult to

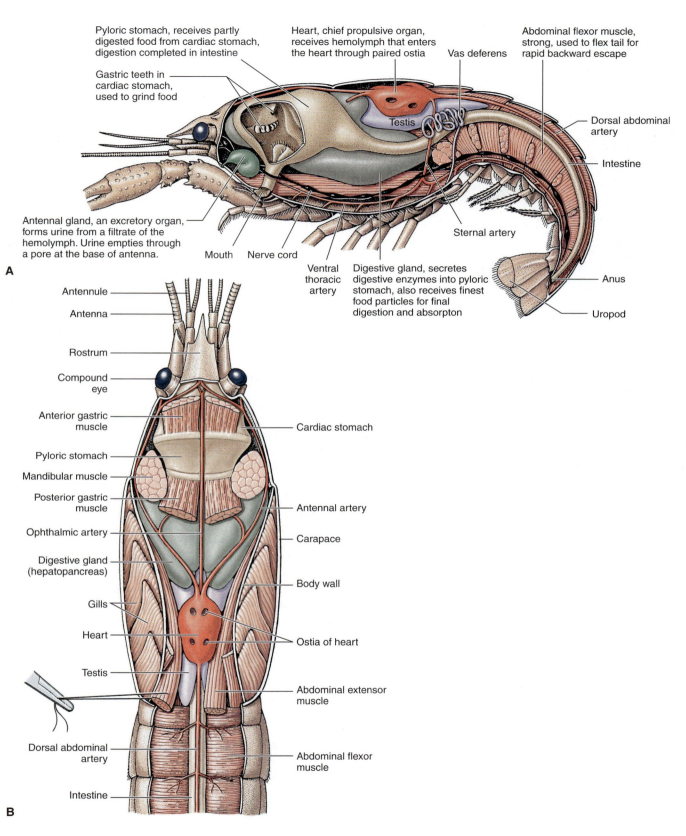

Pyloric stomach, receives partly digested food from cardiac stomach, digestion completed in intestine

Gastric teeth in cardiac stomach, used to grind food

Heart, chief propulsive organ, receives hemolymph that enters the heart through paired ostia

Vas deferens

Abdominal flexor muscle, strong, used to flex tail for rapid backward escape

Testis

Dorsal abdominal artery

Intestine

Antennal gland, an excretory organ, forms urine from a filtrate of the hemolymph. Urine empties through a pore at the base of antenna.

Mouth   Nerve cord

Sternal artery

A

Ventral thoracic artery

Digestive gland, secretes digestive enzymes into pyloric stomach, also receives finest food particles for final digestion and absorpton

Anus

Uropod

Antennule

Antenna

Rostrum

Compound eye

Anterior gastric muscle

Pyloric stomach

Mandibular muscle

Posterior gastric muscle

Ophthalmic artery

Digestive gland (hepatopancreas)

Gills

Heart

Testis

Dorsal abdominal artery

Intestine

B

Cardiac stomach

Antennal artery

Carapace

Body wall

Ostia of heart

Abdominal extensor muscle

Abdominal flexor muscle

Figure 12.4

Internal structure of a male crayfish. **A,** Lateral view. **B,** Dorsal view. What median artery extends forward from the heart to supply the cardiac stomach, esophagus, and head? _____ What pair of arteries pass diagonally forward and downward over the digestive gland to supply the stomach, antennae, antennal glands, and part of the head? _____ What artery leaves the posterior of the heart? _____

distinguish from the digestive glands. The gonads are very slender organs, usually slightly different in color from the digestive glands, lying along the medial line between and slightly above the glands.

In the female, the gonads, or **ovaries,** are slender and pinkish, lying side by side, with the anterior ends slightly raised. In some seasons, the ovaries may be swollen and greatly distended with eggs, appearing orange in color. A pair of **oviducts** leaves the ovaries and passes laterally over the digestive glands to the genital openings on the third walking legs.

The male gonads, or **testes,** are white and delicate. The **vas deferens** pass diagonally over the digestive glands and back to the openings in the fifth walking legs (Figure 12.4A).

Digestive System. The **stomach** is a large, thin-walled organ, lying just back of the rostrum. It is made up of two parts—anteriorly, a large firm **cardiac chamber** and posteriorly a smaller soft **pyloric chamber**—which will be examined later.

Sometimes a mass of calcareous crystals **(gastroliths)** is attached to each side of the cardiac chamber near the time of molting. These limy masses are thought to have been recovered from the old exoskeleton by the blood and used to make the new exoskeleton.

The **intestine,** small and inconspicuous, leaves the pyloric chamber, bends down to pass under the heart, and then rises posteriorly to run along the abdominal length above the large flexor muscles. The intestine ends at the **anus** on the telson.

The large **hepatopancreas,** also called **liver** or **digestive glands** (Figure 12.4B), furnishes digestive secretions that are poured through hepatic ducts into the pyloric chamber. The hepatopancreas is also the chief site of absorption and serves for storage of food reserves.

☞ To see the whole length of the digestive system, remove the left lobe of the hepatopancreas and gonad and push aside the left mandibular muscle.

You can now see the **esophagus,** connecting the stomach and mouth, and the intestine, arising from the stomach and running posteriorly to the abdomen where it lies just ventral to the dorsal abdominal artery. You can also see the **sternal artery** descending ventrally.

☞ Now remove the stomach by severing it from the esophagus and intestine, turn it ventral side up, and open it longitudinally. Wash out the contents, if necessary.

The cardiac chamber contains a **gastric mill,** which consists of a set of three chitinous teeth, one dorsomedial and two lateral, that are used for grinding food. They are held by a framework of ossicles and bars in the stomach and operated by the gastric muscles.

In the cardiac stomach, the food is ground up and partially digested by enzymes from the hepatopancreas before it is filtered into the smaller pyloric stomach in liquid form. Large particles must be egested through the esophagus. Rows of setae and folds of the stomach lining strain the finest particles and pass them from the pyloric stomach into the hepatopancreas or into the intestine where digestion is completed.

Excretory System. In the head region anterior to the digestive glands and lying against the anterior body wall are a pair of **antennal glands** (also called **green glands,** although they will not appear green in the preserved material) (Figure 12.4). They are round and cushion-shaped. In the crayfish, each gland contains an **end sac** that is connected by an excretory tubule to a bladder. Fluid is filtered into the end sac by hydrostatic pressure in the hemocoel. As the filtrate passes through the excretory tubule, reabsorption of salts and water occurs, leaving the urine to be excreted. A duct from the bladder empties through a renal pore at the base of each antenna.

How would urine production differ between a freshwater crayfish and a marine lobster? _____ Why? _____ The role of the antennal glands seems to be largely the regulation of the ionic and osmotic composition of the body fluids.

Excretion of nitrogenous wastes (mostly ammonia) occurs by diffusion in the gills and across thin areas of the cuticle.

Nervous System

☞ Carefully remove all of the viscera, leaving the esophagus and sternal artery in place. The brain is a pair of **supraesophageal ganglia** that lie against the anterior body wall between the antennal glands. Can you distinguish the three pairs of nerves running from the ganglia to the antennae, eyes, and antennules?

From the brain, two connectives pass around the esophagus, one on each side, and unite at the **subesophageal ganglion** on the floor of the cephalothorax.

☞ Chip away the calcified plates that cover and conceal the double ventral nerve cord in the thorax, and follow the cord posteriorly. By removing the big flexor muscles in the abdomen, you can trace the cord for the length of the body. Note the ganglia, which appear as enlargements of the cord at intervals. Observe where the nerve cord divides to pass on either side of the sternal artery. Note the small lateral nerves arising from the cord.

In the annelids there is a ganglion in each segment, but in the arthropods there is some fusion of ganglia. The brain is formed by the fusion of three pairs of head ganglia, and the subesophageal ganglion is formed by the fusion of at least five pairs.

Sense Organs. The crayfish has many sense organs: **tactile hairs** over many parts of the body, **statocysts, antennae, antennules,** and **compound eyes.** The tactile hairs are variously specialized for touch reception, detection of water currents, and orientation. A number of chemoreceptors are present on the antennae, antennules, and mouthparts. The eyes have already been observed.

A **statocyst** is located in the basal segment of each antennule. The pressure of sand grains against sensory hairs in the statocyst gives the crayfish a sense of equilibrium.

Oral Report

Be prepared to demonstrate any phase of your dissection to your instructor. (1) Locate on the dissection any structure mentioned in the exercise, and (2) explain its function. (3) Explain how the appendages of the crayfish or lobster illustrate the principle of serial homology.

Other Crustaceans

Class Branchiopoda

Fairy Shrimp, *Eubranchipus* or *Brachinecta* (Order Anostraca). Fairy shrimp have 11 pairs of basically similar appendages used for locomotion, respiration, and egg carrying. They swim ventral side up. They have no carapace. Note the dark eyes borne on unsegmented stalks. The females carry their eggs in a ventral brood sac that can usually be seen when the animal is moving (Figure 12.5).

Water Flea, *Daphnia* (Order Cladocera). Water fleas (Figure 12.5) are common in pond water. Mount a living *Daphnia* on a slide, using enough water to prevent crushing but not enough to allow free swimming. *Daphnia*

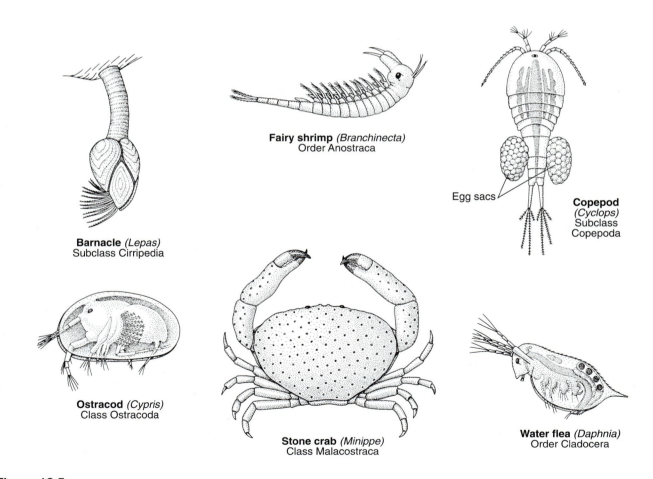

Barnacle *(Lepas)*
Subclass Cirripedia

Fairy shrimp *(Branchinecta)*
Order Anostraca

Egg sacs

Copepod
(Cyclops)
Subclass
Copepoda

Ostracod *(Cypris)*
Class Ostracoda

Stone crab *(Minippe)*
Class Malacostraca

Water flea *(Daphnia)*
Order Cladocera

Figure 12.5
Some representative crustaceans.

is 1 to 3 mm long and, except for the head, is covered by a thin, transparent carapace. The large biramous second antennae are the chief organs of locomotion. The five pairs of small, leaflike swimmerets on the thorax are used to filter microscopic algae from the water for food. The blood in the open circulatory system is red due to the presence of hemoglobin. The paired eyes are fused. The brood pouch in the female is large and posterior to the abdomen. *Daphnia* can reproduce parthenogenetically. (See p. 155 for a demonstration of a heartbeat.)

The Experimenting in Zoology feature at the end of this chapter describes a study of the phototactic behavior of *Daphnia*.

Class Maxillopoda

Subclass Ostracoda. Ostracods, with their transparent bivalved carapace, resemble tiny clams (1 to 2 mm). They have a median eye and seven pairs of appendages, including the large antennules and antennae (Figure 12.5).

Subclass Copepoda. *Cyclops* and other copepods are found everywhere in fresh and brackish water. *Cyclops* has a median eye near the base of the rostrum and long antennae, modified in the male. The cephalothorax bears appendages; the abdomen has none. The sixth thoracic segment in the female carries large, pendulous egg sacs. The last abdominal segment bears a pair of caudal projections covered with setae (Figure 12.5).

Subclass Cirripedia. The barnacles (all marine) might be mistaken for molluscs because they are enclosed in a calcareous shell. *Balanus,* the acorn barnacle, has a six-piece shell that surrounds the animal like a parapet. The animal protrudes its six pairs of biramous appendages through the opening at the top to create water currents and sweep in plankton and particles of detritus.

The gooseneck barnacle, *Lepas* (Figure 12.5), attaches to floating objects by a long stalk, actually the drawnout front part of the head. The body proper is enclosed in a bivalve carapace strengthened by calcareous plates. Six pairs of delicate filamentous and biramous appendages can be protruded from the carapace. Feathery with long setae, the appendages form an effective net to strain food particles from the water.

Class Malacostraca

Class Malacostraca is a large group. It includes the Isopoda (sow bugs, pill bugs, wood lice, and others with a dorsoventrally flattened body), the Amphipoda (beach fleas, the freshwater *Gammarus,* and others that are laterally compressed), and the Decapoda (shrimps, crabs, crayfish, and lobsters).

☞ Examine a crab (Figure 12.5) and compare its structure and appendages with those of the crayfish.

Crustacean Development, Exemplified by the Brine Shrimp

The brine shrimp, *Artemia salina* (Gr. *Artemis,* a goddess of mythology), which is not really a true shrimp but a member of the more primitive order Anostraca, has a pattern of development typical of most marine crustaceans. If time permits, your instructor may have prepared a series of cultures, started 10 days before the laboratory period and every 2 days thereafter, that provides you with five larval stages for study. The larvae hatch 24 to 48 hours after the dry eggs are placed in natural or artificial seawater. The stages are classified on the basis of the number of body segments and the number of appendages present. Since growth cannot continue without molting (ecdysis), the larvae will go through a number of molts before reaching adult size. You may find some of the shed exoskeletons in the cultures.

The animals can be narcotized or killed by adding a few drops of ether or chloroform per 10 ml of culture water.

☞ To observe the larvae, mount a drop of the culture containing a few larvae on a slide and cover with a coverslip. Try to identify the stages listed here:

Nauplius. The newly hatched larva, called the nauplius, has a single median ocellus and three pairs of appendages (two pairs of antennae and one pair of mandibles), and the trunk is still unsegmented.

Metanauplius. The first and second maxillae have developed, and there is some thoracic segmentation.

Protozoea. There are now seven pairs of appendages because the first and second pairs of maxillipeds have been added. The compound eyes are now developing.

Zoea. The third pair of maxillipeds have now appeared. The eyes are complete and there are several thoracic segments.

Mysis. Most or all of the 19 body segments and 11 pairs of appendages found in the adult are now present. In these transparent forms, you should be able to see the digestive tract and, in live specimens, its peristaltic movements.

Brine shrimp attain adulthood at about 3 weeks and can be maintained on a diet of yeast, fed sparingly. The adults closely resemble their much larger freshwater cousins, the fairy shrimp, *Branchinecta* (Figure 12.5).

Drawings

✎ Use separate paper for sketching crustaceans other than the crayfish, or the developmental stages of crustacean larvae, as seen in the brine shrimp.

Projects and Demonstrations

1. *Keeping crayfish.* Crayfish are easily maintained in an aquarium tank at room temperature. They will thrive if fed sparingly on raw meat, such as small pieces of fish, shrimp, frog, or beef. Crayfish need plenty of oxygen. Keep the water in the aquarium shallow and provide rocks that can serve for cover. Rocks will also enable the crayfish to crawl out of the water. Aquatic plants added to the aquarium will enhance its appearance and will also serve as food.

2. *Heartbeat of crayfish.* Anesthetize the crayfish in a suitable-sized dish by immersing in 15% ethanol or freshly opened club soda. Remove the animal from the anesthetic when it fails to respond to prodding; then remove the carapace and observe the beating heart. The crayfish should survive several hours under these conditions.

3. *Heartbeat of* Daphnia. Place a small drop of petrolatum (Vaseline) in the center of a Syracuse watch glass. Fasten *Daphnia* by one valve of its carapace to the petrolatum, but be careful that water circulates between the valves. Observe the heartbeat. Start with water at 0°C. As the water slowly rises to room temperature, record the heartbeats at the different temperature. Determine the number of heartbeats in 15 seconds at each 2° or 3° rise in temperature.

EXPERIMENTING IN ZOOLOGY
The Phototactic Behavior of *Daphnia*

Many species of freshwater zooplankton move toward light (positive phototaxis). In nature, movement toward light often means movement toward shallower, warmer water that is rich in algae and microorganisms. However, most zooplankton have few defense mechanisms and are very vulnerable to predators. Movement toward light would likely make them more visible to potential predators. Thus, natural selection might favor one type of response to light when no predation risk is apparent, and a different response to light when predation risk exists.

Getting Ready

Work with a partner for this exercise. Try to work in a room that can be darkened. Obtain an open-ended glass tube 2 to 3 cm in diameter and 25 to 30 cm long (Figure 12.6). Wash the tube and place a black rubber stopper tightly into one end. Use a wax pencil to mark off three equal sections of tube: left, middle, and right. Fill the tube to within 3 to 4 cm of the top with bottled water or dechlorinated tap water. Have a second rubber stopper available for the open end. Have a stopwatch available for timing the experiment.

How to Proceed

Use a pipette to count out 20 *Daphnia* and add them to the tube. Place the second stopper firmly into the open end. Clamp the tube to the ring stand so that it is horizontal.

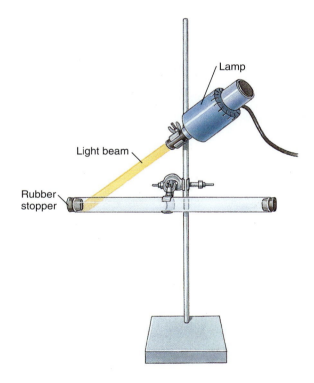

Figure 12.6
Setup for observing the phototactic behavior of *Daphnia*.

Turn out the room lights and allow the *Daphnia* to acclimate to the tube for the next 10 minutes. At the end of each minute for the next 10 minutes, you will record where the

Daphnia are distributed in the tube (left, middle, right). If light from the windows is insufficient to see the *Daphnia*, use diffuse lighting from a lamp some distance from the tube. Turn the lamp on just long enough to count the *Daphnia* and then turn it back off until the next observation. After the tenth observation, calculate the mean percentage of *Daphnia* in each section of the tube over the 10 minutes. Wash out the tube and fill it with clean water and 20 new *Daphnia;* repeat the experiment. Calculate the mean distribution of *Daphnia* for the two trials.

After two trials with the *Daphnia* in the dark, repeat the experiment with a dissecting scope illuminator (see Figure 12.6) clamped at an angle so that it illuminates the end of one stopper making it visible to the *Daphnia*. Clamp it 4 to 5 cm away from the tube so that it does not heat the water. Again, record the distribution of *Daphnia* in the tube over the next 10 minutes and repeat the experiment. Calculate the mean distribution of *Daphnia* for the two trials.

Compile the data from the entire class and calculate overall means for all control trials and all lighted trials.

Do the *Daphnia* appear to respond to light? Do they exhibit negative or positive phototaxis? In the light trials, do you notice any change in distribution over the 10 minutes?

Questions for Independent Investigation

Recent studies on vertical migrations in zooplankton have found that stimuli from predators can influence phototaxis. Aquatic organisms like *Daphnia* are often very sensitive to chemical stimuli in the water.

1. Design an experiment to test how fish odor might affect phototaxis in *Daphnia*. We suggest adding a small amount of water (40 to 50 ml) from a tank that contains a bass or sunfish to your experimental tube containing the *Daphnia*. What would be a good control for this experiment?

2. Do *Daphnia* respond the same to chemical cues from bass or sunfish as they do to cues from frog tadpoles? Why might there be a difference?

3. *Daphnia* also have many invertebrate predators (e.g., hydra, midge larvae—*Chaoborus*). Do chemical cues from invertebrate predators have an effect on their phototactic behavior?

4. How might the effects of gravity impact phototaxis? Design an experiment to examine both gravity and light on *Daphnia* behavior.

References

DeMeester, L. 1991. An analysis of the phototactic behaviour of *Daphnia magna* clones and their sexual descendants. Hydrobiologia **225:**217–227.

DeMeester, L. 1993. Genotype, fish-mediated chemicals, and phototactic behavior in *Daphnia magna*. Ecology **74:**1467–1474.

Neill, W. 1990. Induced vertical migration in copepods as a defense against invertebrate predators. Nature **345:**524–525.

Parejko, K., and S. Dodson. 1991. The evolutionary ecology of an antipredator reaction norm: *Daphnia pulex* and *Chaoborus americanus*. Evolution **45:**1665–1674.

The Arthropods
Myriapods and Insects

EXERCISE 13A
Myriapods—Centipedes and Millipedes

The subphylum Myriapoda, meaning "many footed," consists of four classes (Chilopoda, Diplopoda, Pauropoda, and Symphyla) that have in commom a body of two tagmata—head and trunk—and paired appendages on most or all trunk segments. Of these four many-legged groups, the centipedes (Chilopoda) and millipedes (Diplopoda) comprise the majority of myriapod species.

Core Study

Class Chilopoda—Centipedes

Phylum Arthropoda
   Subphylum Myriapoda
      Class Chilopoda
         Order Scolopendromorph
            Genus *Scolopendra* or *Lithobius*

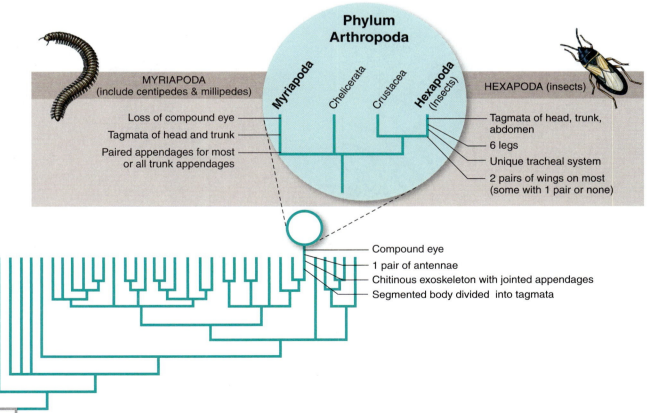

Where Found

Centipedes, or "hundred-leggers," are active predators that live in moist places, such as under logs, stones, and bark, where they feed on worms, larvae, and insects.

 If living examples are available, watch their locomotion and note the agile use of the body and legs.

Note the general shape of the body and the arrangement of segments and appendages. Is the body circular or flattened in cross section? _____

 On a preserved specimen of *Scolopendra* (Gr., a kind of centipede) or *Lithobius* (Gr. *lithos,* stone, + *bios,* life) (Figures 13.1 and 13.2), examine the head.

Some species have simple **ocelli;** others have large faceted **eyes** resembling the compound eyes of insects. Which does your specimen have? _____

 Pull aside the large poison fangs to uncover the mouth area, and examine it under a dissecting microscope.

Find the **antennae;** the **labrum,** anterior to the mouth; the **mandibles** and **first maxillae,** lateral to the mouth; and the **second maxillae,** bearing a long palp and a short labial portion just posterior to the mouth.

The first trunk appendages are the prehensile **maxillipeds,** each bearing a terminal **poison fang.** Do the rest of the trunk appendages have legs? _____ The last segment bears the **gonopores,** and the **anus** is located on a short telson. Find the **spiracles.** In *Lithobius* they are near the bases of the legs; in *Scolopendra* they are located more dorsally and are present on alternate segments.

Class Diplopoda—Millipedes

Phylum Arthropoda
   Subphylum Myriapoda
      Class Diplopoda
         Genus *Julus* or *Spirobolus*

Where Found

Millipedes, or "thousand-foot worms," are found throughout the world, usually hiding in damp woods

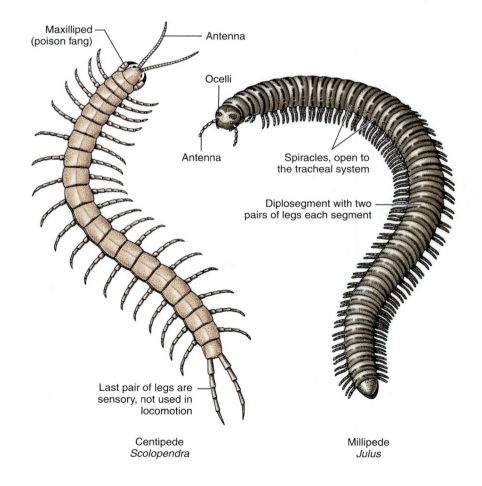

Figure 13.1
Examples of myriapods—centipede and millipede.

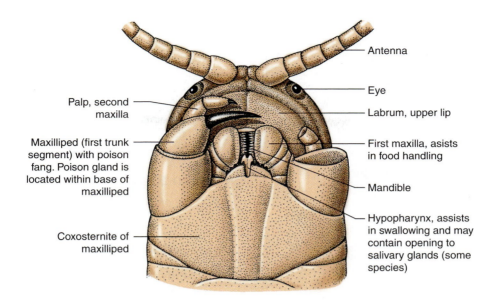

Figure 13.2
Mouthparts of *Scolopendra,* ventral view.

The following labels appear in the figure:
- Antenna
- Eye
- Labrum, upper lip
- First maxilla, asists in food handling
- Mandible
- Hypopharynx, assists in swallowing and may contain opening to salivary glands (some species)
- Palp, second maxilla
- Maxilliped (first trunk segment) with poison fang. Poison gland is located within base of maxilliped
- Coxosternite of maxilliped

under bark, leaves, rocks, or logs. They are herbivorous, feeding on decaying wood or leaves. Their most distinguishing feature is the presence of diplosegments, which are double trunk segments probably derived from the fusion of two single segments and each bearing two pairs of legs. This double-legged characteristic is recorded in their name, Diplopoda, which derives from the Greek, *diploos,* double, and *pous, podos,* foot.

☞ If living millipedes such as *Spirobolus* (L. *spira,* coil, + Gr. *bōlos,* lump) or *Julus* (Gr. *iulus,* plant-down) are available, note the shape of the body, use of the antennae and the legs, and the animal's ability to roll up.

Notice the rhythm of the power and recovery strokes of the legs, with the opposite sides of each segment exactly out of phase, similar to the strokes seen in the polychaete *Nereis.* The movements of the legs pass in regular successive waves toward the anterior end. If a specimen has been chilled, you should be able to determine the number of legs working together in each wave and the number of segments between waves.

☞ Study a preserved specimen, and locate on the head the **ocelli, antennae, labrum, mandibles,** and **labium** (fused second maxillae).

Are there appendages on the first trunk segments? _____ How many pairs are on the next three segments? _____ The **gonopores** open on the third trunk segment at the bases of the legs. How does this compare with the centipede? _____ The

dorsal overlapping of the exoskeletal plates provides full protection for the animal, even when rolled into a ball. Notice that the diplosegments each have two pairs of **spiracles.**

EXERCISE 13B
Insects—Grasshopper and Honeybee

Insects are hexapods, so named for the presence of six legs, all of which are uniramous (i.e., unbranched). Other distinguishing characteristics of insects are shown in the cladogram on p. 157. The enormous success of insects is evident in their numbers: many millions of species (approximately 1.1 million have been described—more than all other animal species combined). They are the most diverse of arthropods, outnumbering all other arthropod groups by more than 8 to 1.

Among insects' chief characteristics are **three pairs of walking legs, one pair of antennae,** a body typically divided into **head, thorax,** and **abdomen,** and a respiratory system composed of **tracheal tubes.** Most insects also have one or two pairs of **wings.** Their sense organs are often specialized and perhaps account for much of their success in the competition for ecological niches.

Most insects are less than 2.5 cm long, but they range from 1 mm to 20 cm, with the largest insects usually living in tropical areas.

A grasshopper has a fairly typical insect body plan. A honeybee, on the other hand, has become specialized for particular conditions. Not only is its morphology modified for special functions and adaptations, but it is a social insect whose patterns of group organization involve different types of individuals and division of labor.

Core Study

Romalea, Lubber Grasshopper

Phylum Arthropoda
    Subphylum Hexapoda
        Class Insecta
            Order Orthoptera
                Genus *Romalea*

Behavior of a Grasshopper

👉 If living grasshoppers are available, observe them in a terrarium or place one in a glass jar with moist paper toweling in the bottom and a stick or other object to perch on.

Is the color adaptive, given the grasshopper's natural habitat? _____ Can the insect move its head? _____

Observe how the grasshopper moves and how it uses its legs. How does it crawl up a stick? _____ How does it use its claws? _____ What position does it assume when quiescent in the jar? _____ How does it jump? _____ How are its legs adapted for jumping? _____ Does it use its wings? _____.

Note movements of the body while the grasshopper is at rest. Are the movements related to breathing? _____ How does the grasshopper breathe? _____ How does it get air into its body? What is the common food of grasshoppers? _____ Observe how one eats a piece of lettuce leaf. Watch how it moves its mouthparts.

Take a specimen from the jar. Note that it will regurgitate its greenish digestive juices and food on a glass plate. Is this a defensive adaptation? _____

Written Report

✍ Record your observations on separate paper.

External Structure

👉 Study a preserved specimen of a grasshopper. This insect is a model of compactness, so use a dissecting microscope or hand lens to observe the smaller structures.

Note the division of the body into three tagmata: **head, thorax,** and **abdomen.** Is the grasshopper segmented throughout, or is segmentation more apparent in certain regions of the body? _____

The chitinous **exoskeleton** is secreted by the underlying epidermis. It is composed of hard plates, called **sclerites,** that are bounded by sutures of soft cuticle.

Head of a Grasshopper. A grasshopper's head is freely movable. Notice the **compound eyes,** the **antennae,** and three **ocelli,** one dorsal to the base of each antenna and one in the groove between them. Lift the movable bilobed upper lip, or **labrum** (L., lip) (Figure 13.3)

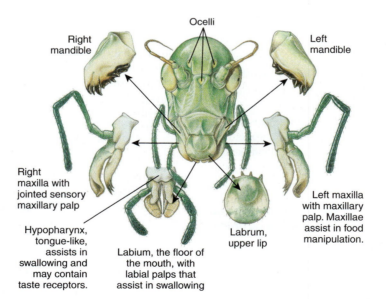

Figure 13.3
Head and mouthparts of a grasshopper.

and observe the toothed **mandibles.** The mouth contains a membranous **hypopharynx** for tasting food. The bilobed lower lip, or **labium** (L., lip), is the result of the fusion of the second maxillae. The labium bears on each side a three-jointed labial palp. Between the mandible and the labium are the paired **maxillae** (L., jawbone), each with a maxillary palp, a flat lobe, and a toothed jaw. Note how the mouthparts are adapted for biting and chewing.

In summary, the insect head bears four pairs of true appendages: the antennae, the mandibles, the maxillae, and the labium (fused second maxillae). There are at least six segments in the head region, although some of them are apparent only in the insect embryo.

☞ After you study the rest of the external features, if time permits, your instructor may ask you to use forceps and teasing needles to carefully remove all the mouthparts and arrange them in their relative positions on a sheet of paper. Permanent demonstration mounts can be made very easily by a method given in Appendix A, pp. 286–287.

Thorax of a Grasshopper. The thorax is made up of three segments: **prothorax, mesothorax,** and **metathorax,** each bearing a pair of legs (Figure 13.4).

The mesothorax and metathorax also bear a pair of wings. **Spiracles** (external openings of the insect's tracheal system) are located above the legs in the mesothorax and metathorax. Note the leathery **forewings** (on the mesothorax) and the membranous **hindwings** (on the metathorax). Which is more useful for flight? _____ What appears to be the chief function of the forewings? _____
The small **veins** in the wings, which are really tracheal tubes, are used by entomologists to identify and classify insects.

Examine the legs of the grasshopper and identify the basal **coxa** (L., hip), small **trochanter** (Gr., ball of hip joint), large **femur,** slender, spiny **tibia,** and five-jointed **tarsus** with two **claws** and a terminal pad, the **arolium.** Which pair of legs is most specialized, and for what function? _____

Abdomen of a Grasshopper. There are 11 segments in the abdomen of a grasshopper. Notice the large **tympanum,** the organ of hearing, one located on each side of the first abdominal segment. On which of the abdominal segments are the paired spiracles located? _____ In both sexes, segments 2 to 8 are similar and unmodified, and segments 9 and 10 are partially fused.

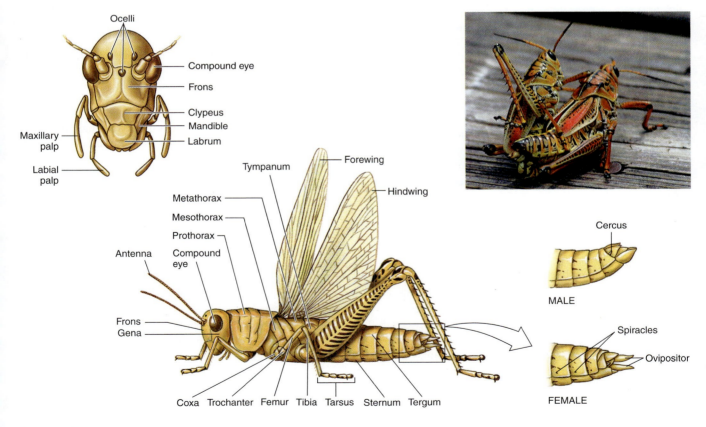

Figure 13.4
External features of a female grasshopper. The terminal segments of a male with external genitalia are shown in the inset.

The eleventh segment forms the genitalia (secondary sex organs). On each side behind the tenth segment is a projection, the **cercus** (pl., **cerci;** Gr. *kerkos,* tail). In the female, the posterior end of the abdomen is pointed and consists of two pairs of plates with a smaller pair between, with the whole forming the **ovipositor.** Between the plates is the opening of the **oviduct.** The end of the abdomen in the male is rounded. What is the sex of your specimen?

Drawing

Label the external view of the grasshopper in the lab report on p. 165.

Further Study

Apis, Honeybee

Phylum Arthropoda
   Subphylum Hexapoda
      Class Insecta
         Order Hymenoptera
           Genus *Apis*
              Species *Apis mellifera*

External Structure

Examine a preserved or freshly killed honeybee. Use a dissecting microscope or hand lens to study smaller structures.

The body of the honeybee, like that of the grasshopper, is divided into three tagmata: **head, thorax,** and **abdomen.**

Head of a Honeybee. Identify **antennae, compound eyes,** and the three **ocelli,** located dorsally between the compound eyes (Figure 13.5). The mouthparts can be used for both chewing and sucking. Observe the narrow upper lip, or **labrum,** with a row of bristles on its free margin. Below the labrum are the **mandibles,** which are brownish in color. From the honeybee's mouth projects a sucking apparatus made up of the long, slender, hairy **tongue** (or **labium**), paired **labial palps,** one on each side of the tongue, and paired, broad **maxillae,** one on each side of the labial palps.

Thorax of a Honeybee. The thorax is composed of three segments: **prothorax, mesothorax,** and **metathorax,** but the lines of division between the segments are less distinguishable than on a grasshopper. The honeybee's first abdominal segment is also a part of the thorax, lying anterior to its narrow "waist." Each segment bears a pair of legs, and the mesothorax and metathorax each bear a pair of wings. Note that both pairs of wings are thin and membranous, unlike the grasshopper in which only the hindwings are membranous. How would this difference affect the wings' function?_____

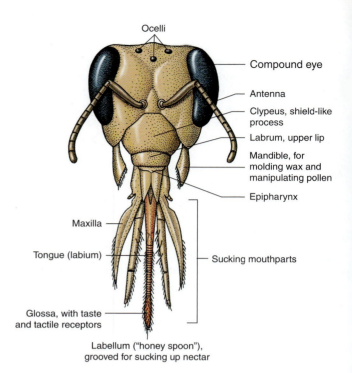

Figure 13.5
Head and mouthparts of a honeybee.

Notice small hooks on the front margin of the hindwing. During flight, the hooks catch hold of a groove near the margin of the forewing. The wings of a bee may vibrate 400 times or more per second during flight.

The legs of the honeybee have the same segments as those of the grasshopper but are highly adapted for specific functions.

Use your hand lens, or remove the legs from one side of the bee and examine them under a dissecting microscope.

Note the hairs on the **foreleg.** Are they branched? _____ These hairs collect pollen. The **pollen brush** (Figure 13.6) consists of long hairs on the proximal end of the tarsus. Pollen brushes on the forelegs and midlegs brush pollen off the body hairs and deposit it on the pollen combs of the hindlegs (Figure 13.6).

At the distal end of the tibia is a movable spine, the **velum** (L., covering). The velum covers a **semicircular notch** that bears a row of stiff bristles, the **antenna comb.** The velum, notch, and comb together make up the **antenna cleaner.** The antenna is freed from pollen as it is drawn through this antenna cleaner.

The **midleg** has a long, sharp **spur** projecting from the end of the tibia that is used to remove pollen from the pollen basket of the leg behind. The midleg also bears a pollen brush.

The **hindleg** is the largest and most specialized of the legs. One of its striking adaptations is the **pollen**

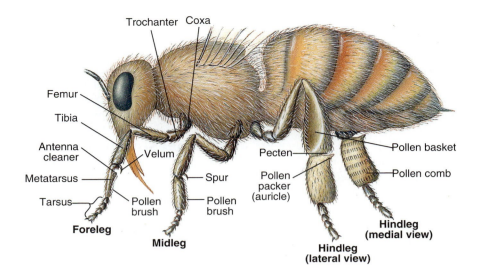

Figure 13.6
Adaptive legs from the left side of a worker honeybee.

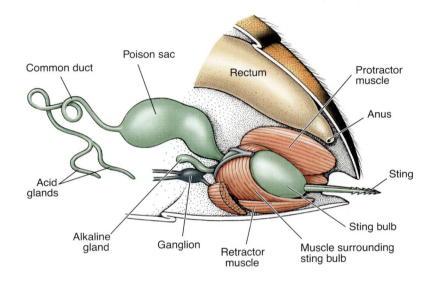

Figure 13.7
The sting and associated structures in a worker honeybee.

basket, a wide groove with bristles on the outer surface of the tibia. By keeping these bristles moist with mouth secretions, the bee can use the basket for carrying pollen. On the inner surface of the metatarsus are **pollen combs,** which are composed of rows of stout spines. The large spines found along the distal end of the tibia and the proximal end of the metatarsus make up the **pollen packer** (auricle). The **pecten** (pollen rake) removes the pollen from the pollen brush of the opposite leg, and then, when the leg is bent, the auricle packs it into the pollen basket. The bee carries her baskets full of protein-rich pollen back to the hive and pushes the pollen into a cell, where it will be cared for by other workers.

Abdomen of a Honeybee. The abdomen of a bee has 10 segments, the first of which is really part of the thorax, as mentioned before. The last three segments are modified and hidden within the seventh segment. Can you identify five pairs of spiracles on the abdomen?

Honeybee Sting. The amazingly intricate **sting** is a modified ovipositor that has evolved for defense. It consists of a large **poison sac** that receives secretions (a mixture of proteins, peptides, and other compounds) from two acid glands by way of a common duct (Figure 13.7). The poison sac discharges into the cavity of a sting bulb; from there the poison passes through the canal of the sting shaft and into the wound. The sting bulb also receives a short alkaline gland of uncertain function.

When a honeybee stings, it thrusts the barbs of the sting into the victim's flesh. The entire sting apparatus is torn out of the bee's abdomen, fatally wounding the bee. Each sting actually consists of a sting shaft together with two barb lancets that are moved rapidly back and forth on the stylet by the action of protractor and retractor muscles. Therefore, even after the bee has hurriedly departed, the self-sacrificial sting continues to work deeper into the flesh, and poison continues to be injected for 30 to 60 seconds.

The sting of the queen bee is used only against rival queens and can be withdrawn and reused many times. It is much more firmly attached within the queen's abdomen, and the lancets have fewer and smaller barbs.

Honeybees as Social Animals. In the hive, the honeybees are of three **castes:** workers, queen, and drones. **Workers**, which are sexually inactive genetic females, make up most of the society. They do most of the work of the hive, except lay eggs. They collect food, clean out the hive, make honey and wax, care for the young, guard the hive, ventilate the hive, and so on.

The single **queen** is a sexually mature female that lays the eggs, which may or may not be fertilized by sperm stored in her spermatheca.

Drones are males; all develop from unfertilized eggs and consequently are haploid (workers and the queen are diploid). There are usually only a few hundred drones in a hive. Their main duty is to fertilize the queen during the nuptial flight, although usually only one is required to provide the queen with enough sperm to last her lifetime.

Internal Structure of an Insect

Studying the internal structures of a grasshopper is not always satisfactory because the various organs appear somewhat poorly defined. We suggest that internal structures are more easily studied in an anesthetized cricket, instructions for which are given in Exercise 13C.

Oral Report

Be prepared to (1) identify the major external features of the grasshopper and/or honeybee, (2) explain the adaptations of the honeybee worker appendages, and (3) explain the sting of the worker honeybee.

Phylum _____

Subphylum _____

Genus _____

N a m e _____

D a t e _____

S e c t i o n _____

Lubber Grasshopper

External view of female (male abdomen at bottom left)

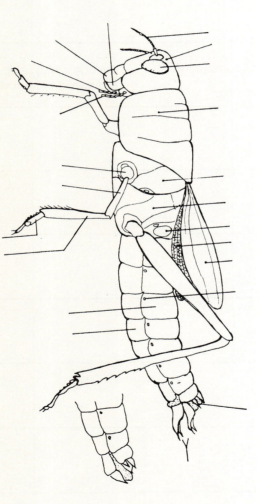

Name_____

Date_____

Section_____

Comparing Characteristics of Annelida and Arthropoda

	Annelida	Arthropoda
Segmentation		
Body covering		
Main body cavity		
Appendages		
Circulatory system		
Respiration		
Excretory organs		
Sense organs		
Method of growth		
Division of animal kingdom (protostome/deuterostome)		

EXERCISE 13C
Insects—House Cricket[1]

Acheta domesticus, House Cricket

Phylum Arthropoda
  Subphylum Hexapoda
    Class Insecta
      Order Orthoptera
        Genus *Acheta*
          Species *Acheta domesticus*

Where Found

The house cricket, or gray cricket, commercially available as fish bait, is a European import and is not native to the United States. Although crickets have functional wings, the flight muscles do not completely develop and crickets never fly. Adult house crickets live an average of two months, during which the female may deposit up to 2000 eggs. Only the adult male sings, and one of the functions of his singing is to attract a female. At a rearing temperature of 30°C, the eggs hatch in 13 days. Crickets undergo **gradual metamorphosis.** The animal grows by successive molts (ecdysis), with each stage called a larval instar; after the last molt, the insect is called an adult. Eight larval instars occur before the final ecdysis to the adult 48 days later.

External Structure

☞ Examine several alcohol-preserved crickets. Determine the sex of the crickets (Figures 13.8E and 13.9C), and determine which are larval instars and which are adults.

Only adult crickets have fully developed wings. The last two or three larval instars of crickets have external wing pads (buds).

☞ Pin an adult male or female preserved cricket lateral side up in the dissecting dish.

Note that the basic 18 segments of the insect are functionally organized into three body regions, or tagmata: **head** (five segments), **thorax** (three segments), and **abdomen** (10 segments). Feeding and sensory organs are on the head; locomotory organs (wings and legs) are on the thorax; and digestive and reproductive organs are in the abdomen. All head segments are fused into a unit head capsule, and the mouthparts represent the modified appendages. The three thoracic segments are the **prothorax, mesothorax,** and **metathorax.** The abdominal segments are simply numbered 1 through 9

---

[1]Exercise contributed by J. P. Woodring of Louisiana State University.

(the cerci represent segment 10). The entire dorsum is called the **tergum** (L., back) (or notum), and any single segment of the dorsal plate is called a **tergite.** The lateral body surface is the **pleuron** (Gr., side) (pleurite for one segment), and the ventral body surface is the **sternum** (L., breastbone) (sternite for one segment). Find and identify all of these parts.

☞ Cut off a prothoracic leg, and identify all of the segments and the tympanic membrane (Figure 13.8). Cut off the mesothoracic wing of a male cricket (Figure 13.8A), and determine how male crickets are able to sing. Place a cricket ventral side up, bend the head back, fold out the labrum, and pin the head in this position (Figure 13.8C). Identify all the mouthparts by moving each with fine forceps. Carefully remove one maxilla and the labium to see the **hypopharynx.**

The hypopharynx acts as a tongue and bears the openings of the **salivary glands.**

Functional Observations

☞ Work in pairs. Anesthetize two crickets with carbon dioxide. Immobilize one cricket *ventral side up* by crossing insect pins over the insect (do not stick the pins through the cricket). When the cricket recovers from the anesthetic, feed it some colored, moistened food and observe the action of the mouthparts under the dissecting microscope.

The heart is an almost transparent tube visible through the intersegmental membranes along the middorsal line. Cut the wings off the other cricket and immobilize it *dorsal side up* with insect pins crossed over the body. When the cricket recovers from the anesthetic, determine the ventilation rate (abdominal contractions) and the heart rate in ventilations and beats per minute.

Internal Anatomy

☞ If necessary, reanesthetize the cricket from which the wings have been removed, and pin the animal *dorsal side up* through the head and epiproct (Gr., *epi,* upon, + *proktos,* anus; see Figure 13.9C, D) onto the wax in the dish. Flood the entire animal with saline and cut the cricket open with fine-tipped scissors. Pin open the cricket with insect pins.

Observe **peristalsis** of the gut and movement of the **malpighian tubules.** The cricket will remain alive and the organ systems will function normally for several hours under these conditions. The chalky white

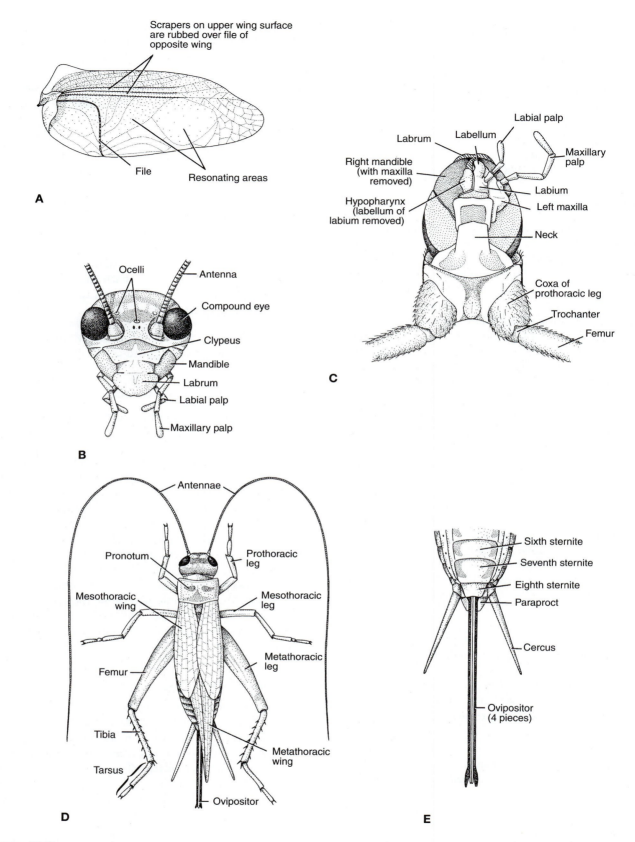

Figure 13.8

External structure of a cricket. **A,** Undersurface of the mesothoracic wing of a male. **B,** Frontal and **C,** ventral views of the head. **D,** Dorsal aspect of an adult female. **E,** Ventral aspect of the adult female abdomen.

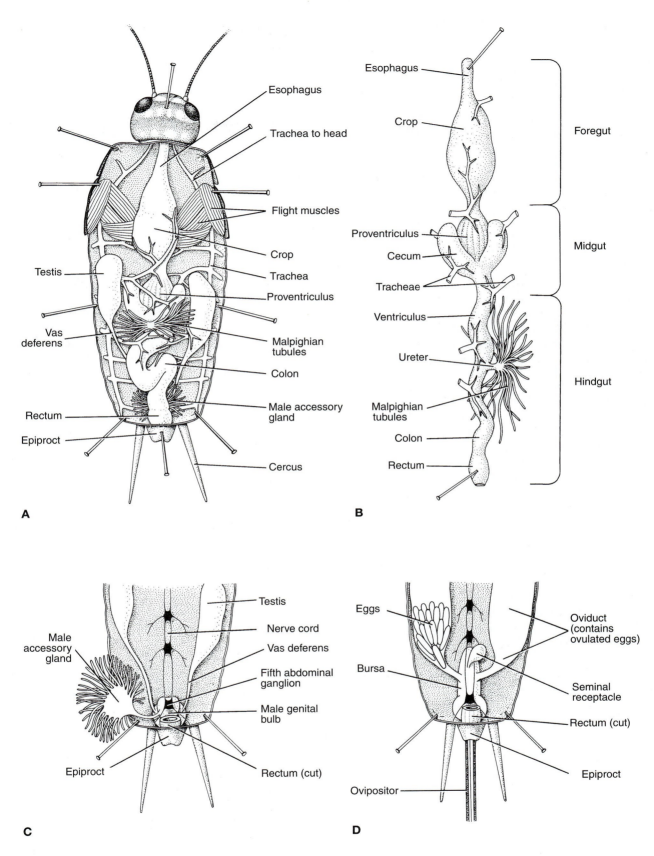

Figure 13.9

A, Internal anatomy of a cricket (fat bodies not shown). **B,** Isolated gut. **C,** Male reproductive anatomy. **D,** Female reproductive anatomy.

fat body, which functions as a liver, is spread through-out the body and will vary greatly in amount and location according to the age and diet of the cricket. Note the **tracheal tubes,** which appear silvery because of contained air. These branch throughout the body.

☞ Sever the esophagus and rectum. Grasp the esophagus with forceps and gradually lift the entire digestive tract out by carefully cutting each tracheal connection to the gut (use the dissecting microscope). Stretch and straighten out the gut, and pin it to one side of the dish as illustrated (Figure 13.9B).

Identify all the structures labeled in Figure 13.9. Note that the **ureter** arises in the middle of the hindgut.

Both the male and female reproductive systems are best seen after the digestive tract has been removed. Identify the gonads **(testes** or **ovaries),** gonoducts **(vas deferens** or **oviducts), accessory gland, bursa** or **genital bulb,** and the genitalia. The genitalia are the **aedeagus** (e-de′a-gus; Gr. *aidoia,* genitals) of the male and the **ovipositor** (L. *ovum,* egg, + *ponere,* to place) of the female.

☞ Pour off the saline from your specimen, but leave it pinned in place. Rinse out the debris and fat body with a mild jet of water from a plastic squeeze bottle. Pour off all the water and add several crops of methylene blue dye to the tissues to stain the nerve cord and ganglia. Wait 1 minute; then cover the specimen with water. Pick away tissue covering parts of the nervous system. Rinse and restain if necessary.

A double ventral **nerve cord** passes back through the body. In the thorax are three pairs of large, fused **segmental ganglia.** The nerve cords are widely separated. In the abdomen are five smaller ganglia; the fifth ganglion (overlying the genital bulb in the male and the bursa in the female) is larger than the rest and supplies all the posterior end of the body (Figure 13.9C, D).

Two cerebral ganglia lie in the head, but are not visible without dissecting the head.

Oral Report

Be prepared to locate on your dissection any structure mentioned in the exercise and state its function.

EXERCISE 13D
Collection and Classification of Insects

Where to Collect Insects

1. Using a sweep net, sweep over grass, alfalfa, or weed patches.

2. Spread a cloth, newspaper, or inverted umbrella under a bush or shrub; beat or shake the plant vigorously.

3. Overturn stones, logs, bark, leaf mold, and rubbish; look under dung in pastures.

4. Watch for butterflies to alight, and then drop a net over them.

5. Look around outdoor lights at night.

6. At night, suspend a sheet from a tree limb or clothesline, with the lower part of the sheet spread on the ground. Direct automobile headlights or a spotlight on the sheet. The insects will be attracted to the white light, hit against the sheet, and drop onto the cloth below.

7. Moths may be baited by daubing a mixture of crushed banana or peach and molasses or sugar on the bark of trees; visit the trees at night with a flashlight to collect the moths.

8. Locate soil insects by placing humus and leaf matter in a Berlese funnel.

9. Aquatic insects may be seined with a water net. Their larvae may be found in either quiet or running water, attached to plants and leaves, under stones, or in bottom mud, from which they may be sieved.

10. In early spring, the sap exuding from stumps or tree trunks attracts various insects.

Key to the Principal Orders of Insects[2]

This key will enable you to place your insects in the correct order. The use of a two-choice (dichotomous) key such as this is explained in Exercise 2. This key is designed for use with adult (final instar) insects and is therefore not suitable for identifying insect larvae. If you wish to identify insects to the family, genus, or species level, consult one of the references listed on page 176.

Terms Used in the Key

Cercus (pl., cerci) One of a pair of jointed anal appendages.

Tarsus (pl., tarsi) The leg segment distal to the tibia, consisting of one or more segments or subdivisions.

Lepidoptera

1   With functional wings 2
Without functional wings, or with forewings thickened and concealing membranous hindwings 15

2 (1)   Wings covered with minute scales; mouthparts usually a coiled tube (butterflies, moths) . . . . **Lepidoptera**
Wings usually clear, not covered with scales; mouthparts not a coiled tube 3

Diptera

3 (2)   With one pair of wings (true flies) . . . . **Diptera**
With two pairs of wings 4

4 (3)   Wings long, narrow, fringed with long hairs, body length 5 mm or less (thrips) . . . . **Thysanoptera**
Wings not narrow and fringed, body usually longer than 5 mm 5

Thysanoptera

5 (4)   Abdomen with two or three threadlike "tails"; hindwings small (mayflies) . . . . **Ephemeroptera**
Abdomen with only short filaments or none; hindwings larger 6

6 (5)   Forewings clearly longer and with greater area than hindwings 7
Forewings not longer, or only slightly longer than hindwings, and with same or less area than hindwings 9

Ephemeroptera

7 (6)   Forewings noticeably hairy; antennae as long or longer than body (caddisflies) . . . . **Trichoptera**
Wings transparent or translucent, not hairy; antennae shorter than body 8

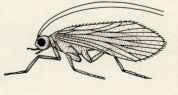
Trichoptera

continued

[2]Key adapted from Borror, D. J., and White, R. E. 1970. A field guide to the insects of America north of Mexico. Boston, Houghton Mifflin Co.

Key to the Principal Orders of Insects—*continued*

8 (7)   Tarsi 2-segmented or 3-segmented; body
        not wasplike or beelike 14
        Tarsi 5-segmented; usually wasplike or
        beelike (sawflies, ichneumons, winged
        ants, wasps, bees) . . . . **Hymenoptera**

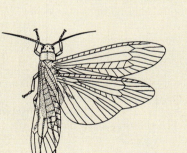

Tarsi

Hymenoptera

9 (6)   Head prolonged ventrally into a beaklike
        structure (scorpionflies) . . . . **Mecoptera**
        Head not prolonged ventrally 10

Mecoptera

10 (9) Antennae very short and bristlelike; eyes
        large; abdomen long and slender
        (dragonflies, damselflies) . . . . **Odonata**
        Antennae not short and bristlelike; eyes
        moderate to small 11

Odonata

11 (10) Hindwings broader than forewings; cerci
        present (stoneflies) . . . . **Plecoptera**
        Hindwings little, if any, broader than
        forewings; cerci absent 12

Plecoptera

12 (11) Mothlike; wings noticeably hairy and
        opaque; antennae as long or longer than
        body (caddisflies) . . . . **Trichoptera**
        Not mothlike; wings not noticeably hairy,
        usually clear; antennae shorter than
        body 13

Trichoptera

13 (12) Wings with few cross veins; tarsi
        4-segmented; length to 8 mm
        (termites) . . . . **Isoptera**

Isoptera

Wings with numerous cross veins;
tarsi 5-segmented; length to 75 mm
(fishflies, dobsonflies, lacewings,
ant lions) . . . . **Neuroptera**

Neuroptera

14 (8) Mouthparts sucking, beak arising from
rear of head (cicadas, hoppers,
aphids) . . . . **Hemiptera, suborder
Auchenorrhyncha**
Mouthparts chewing, beak absent; body length
less than 7 mm (book lice, bar lice)
. . . . **Pscoptera**

15 (1) Wings entirely absent 16
Wings modified, forewings hard and
leathery and covering hindwings 27

Hemiptera

16 (15) Narrow-waisted, antlike (ants, wingless
wasps) . . . . **Hymenoptera**
Not narrow-waisted or antlike 17

Hymenoptera

17 (16) Body rarely flattened laterally; usually do
not jump 18
Body flattened laterally; small jumping
insects (fleas) . . . . **Siphonaptera**

18 (17) Parasites of birds and mammals; body
nearly always flattened dorsoventrally 19
Never parasitic; body usually not flattened . . 20

Siphonaptera

19 (18) Head as wide as or wider than thorax
(chewing lice). . . . . **Phthiraptera**

Phthiraptera

Head narrower than thorax
(sucking lice) . . . . **Phthiraptera suborder
Anoplura**

Anoplura

continued

20 (18)   Abdomen with stylelike appendages
or threadlike tails (silverfish,
bristletails) . . . . **Thysanura**
Abdomen with neither styles nor tails 21

Thysanura

21 (20)   Abdomen with a forked, tail-like jumping
mechanism (springtails) . . . . **Collembola**
Abdomen lacking a jumping
mechanism 22

Collembola

22 (21)   Abdomen usually with two short tubes;
small, plump, soft-bodied (aphids,
others) . . . . **Hemiptera,
suborder Sternorrhyncha**
Abdomen without tubes; usually not
plump and soft-bodied 23

23 (22)   Lacking pigment, whitish; soft-bodied 24
Distinctly pigmented; usually
hard-bodied 25

Sternorrhyncha

24 (23)   Antennae long, hairlike; tarsi 2-segmented
or 3-segmented (psocids) . . . . **Psocoptera**
Antennae short, threadlike; tarsi
4-segmented (termites) . . . . **Isoptera**

25 (23)   Body shape variable; length over
5 mm 26
Body narrow; length less than 5 mm
(thrips) . . . . **Thysanoptera**

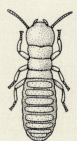

Isoptera

26 (25)   Antennae 4-segmented or 5-segmented;
mouthparts sucking
(wingless bugs) . . . . **Hemiptera, suborder
Heteroptera**

Antennae many-segmented; mouthparts
chewing 31

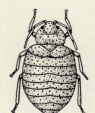

Heteroptera

27 (15)   Abdomen with forcepslike cerci
          (earwigs) . . . . **Dermaptera**
          Abdomen lacks forcepslike cerci 28

Dermaptera

28 (27)   Mouthparts sucking; beak usually
          elongate 29
          Mouthparts chewing 30

29 (28)   Forewings nearly always thickened at
          base, membranous at tip; beak rises from
          front or bottom of head (true bugs)
          . . . . **Hemiptera, suborder Heteroptera**

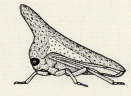

Heteroptera

          Forewings of uniform texture throughout;
          beak arises from hind part of head
          (hoppers) . . . . **Hemiptera, suborder
          Auchenorrhyncha**

Auchenorrhyncha

30 (28)   Forewings with veins, at rest held rooflike
          over abdomen or overlapping
          (grasshoppers, crickets, cockroaches,
          mantids) . . . . **Orthoptera**

Orthoptera

          Forewings without veins, meeting in a
          straight line down back
          (beetles) . . . . **Coleoptera**

Coleoptera

31 (26)   Body oval and flattened (cockroaches)
          . . . . **Blattodae**
          Body slender and elongate 32

32 (31)   Sticklike body (walking sticks) . . . **Phasmatodea**
          Elongate body with raptorial
          front legs (mantids) . . . . **Mantodea**

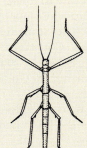

Phasmatodea

References

Note that, in addition to the following general guide-books, regional guides are often useful for surrounding areas.

Arnett, R. H. 2001. American insects: a handbook of the insects of America and Mexico, ed. 2. Boca Raton, Florida, CRC Press.

Arnett, R. H., Jr., and M. C. Thomas, eds. 2000. American beetles, vol. 1. Boca Raton, Florida, CRC Press.

Borror, D. J., and R. E. White. 1998. A field guide to the insects: America north of Mexico. Boston, Houghton Mifflin Harcourt.

Chu, H. F. 1949. How to know the immature insects. Dubuque, Iowa, Wm. C. Brown Company.

Claassen, P. W. 1931. Plecoptera nymphs of America (north of Mexico). Springfield, Ill., Pub. of the Thomas Say Foundation, by Charles C. Thomas.

Covell, C. V., Jr. 2005. A field guide to the moths of eastern North America, ed. 2. Martinsville, VA, Virginia Museum of Natural History.

Dillon, E. S., and L. S. Dillon. 1972. A manual of common beetles of eastern North America. New York, Dover Publications, Inc.

Edmunds, G. F., Jr., S. L. Jensen, and L. Berner. 1976. The mayflies of North and Central America. Minneapolis, University of Minnesota Press.

Ehrlich, P. R., and A. H. Ehrlich. 1961. How to know the butterflies. Dubuque, Iowa, Wm. C. Brown Company.

Fisher, B. L., and Stepfan P. Cover. 2007. Ants of North America: A guide to the genera. Berkeley, University of California Press.

Harris, J. R. 1952. An angler's entomology. New York, F. A. Praeger.

Hogue, C. L. 1993. Insects of the Los Angeles Basin. Los Angeles, Natural History Museum of L. A. County.

Holland, W. J. 1968. The moth book: a popular guide to a knowledge of the moths of North America. New York, Dover Publications, Inc.

Jaques, H. E. 1947. How to know the insects, ed. 2. Dubuque, Iowa, Wm. C. Brown Company.

Jaques, H. E. 1951. How to know the beetles. Dubuque, Iowa, Wm. C. Brown Company.

Jewett, S. G. 1959. The stoneflies (Plecoptera) of the Pacific Northwest. Corvallis, Oregon State College.

Kaufman, K., and F. P. Brock. 2006. Butterflies of North America (Kaufman Field Guides). Boston, Houghton Mifflin Harcourt.

Klots, A. B. 1951. A field guide to the butterflies of North America east of the great plains. Boston, Houghton Mifflin Company.

LaFontaine, G. 1981. Caddisflies. New York, Lyons & Burford.

Lehmkahl, D. M. 1979. How to know the aquatic insects. The pictured key nature series. Dubuque, Iowa, Wm. C. Brown Company.

McPherson, J. E. 1982. The Pentatomoidea (Hemiptera) of northeastern North America. Carbondale, Ill., Southern Illinois University Press.

Merritt, R. W., and K. W. Cummins. 1996. An introduction to the aquatic insects of North America, ed. 3. Dubuque, Iowa, Kendall/Hunt Publishing Company.

Miller, P. L. 1984. Dragonflies. New York, Cambridge University Press.

National Audubon Society. 1980. National Audubon Society field guide to North American insects and spiders. New York, Alfred A Knopf, Inc.

Needham, J. G., and M. J. Westfall, Jr. 1975. A manual of the dragonflies of North America (Anisoptera): including the Greater Antilles and the provinces of the Mexican border. Berkeley, University of California Press.

Opler, P. A. 1998. A field guide to eastern butterflies, ed. 2. The Peterson field guide series. Boston, Houghton Mifflin Harcourt.

Otte, D. 1981–84. The North American grasshoppers. Vol. 1: Acrididae (Gomphocerinae and Acridinae); Vol. 2: Acrididae (Oedipodinae). Cambridge, Mass., Harvard University Press.

Pyle, R. M. 1981. The Audubon Society field guide to North American butterflies. New York, Alfred A. Knopf, Inc.

Sborboni, V., and S. Forestiero. 1998. Butterflies of the world. Buffalo, N.Y., Firefly Books Inc.

Schuh, R. T., and J. A. Slater. 1995. True bugs of the world. Ithaca, N.Y., Cornell University Press.

Scott, J. A. 1986. The butterflies of North America: a natural history and field guide. Stanford, Calif., Stanford University Press.

Smart, P. 1975. The international butterfly book. New York, Crowell Company.

Stehr, F. W. (ed.). 1991. Immature insects. Vols. 1 & 2. Dubuque, Iowa, Kendall/Hunt Publishing Company.

Swan, L. A., and C. S. Papp. 1972. The common insects of North America. New York, Harper & Row, Publishers.

Triplehorn, C. A., and N. F. Johnson. 2005. Borror and DeLong's introduction to the study of insects, ed. 7. Belmont, Calif., Brooks/Cole.

Westfall, M. J., Jr., and J. L. May. 1996. Damselflies of North America. Jodhpur, India, Scientific Publishers.

White, R. E. 1983. A field guide to the beetles of North America. Boston, Houghton Mifflin Company.

Wiggins, G. B. 1977. Larvae of the North American caddisfly genera (Trichoptera). Toronto, University of Toronto Press.

The Echinoderms
Phylum Echinodermata

E chinoderms are an all-marine phylum that comprises sea lilies, sea stars, brittle stars, sea urchins, sand dollars, and sea cucumbers. They are a strange group, strikingly different from any other animal phylum, a group that abandoned the adaptive advantages of bilateral symmetry to become radial.

Echinoderms are deuterostomes, thus sharing with Hemichordata and Chordata several embryological features that set them apart from all the rest of the animal kingdom: anus developing from or near the blastopore and mouth developing elsewhere, enterocoelous coelom, radial and regulative cleavage, and mesoderm derived from enterocoelous pouches. Thus, all three phyla are presumably derived from a common ancestor.

A key characteristic of echinoderms is their **dermal endoskeleton** composed of calcareous plates and spines, often fused into an investing armor. This spiny surface provides the group with its name (Gr. *echinos,* prickly, + *derma,* skin). Echinoderms also have a **water-vascular system** that powers a multitude of tiny tube feet used for locomotion and food gathering. Many are invested with pincerlike **pedicellariae** that snap at creatures that would settle on them. To breathe, many echinoderms rely on numerous **dermal branchiae** (skin gills) that project delicately through spaces in their skeletal armor. As if to emphasize their uniqueness, echinoderms lack a definite head, their nervous system and sense organs are primitive, locomotion is slow, and they lack segmentation. Of all their distinguishing features, however, none delineates the group more conspicuously than its **pentaradiate symmetry:** body parts always arranged radially in five or multiples of five, no matter how the body plan has become adapted to different feeding strategies. This radial symmetry is secondarily acquired because echinoderm larvae are unmistakably bilaterally symmetrical (Figure 14.1).

EXERCISE 14A
Class Asteroidea—Sea Stars
Asterias

EXERCISE 14B
Class Ophiuroidea—Brittle Stars
Brittle Stars

EXERCISE 14C
Class Echinoidea—Sea Urchins
Arbacia

EXERCISE 14D
Class Holothuroidea—Sea Cucumbers
Sea Cucumber

Classification: Phylum Echinodermata

EXERCISE 14A
Class Asteroidea—Sea Stars

Core Study

Asterias
Phylum Echinodermata
   Subphylum Eleutherozoa
      Class Asteroidea
         Order Forcipulatida
            Genus *Asterias*

Where Found

Sea stars are the "prima donnas" of echinoderms, familiar to many people as beautifully symmetrical symbols of marine life. Those commonly seen are intertidal species, especially along rocky coastlines. Many live on high-energy beaches that receive the full force of the surf. Other species inhabit a variety of benthic habitats, often at great depths in

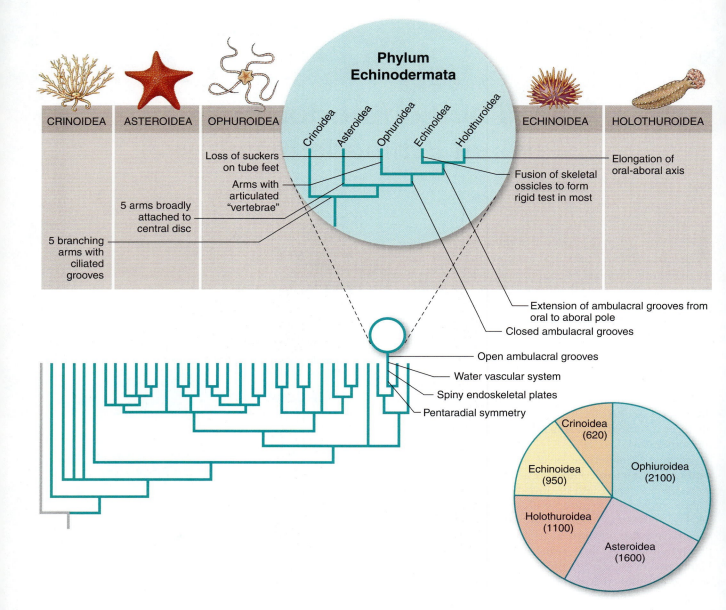

| CRINOIDEA | ASTEROIDEA | OPHUROIDEA | | ECHINOIDEA | HOLOTHUROIDEA |

Phylum Echinodermata

Crinoidea   Asteroidea   Ophuroidea   Echinoidea   Holothuroidea

Loss of suckers on tube feet

Arms with articulated "vertebrae"

5 arms broadly attached to central disc

5 branching arms with ciliated grooves

Fusion of skeletal ossicles to form rigid test in most

Elongation of oral-aboral axis

Extension of ambulacral grooves from oral to aboral pole

Closed ambulacral grooves

Open ambulacral grooves

Water vascular system

Spiny endoskeletal plates

Pentaradial symmetry

Crinoidea (620)
Ophiuroidea (2100)
Echinoidea (950)
Holothuroidea (1100)
Asteroidea (1600)

the ocean. Sea stars, like other echinoderms, are strictly bottom-dwellers. Some are particle feeders, but most are predators of slow-moving prey, such as molluscs (their favorite food), crabs, corals, and worms, since sea stars are themselves slow-moving animals. Adult sea stars seem to have few enemies, suggesting that their bodies produce some substance that discourages potential predators.

Behavior[1]

☞ Examine a living sea star in a dish of seawater. Using the following suggestions, observe its behavior.

Note the general body plan of the star with its **pentaradial** (Gr. *pente,* five, + L. *radius,* ray) symmetry,

its five **arms,** or **rays,** its **oral-aboral flattening,** and its **mouth** on the underside. Lift up the dish and look at this oral surface. Notice the rows of **tube feet.** How are they used? _____ How are the ends of the tube feet shaped? _____ What is the sequence of action of a single tube foot as the animal is moving? _____ Tube feet are muscular and filled with fluid, providing the necessary components for a hydraulic skeleton. When a foot contracts, water flows into a bulblike **ampulla** inside the arm. Tube feet and ampullae are parts of the **water-vascular system.**

Tilt the dish to one side (pour out a little water if necessary) and watch the animal's reaction. Does it move up or down the inclined plane? _____ Now tilt the dish in the opposite direction. Does the animal change direction? Is it

[1]Suggestions for demonstrations with living sea stars are found in Appendix A, pp. 287–288.

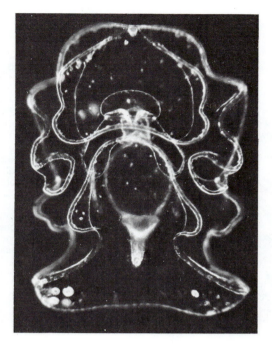

Figure 14.1
Bilaterally symmetrical *Auricularia* larva of the sea cucumber (×100).

Sensory tentacles, modifed tube feet, tactile and chemosensory

Tube foot

Eyespot, composed of retinal cells, each with a lens and separated by red pigment

Spine surrounded by pedicellariae

Skin gills (dermal branchiae) for gas exchange

Figure 14.2
Distal portion of the ray of a living sea star (*Asterias*).

positively geotactic (moves toward the earth) or negatively geotactic? _____

Place a piece of fresh seafood (oyster, fish, or shrimp) near one of the arms. Is there any reaction? _____ Are the arms flexible? _____ If the sea star makes no move toward the food, touch the tip of an arm with it, or slip the food under the end of an arm. Hold the dish up and look underneath. How is the food grasped? _____ How is it moved toward the mouth? _____ What position does the animal assume when feeding?_____

Examine the aboral surface with a hand lens or dissecting microscope. The **epidermis** is ciliated. Place a drop of carmine suspension (in seawater) on the exposed surface and see which direction the ciliary currents take. Notice the calcareous **spines** protruding through the skin (Figure 14.2). These are extensions of skeletal ossicles. Do the spines move? _____ Small, fingerlike bulges in the epidermis are **skin gills** [also called **dermal branchiae** or **papulae** (sing., **papula;** L., pimple)], concerned with gas exchange. Around the spines you will see small **pedicellaria** (L., *pediculus,* little foot, + *aria,* like) (Figure 14.2). These are calcareous, two-jawed pincers that are modified spines and are concerned with capturing tiny prey and protecting the dermal branchiae from collecting sediment or small parasites. Touch a pedicellaria with a fine artist's brush and observe the pincer action.

At the tip of each arm, note a small red **eyespot** and elongate tube feet modified as **sensory tentacles** (Figure 14.2).

Written Report

 Record your observations on separate paper.

External Structure

☞ Place a preserved sea star in a dissecting pan and cover with water. If observing a living sea star, place the animal in a clean pan or culture dish and cover with seawater.

The star-shaped body is composed of a **central disc** and five **rays** (arms). Are all the rays alike? What would account for some of the rays being shorter than others? _____ Compare your specimen with those of your neighbors. Preserved specimens may seem rigid, but live stars can bend their arms by means of muscles.

Aboral Surface. The central disc bears a small, porous **madreporite plate** (Fr. *madrēpore,* reef-building coral, + *ite,* suffix for body part) composed of calcium carbonate (Figure 14.3). It allows seawater to seep into an intricate **water-vascular system,** which provides the means of locomotion. The rays on each side of the madreporite are called the **bivium;** the other three are the **trivium.** The **anus** opens in the center of the central disc, but it is probably too small to see.

☞ Submerge one of the arms in water and examine the dorsal body wall under a dissecting microscope. Compare with a piece of dried-up test.

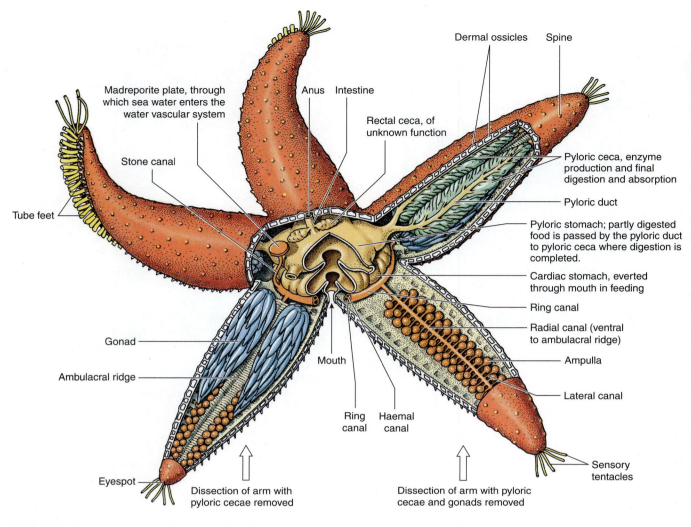

Figure 14.3
Anatomy of the sea star. Gonad size varies greatly with season when specimen was collected.

The body is covered with a thin, ciliated epidermis, through which white calcareous spines extend from the endoskeleton beneath. Are the spines movable? _____ Surrounding the base of each spine is a raised ring of skin bearing tiny, calcareous, pincerlike pedicellariae. Some are also found between the spines. Sometimes pedicellariae can be seen more easily on a dried-up piece of test. Some pedicellariae have straight jaws and others have curved jaws. They are moved by tiny muscles. What is the function of the pedicellariae? _____ Between the spines are soft, transparent, fingerlike projections, the skin gills, or dermal branchiae. These are hollow evaginations of the coelomic cavity. What is the function of the dermal branchiae? _____

Drawings

✐ **Sketch some pedicellariae in your notebook.**

A small, pigmented eyespot is located at the tip of each ray.

Oral Surface. The **ambulacral** (L. *ambulare,* to walk) groove of each ray contains rows of **tube feet (podia).** The **ambulacral spines** bordering the groove are movable and can interlock when the groove is contracted to protect the tube feet. Note the size, shape, number, and arrangement of the tube feet. How many rows of tube feet are there? _____

☞ Scrape away some tube feet and note the arrangement of the pores through which they extend. If you are making observations on a living sea star, do not scrape off the tube feet but instead brush them with a fine artist's brush, noting the responses of the tube feet and the ambulacral ossicles.

The tube feet are part of the water-vascular system. They are hollow, and their tips form suction discs for

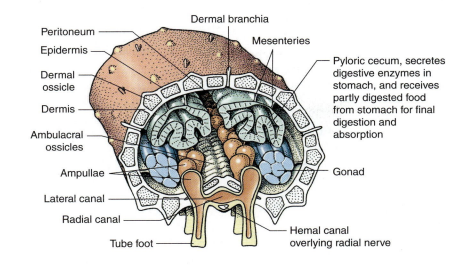

Figure 14.4
Cross section of a sea star arm.

Labels on figure:
Peritoneum
Epidermis
Dermal ossicle
Dermis
Ambulacral ossicles
Ampullae
Lateral canal
Radial canal
Tube foot
Dermal branchia
Mesenteries
Pyloric cecum, secretes digestive enzymes in stomach, and receives partly digested food from stomach for final digestion and absorption
Gonad
Hemal canal overlying radial nerve

attachment (Figures 14.3 and 14.4). These are effective not only in locomotion but also in opening bivalves for food.

The central **mouth** is surrounded by five pairs of movable spines. Push the spines outward, bend the arms back slightly, and note the thin **peristomial membrane** that surrounds the mouth. Sometimes the mouth is filled with part of the everted **stomach.**

Endoskeleton

Echinoderms are the first of the invertebrates to have a mesodermal endoskeleton. It is formed of calcareous plates, or **ossicles,** bound together by connective tissue.

☞ If you are studying a preserved sea star, cut off part of one of the rays of the bivium, remove the aboral wall from the cut-off piece, and study its inner surface. Compare with a piece of dried body wall, which is excellent for examining the arrangement of ossicles.

Note the skeletal network of irregular ossicles. Now look at the cut edge of the body wall and identify the outer layer of **epidermis,** the thicker layer of **dermis,** or connective tissue in which the ossicles are embedded, and the thin inner layer of ciliated **peritoneum.** Do sea stars have a true coelom? _____ The dermis and peritoneum are mesodermal in origin. Are the spines part of the endoskeleton? _____ Hold the piece up to the light. The thin places you see between the ossicles are where the dermal branchiae extend through the connective tissue.

Look at the oral surface of the cut-off ray. Note how the **tube feet** extend up between the **ambulacral ossicles,** emerging on the inside surface of the wall as bulblike **ampullae** (am-pool′ee; sing., **ampulla;** L., flask). Compress some of the ampullae and note the effect on the tube feet. Press on the tube feet and see the effect on the ampullae. Both are muscular and can regulate the water pressure by contraction. Scrape away some of the ampullae and tube feet, and examine the shape of the ossicles in the ambulacral groove. How do they differ from the ossicles elsewhere? _____ Note the alternating arrangement of the **ambulacral pores,** through which the tube feet extend. Whereas the ambulacral ossicles form a groove on the oral surface of each arm, they form an **ambulacral ridge** on the inner surface.

Drawings

✏ In your notebook, sketch a series of ambulacral ossicles, showing the arrangement of the openings for the tube feet.

Further Study

Internal Structure (Dissection)[2]

☞ Place the preserved specimen aboral side up in a dissecting pan and cover with water. Select the three rays of the trivium and snip off their distal ends. Insert a scissors point under the body wall at the cut end of one of the rays. Carefully cut along the dorsolateral margins of each ray to the central disc. Lift up the loosened wall and carefully free any clinging organs. Uncover the central disc, but cut around the madreporite plate, leaving it in place. Be careful not to injure the delicate tissue underneath. The tissue around the anal opening in the center of the disc will cling. Cut the very short intestine close to the aboral wall before lifting off the body wall.

---

[2]Directions for anesthetizing living sea stars for this study are found in Appendix A, p. 288.

The **coelomic cavity** inside the rays and disc contains **coelomic fluid,** which bathes the visceral organs.

Digestive System. A pentagonal **pyloric stomach** (Gr. *pylōros,* gatekeeper) (see Figure 14.3) lies in the central disc, and from it a **pyloric duct** extends into each arm, where it divides to connect with a pair of large, much-lobulated **pyloric ceca** (digestive glands). A very short **intestine** leads up from the center of the stomach to the anus in the center of the disc. Attached to the intestine are two **rectal ceca,** small, branched sacs of uncertain function. Below (ventral to) the pyloric stomach is the larger, five-lobed **cardiac stomach,** which fills most of the central disc (see Figure 14.3). Each lobe of the stomach is attached to the ambulacral ridge of one of the arms by a pair of **gastric ligaments,** which prevent too much eversion of the stomach.

When a sea star feeds on a bivalve, it folds itself around the animal, attaches its tube feet to the valves, and exerts enough pull to cause the shell to gape a little. Then, by contracting its body walls (to increase coelomic fluid pressure), it everts its stomach and inserts it into the slightly opened clam shell. There it digests the soft parts of the clam with juices from the pyloric ceca. Partly digested material is drawn up into the stomach and pyloric ceca, where digestion is completed. There is little waste fecal matter. When the sea star is finished feeding, the stomach withdraws into the coelom by contraction of stomach muscles and relaxation of the body wall, which allows coelomic fluid to flow back into the arms.

Many stars feed on small bivalves by engulfing the entire animal, digesting out its contents, and then casting the shell out through the mouth.

Reproductive System

Sexes are separate (dioecious) in sea stars.

☞ Remove the pyloric ceca from one arm to find the gonads attached to the sides of the arm where the arm joins the disc (see Figure 14.3).

Each gonad opens aborally by a very small **reproductive duct** and **gential pore.** Even during the breeding season when the gonads are enlarged, microscopic examination usually is required to distinguish female gonads from male gonads.

☞ Make a wet mount of a mashed bit of gonad and examine with a microscope.

In the ovary, there are eggs that have large nuclei; in the testes, many small sperm are to be found. In early summer, sea stars shed large streams of eggs and sperm into the water, where fertilization occurs externally.

Nervous System. The nervous system of a sea star consists of three interrelated systems. Foremost is the **oral** system, consisting of a **nerve ring** around the mouth in the peristomial membrane, and a **radial nerve** to each arm running along the ambulacral groove to the eyespot.

☞ To find the nerve ring, remove the tube feet and movable spines around the mouth and expose the peristomial membrane.

The nerve ring is a whitish thickening on the outer margin of this membrane. To see one of the **radial nerves,** bend an arm aborally and look along the oral surface of the ambulacral groove for a whitish cord (Figure 14.4). Trace the nerve from the ring to its termination in the arm.

An **epidermal nerve plexus** lying just beneath the epidermis coordinates responses of the dermal branchiae to tactile stimulation.

Sense Organs. Chemoreceptors and cells sensitive to touch are found all over the surface. Each pigmented **eyespot** consists of a number of light-sensitive ocelli.

Water-Vascular System. The water-vascular system is found only in echinoderms, which use it for locomotion and, in the case of sea stars, for opening clam shells. If this system in your specimen has been injected with a colored injection mass, its features can be studied to greater advantage.

☞ Carefully remove the stomach from the central disc.

The **madreporite plate** (orange in life) on the aboral surface (see Figure 14.3) contains ciliated grooves and pores. From it, a somewhat curved **stone canal** (yellow in life) leads to a **ring canal** (see Figure 14.3), which is found around the outer edge of the peristomial membrane next to the skeletal region of the central disc. The ring canal may be difficult to find if not injected.

Five **radial canals,** one in each arm, radiate out from the ring canal, running along the apex of the ambulacral groove just below the ambulacral ossicles and above the radial nerve. The position of the radial canal is best seen in a cross section of one of the arms (Figure 14.4). Short **lateral canals,** each with a valve, connect the radial canal with each of the tube feet. Now look on the inside of an arm and study the alternating arrangement of the ampullae. Note how each ampulla connects with a tube foot through a **pore** between the ambulacral plates.

Seawater that enters the madreporite passes down the stone canal to the ring canals, from there to the radial canals, and finally through the lateral canals to the ampullae and tube feet.

Tube feet have longitudinal muscles; ampullae have circular muscles. When the tube feet are contracted, most of the water is held in the ampullae (Figure 14.4). When the ampullae contract, water is forced into the elastic tube feet, which elongate because of the hydrostatic pressure within them. When the cuplike ends of the extended tube feet contact a hard surface, they attach with a suction force and then contract, pushing the water back into the ampullae and pulling the animal forward. Valves in the lateral canals prevent the backflow of water into the radial canals.

Although a single foot is not very strong, hundreds of them working together can move the animal along slowly and can create a tremendous pull on the shell of a mussel. Suckers are of little use on a sandy surface, where the tube feet serve as tiny legs. Some species have no suckers but use the stiff tube feet like little legs to "walk." Sea stars can travel about 15 cm per minute.

Cross Section of the Arm of a Sea Star (Microslide)

Identify the **epidermis** covering the entire animal, the **dermis** containing the **ossicles,** muscular tissue, the **peritoneum,** and the **coelomic cavity.** The spines have not yet erupted, for these are young sea stars. Observe the **dermal branchiae** projecting from the coelom through the body wall. **Pyloric ceca** hang from the aboral wall by **mesenteries.** Notice the ampullae in the coelom and their connection to the **tube feet** and **lateral canals.** The canals may not always be visible. Why? _____

Locate the **ambulacral ossicles** with the pores through which tube feet extend. Find the **radial canal,** a small tube under the **ambulacral groove,** and the **radial nerve** beneath the radial canal (Figure 14.4).

EXERCISE 14B
Class Ophiuroidea—Brittle Stars
Core Study

Brittle Stars
Phylum Echinodermata
　　Subphylum Eleutherozoa
　　　　Class Ophiuroidea
　　　　　　Order Ophiurida
　　　　　　　　Available genera

Where Found

Brittle stars, the most agile echinoderms, are widely distributed in all oceans and at all depths. In many habitats they are also the most abundant echinoderms, and yet they are seldom seen by casual observers. Brittle stars are reclusive animals that hide under and between intertidal and subtidal rocks by day to escape predation by fish. Even at night they may extend only their arms from hiding places to feed. If a diver exposes their retreat, they quickly scuttle to safety, using rowing movements of their arms. Should a fish (or a human) catch a brittle star by one arm, the animal usually will simply cast off the arm (autotomize), leaving the predator with a wiggling arm while its erstwhile owner scurries to safety beneath a nearby rock. It is common to find brittle stars with missing or partly regenerated arms.

External Features

General Body Form. Note that the arms of brittle stars are sharply marked off from the central disc—a characteristic of ophiuroids (Figure 14.5). The arms are more flexible than those of the sea stars. Does their appearance give a clue as to why? _____
In both appearance and function, the arms resemble vertebral columns. In fact, the elements of which the skeleton is composed are called **vertebral ossicles** because of a fancied resemblance to the vertebrae of vertebrate animals. Each ossicle is joined to the next by two pairs of muscles. Externally, the arm is encased in a series of aboral, lateral, and oral **plates.**

☞ Pull an arm off a preserved specimen and examine it in cross section. Locate the muscles and note the vertebral articulations.

Are there spines on the arms? _____
How do the spines compare with those of asteroids? _____ Do you find any pedicellariae or skin gills? _____
Oral Surface. On the oral side, note five triangular **jaws** around the mouth. Find the five **oral shields** (also called **buccal shields**), which are oval plates located on the interradial area between the rays (Figure 14.6). One of these is slightly modified as a **madreporite plate,** and its tiny pores connect with a madreporite canal inside. Compare the location of the madreporite plate in asteroids and ophiuroids.

Distal to the oral shields and close to each arm is a pair of grooves, representing the openings of the **bursae.** (In *Ophioderma,* a second pair is located distal to the first.) The bursae, comprising 10 saclike cavities within the disc, are peculiar to ophiuroids. Water is pumped in and out of them for respiratory purposes, and the gonads discharge their products into them. In some species they also serve as brood pouches, but in *Ophioderma* development is external.

The tube feet (podia) are often called tentacles in ophiuroids. They are small, do not have suckers, and project laterally between the skeletal plates. They are largely sensory in function but may also assist in locomotion. Examine the rough spines used to grip the substrate. How do they compare with those of asteroids? _____

Behavior

Avoid rough handling of a brittle star because this may cause it to "freeze," becoming immobile, or even to cast off one or more of its arms.

Locomotion. Brittle stars "walk" using twisting, highly flexible arm movements. Watch their movements carefully. What provides the actual force for forward movement? Do the arms push or pull the brittle star? _____ Do the arms work in pairs?

Figure 14.5
A, The brittle star *Ophiopholis aculeata* from the Gulf of Maine. **B,** A basket star, *Astrophyton muricatum,* on octocoral. Ophiuroids have sharply marked-off arms and are fragile.

_____ How are the tube feet used—or are they used at all in locomotion? _____ Does one arm always lead or follow? _____ Note that locomotory patterns differ somewhat in different species of brittle stars.

Turn the brittle star over and watch its righting response. If placed on a sandy substrate, will the brittle star burrow? _____

Feeding. Most ophiuroids are either active predators or selective deposit feeders. Deposit feeders capture organic material with their tube feet, which secrete copious amounts of mucus. Mucus and food are rolled into a ball by the tube foot, and then transferred to a small scale that lies adjacent to the tube foot. The next tube foot picks up the food and transfers it to the next scale, and so on until the food ball reaches the mouth.

Active predators such as species of the genus *Ophioderma* capture benthic (bottom-dwelling) organisms by curling an arm around the prey and sweeping it into the mouth.

☞ To see the method of feeding, drop very small bits of fish or shrimp into the aquarium near (but not touching) some brittle stars.

Do the animals appear to sense the presence of food before touching it? _____ Is the species you are studying an active predator or a selective deposit feeder? _____ The digestive system is much reduced compared to that of sea stars. The mouth leads by a short esophagus to a stomach, the site of digestion and absorption. There is no intestine, anus, or hepatic ceca.

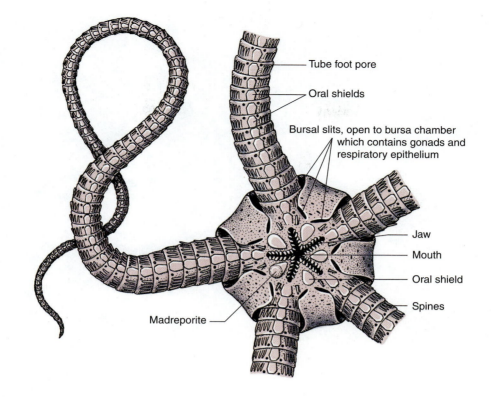

Figure 14.6
Oral view of central disc of a brittle star.

Reactions to Other Stimuli. Note how the animal reacts to mechanical stimulation. Touch the tip of an arm. Does it retreat or advance toward the source of the stimulus? _____ Touch a more proximal part of the arm. Is the reaction the same or different? _____ Stimulate the base of an arm or the central disc. What happens? _____

Can you determine whether ophiuroids—at least the species you are studying—respond positively or negatively to light? _____

The water-vascular, hemal, and nervous systems are organized similarly to those of asteroids.

Written Report

 Record your behavioral observations on separate paper.

Table of Comparison

In the lab report on p. 192, begin completing the comparative table of echinoderms by filling in the characteristics of the external anatomy of the sea star and the brittle star. Include such features as symmetry, shape, integument, ambulacra, tube feet, skin gills, pedicellariae, skeleton, and digestive system. After you have observed the sea urchin and sea cucumber in the next exercise, add their characteristics to the table.

EXERCISE 14C
Class Echinoidea—Sea Urchins

Core Study

Sea urchins, like sea stars, are familiar denizens of seashores. Lacking arms, and with the body enclosed within a globose shell, or test, composed of interlocking plates that bear movable spines, sea urchins have evolved a body design quite unlike that of sea stars and brittle stars. Nevertheless, they bear the typical pentamerous plan of all echinoderms, including a water-vascular system and other characteristics that set the strange echinoderms apart from any other animal phylum. Sea urchins and their kin—heart urchins and sand dollars—all bear a spiny armament suggesting *echinos,* the Greek word for hedgehog, from which the echinoids get their scientific name. Unlike the carnivorous sea stars and brittle stars, sea urchins are herbivores that scrape encrusting algae from rock surfaces, nibble on plants, or trap and eat drifting food. Sea urchins are "regular" echinoids: radially symmetrical, globose in shape, and armed with long spines. Other echinoids, are "irregular": bilaterally symmetrical with bodies variably shaped, such as the flattened sand dollars or the heart-shaped heart urchins.

Figure 14.7
The green sea urchin *Strongylocentrotus drobachiensis*. Note the slender, suckered tube feet. Urchins often attach bits of shell, marine algae, and other debris to themselves for camouflage. Small, stalked, white-tipped pedicellariae can be seen surrounding the bases of spines near the center of the photograph.

Arbacia

Phylum Echinodermata
    Subphylum Eleutherozoa
        Class Echinoidea
            Genus *Arbacia*
                Species *Arbacia punctulata*

Where Found

Like sea stars and brittle stars, echinoids are strictly marine, benthic animals widely distributed in all seas, from intertidal regions to the deep sea. Some favor rocky, high-energy coastlines with pounding surf, but most echinoids are subtidal, grazing in turtle-grass beds or on coral reefs or (especially the irregular echinoids) burying themselves in sandy bottoms where they feed on microscopic organic matter. *Arbacia punctulata* (Gr. *Arbaces*, first king of Media[3]), the purple sea urchin on the East Coast, is found from Cape Cod to Florida and Cuba. Other common species include *Strongylocentrotus drobachiensis* (Gr. *strongylos*, round, + *kentron*, spine), the green sea urchin of both the East and West Coasts of North America (Figure 14.7); *Strongylocentrotus purpuratus*, the purple

urchin of the West Coast; and species of *Lytechinus* (Gr. *lytos*, broken, + *echinos*, urchin) on both coasts.

Behavior and External Structure

The external features of a sea urchin are best observed in a living animal. However, if living forms are not available, submerge a preserved specimen in a bowl of water and use the following suggestions, directed toward live urchins, to identify the external structures.

☞ Place a living sea urchin on a glass plate, submerge in a bowl of seawater, and observe under a dissecting microscope.

Spines. Examine the long, movable **spines,** each attached at a ball-and-socket joint by two sets of ring muscles. Remove a spine (instructor's option) and note that its socket fits over a rounded **tubercle** on the test (Figure 14.8). An inner ring of **cog muscles** holds the spine erect. Hold the tip of a spine and try to move it. Do you feel the locking mechanism of the cog muscles? _____ Of what advantage is such a locking mechanism? _____ Now use a probe to touch the epidermis near a spine. Does the spine move? _____ In which direction? _____ The outer ring of muscles is responsible for directional movement. Are all the spines the same length? _____ Are they all pointed on the distal ends? _____

---

[3]The name of the genus *Arbacia*, bestowed by British zoologist John Edward Gray in 1835, is an example of a "nonsense" name that lacks any descriptive value. Gray apparently chose the name after reading Lord Byron's poem *Sardanapalus* concerning Arbaces, whom legend credits with founding the Median empire (now part of northern Iran) about 830 B.C.

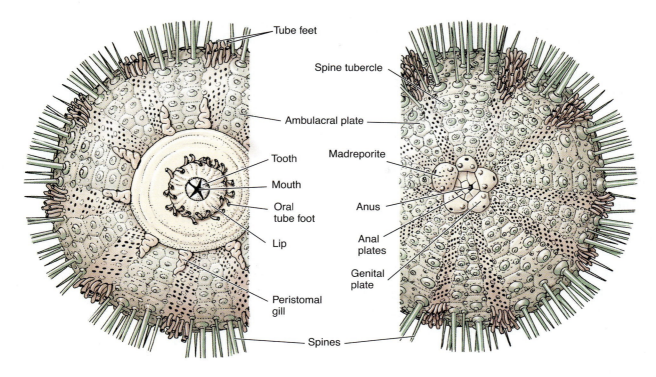

Tube feet

Spine tubercle

Ambulacral plate

Tooth

Madreporite

Mouth

Oral tube foot

Anus

Lip

Anal plates

Genital plate

Peristomal gill

Spines

Figure 14.8

External structure of the sea urchin. The oral (*left*) and aboral (*right*) surfaces are shown with the spines partly removed.

Tube Feet. Notice that the tube feet all originate from rows of perforations in the five **ambulacral regions.** The tube feet can be extended beyond the ends of the spines. Do any of them possess suckers? _____ Are any of them suckerless? _____ Do they move? _____ What happens when you touch one? _____ When you jar the bowl? _____

Mouth and Peristome. Examine the oral side of the urchin and find the **mouth** with its five converging **teeth** and collarlike **lip** (Figure 14.8 *left*). The teeth are part of a complex chewing mechanism called **Aristotle's lantern,** which is operated internally by several sets of muscles.[4] The lip contains circular, or "purse-string," muscles.

The membranous **peristome** surrounding the mouth is perforated by five pairs of large oral tube feet called **buccal podia.** Do these tube feet have suckers? _____ They are probably sensitive to chemical stimuli. The peristome also bears some small spines.

Pedicellariae. Notice on the peristome a number of three-jawed **pedicellariae** on the ends of long, slender stalks. Smaller but more active pedicellariae are located among the spines. Stimulate some of them by touching gently with a fine artist's brush. You may want to pinch off some of the pedicellariae

to examine them more closely on a slide, particularly if you are using a preserved specimen. Their functions are to discourage intruders and to help keep the skin clean. If you are working with a living urchin, drop some sand grains on the surface to see how the pedicellariae respond to clean the animal.

Locomotion. Note how the urchin uses its spines and tube feet in locomotion. Carefully, so as not to injure its tube feet, turn the urchin over (oral side up). Does it use its spines or tube feet to right itself? _____ Notice which ambulacra turned first, and mark that row by removing some of its spines or by marking some of the spines with thread; then turn the animal over again and see if the same ambulacra turn first. Repeat once more.

Tilt the glass plate to determine if the urchin moves up or down. Tilt in the opposite direction and see what happens. Is it positively or negatively geotactic? _____

Some sea urchins are adapted for burrowing into rock or other hard material by using both their spines and their chewing mechanisms. For example, *Strongylocentrotus purpuratus* excavates cup-shaped depressions in stone.

Most echinoids have tiny modified spines called **sphaeridia** (Gr. *sphaira,* sphere, + *idion,* dim. suffix), believed to be organs of equilibrium. In *Arbacia,* these are minute, glassy bodies located one in each ambulacrum close to the peristome. Try to find and remove

---

[4]The curious name for the protrusible chewing mechanism of sea urchins derives from a passage in the writings of Aristotle (384–322 B.C.) where he compared the apparatus to the frame of a lantern.

the sphaeridia to see whether their removal affects the urchin's righting reaction or its geotactic responses.

Direct a beam of bright light toward an urchin. How does it react? _____

Epidermis. The test, tube feet, pedicellariae, and spines are covered with **ciliated epidermis,** although the epidermis may have become worn off from the exposed spines. Drop a little carmine suspension on various parts of the sea urchin and find the direction of the ciliary currents. Of what advantage are such currents to the urchin? _____

Gills. At the outer edge of the peristome, between the ambulacra, find five pairs of branching peristomial **gills,** which open into the coelomic cavity.

Written Report

Record your observations on sea urchin behavior on separate paper.

Further Study

Test (Endoskeleton)

☞ Examine a dried sea urchin test from which the spines have been removed and also a dried sand dollar and/or heart urchin.

The test, or endoskeleton, is composed of calcareous plates (ossicles) that are symmetrically arranged and interlocked or fused so as to be immovable. Note the tubercles to which the spines were attached. Note the arrangement of the plates into 10 meridional double columns—five double rows of **ambulacral plates** alternating with double rows of interambulacral plates. What is the function of the perforations in the ambulacral plates? _____ Recall that in asteroids, the tube feet are extended between the plates rather than through them.

Examine the test of a sand dollar, a heart urchin, or both. Can you find perforations in them similar to those of the sea urchin? _____ Do they have ambulacra, and is their arrangement pentamerous? _____ The ambulacra in echinoids are homologous to the ambulacra of asteroids and ophiuroids.

On the aboral surface, note the area that is free of spines, the **periproct.** The **anus** is centrally located, surrounded by four (sometimes five) valvelike **anal plates** (Figure 14.8). Around the anal plates are five **genital plates,** so called because each bears a **genital pore.** Note that one of the genital plates is larger than the others and has many minute pores. This is the **madreporite plate,** which has the same function in sea urchins as it does in sea stars, for the echinoids have a **water-vascular system** in common with all echinoderms.

The test grows both by the growth of plates and by the production of new plates in the ambulacral area near the periproct.

Table of Comparison ◀

Continue completing the comparative table of echinoderms on p. 192 by filling in the characteristics of the external anatomy of the sea urchin. Include such features as symmetry, shape, integument, ambulacra, tube feet, skin gills, pedicellariae, skeleton, digestive system, and so on. After you have observed the sea cucumber in the next exercise, add its characteristics to the table.

EXERCISE 14D
Class Holothuroidea— Sea Cucumbers

Core Study

Sea Cucumber

Phylum Echinodermata
    Subphylum Eleutherozoa
        Class Holothuroidea
            Order Dendrochirotida
                Genus *Cucumaria* (or *Thyone*)
                or
            Order Aspidochirotida
                Genus *Parastichopus*

Where Found

Sea cucumbers, perhaps the oddest members of a phylum distinguished by strange animals, look remarkably like their vegetable namesake. Their characteristics include an elongate body, leathery body wall with warty surface, absence of arms, and mouth and anus located at opposite poles of the animal. They are benthic (bottom-dwelling), slow-moving animals found in all marine habitats. Two common genera on the East Coast of the United States are *Thyone* and *Cucumaria; Parastichopus* is a familiar genus on the West Coast (Figure 14.9).

Behavior and External Structure

☞ If possible, study living specimens in an aquarium or in a bowl of seawater containing a generous layer of sand, and then examine a preserved specimen. Sea cucumbers are slow to react. They should be left undisturbed for some time before the laboratory period if you are to see them relaxed and feeding.

Note that holothurians, unlike the other echinoderms, are orally-aborally elongated and have a cylindrical body with the **mouth** encircled by **tentacles** at one end and the **anus** at the other.

A more detailed description of the animal and its behavior will depend somewhat on the species you are observing. Notice the tentacles. Are they branched and extensible (as in *Cucumaria* and *Thyone*) or short and shield-shaped (as in *Parastichopus*), or of some other

A

B

Figure 14.9

Two sea cucumbers from the Pacific coast of North America.
A, *Parastichopus californicus* is a deposit feeder that grazes the bottom with its tentacles. **B,** *Cucumaria miniata* is a suspension feeder that traps planktonic organisms on its extended, mucus-coated tentacles. When loaded, the tentacles are pushed one by one into the mouth, and the food is ingested.

type? _____ The tentacles, which are modified tube feet, are hollow and a part of the **water-vascular system;** they are connected internally with the radial canals. The type of tentacle structure is related to feeding habits. *Cucumaria* and *Thyone* are suspension feeders that stretch their mucus-covered tentacles into the water or over the substrate until they are covered with tiny food organisms. Then they thrust the tentacles into the mouth, one by one, to lick off the food. Can you observe these actions? _____ *Parastichopus* and some others are deposit feeders that simply shovel mud and sand into the mouth, digest out organic particles, and void the remainder.

Have you noticed any rhythmic opening and closing of the anus? _____ This is a respiratory movement coordinated with the pumping action of the cloaca, which pumps water into and out of the **respiratory trees** (internal respiratory organs).

Does the animal try to burrow into the sand? _____ Does it cover itself completely or leave the ends exposed? _____ *Thyone* may take 2 to 4 hours to bury its middle by alternate circular and longitudinal muscle contractions. It is likely to be more active in the late afternoon and night than in the morning. If you are watching *Parastichopus* move, what does its muscular action remind you of? _____ Are there waves of contraction? _____

Does the sea cucumber react to mechanical stimulus? _____ Try touching a tentacle. Does one or more than one tentacle react? _____ Touch several tentacles. What happens when you stroke the body or gently pick the animal up? _____ Did it expel water when you picked it up? _____ Having observed its movement and reactions, can you see the advantages of the hydrostatic skeleton? What other phyla used the hydraulic skeleton to advantage? _____

Note the **tube feet.** Are they scattered all over the body or arranged in ambulacral rows? _____ This pattern differs among different species. Are the tube feet all alike—that is, are ventral tube feet any different from dorsal tube feet? _____ Are any of them suckered? _____ If the pentamerous arrangement of ambulacra is evident in your specimen, how many rows make up the ventral **sole?** _____ How many are on the dorsal surface? _____ If you place the animal on a solid surface, do the tube feet attach themselves? _____ Can the animal right itself if turned over? _____ Are the tube feet involved in the righting action? _____ Are muscles involved?

Does the animal show any geotactic reaction if placed on a vertical or sloping surface? _____ Some burrowing forms are positively geotactic and move downward; other species are negatively geotactic and climb upward. *Thyone* gives no geotactic response.

Do you find any pedicellariae or skin gills? _____ Do you feel the presence of a test under the epidermis? That is because the skeleton of the holothurian is usually limited to microscopic ossicles embedded in the tough, leathery body wall.

Written Report

Record your observations on sea cucumber behavior on separate paper.

Further Study

Internal Structure

☞ Locate the five longitudinal ambulacral areas of the sea cucumber. The sole (the ventral side applied to the substrate) has three ambulacra with well-developed tube feet. The dorsal side has two ambulacra with smaller tube feet (in

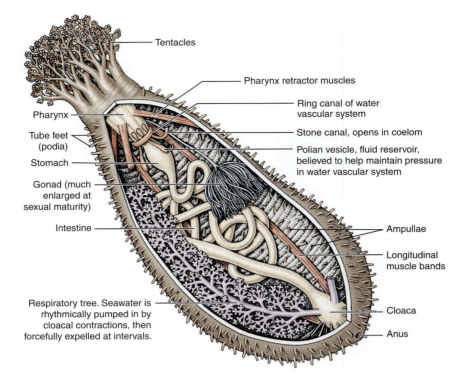

Tentacles

Pharynx retractor muscles

Ring canal of water
vascular system

Pharynx

Stone canal, opens in coelom

Tube feet
(podia)

Polian vesicle, fluid reservoir,
believed to help maintain pressure
in water vascular system

Stomach

Gonad (much
enlarged at
sexual maturity)

Intestine

Ampullae

Longitudinal
muscle bands

Respiratory tree. Seawater is
rhythmically pumped in by
cloacal contractions, then
forcefully expelled at intervals.

Cloaca

Anus

Figure 14.10
Internal anatomy of a sea cucumber.

some species, tube feet are absent on the dorsal surface). With scissors, open the body by making a longitudinal incision on the ventral (sole) side of the animal, between the central and right ambulacra. Pin down the walls and cover with water.

Digestive System. Note the large **coelomic cavity.** Just behind the mouth is the **pharynx,** supported by a ring of calcareous plates (Figure 14.10). It is followed by a short, muscular **stomach** and a long, convoluted **intestine,** held in place by mesenteries and expanding somewhat at the end to form a **cloaca,** which empties at the **anus.**

Respiratory System. Two branched **respiratory trees** are attached to the cloaca; they serve as both respiratory and excretory organs. They are aerated by rhythmic pumping of the cloaca. Several inspirations 1 minute or more apart are followed by a vigorous expiration that expels all the water.

Note the muscles between the cloaca and the body wall. Long **retractile muscles** (how many? ) run from the pharynx to join the longitudinal **muscle bands** in the body wall.

Water-Vascular System. A **ring canal** surrounds the pharynx. One or more rounded or elongated sacs called **polian vesicles** hang from the ring canal into

the coelom and open into the ring canal by a narrow neck. Polian vesicles are believed to function as expansion chambers in maintaining pressure within the water-vascular system. One or more **stone canals** also open into the ring canal from the body cavity. In adult sea cucumbers, the water-vascular system has usually lost contact with the seawater outside. Coelomic fluid, rather than seawater, enters and leaves the system.

Five **radial canals** extend from the ring canal forward along the walls of the pharynx to give off branches to the tentacles, which are actually modified tube feet. From there, the radial canals run back along the inner surface of the ambulacra, where each gives off **lateral canals** to the **tube feet** and **ampullae.** Valves in the lateral canals prevent backflow. Note the ampullae along the ambulacra in the inner body wall.

Reproductive System. The **gonad** consists of numerous tubules united into one or two tufts on the side of the dorsal mesentery. These become quite large at sexual maturity. A **gonoduct** passes anteriorly in the mesentery to the **genital pore.** The sexes are separate, and fertilization is external.

Endoskeleton. The **endoskeleton** consists largely of tiny calcareous ossicles scattered in the dermis. These can be seen on a prepared slide.

Complete the comparative table of the echinoderms in the lab report on p. 192 by filling in the characteristics of the external anatomy of the sea cucumber, mentioning such features as shape, ambulacra, tube feet, skin gills, skeleton, integument, and so on.

Classification: Phylum Echinodermata

The following classification is abbreviated.

Class Crinoidea (cry-noi′de-a; Gr. *krinon,* lily, + *eidos,* form, + *ea,* characterized by). Sea lilies and feather stars. Aboral attachment stalk of dermal ossicles. Anus on oral surface; five branching arms with pinnules; ciliated ambulacral groove on oral surface with tentacle-like tube feet for food collecting; spines, madreporite, and pedicellariae absent. Examples: *Antedon, Florometra.*

Class Asteroidea (as′ter-oi′de-a; Gr. *aster,* star, + *eidos,* form, + *ea,* characterized by). Sea stars. Star-shaped, with arms not sharply marked off from central disc; ambulacral grooves open, with tube feet on oral side; tube feet often with suckers; anus and madreporite aboral; pedicellariae present. Example: *Asterias.*

Class Ophiuroidea (o′fe-u-roi′de-a; Gr. *ophis,* snake, + *oura,* tail, + *eidos,* form). Brittle stars and basket stars. Star-shaped, with arms sharply marked off from central disc; ambulacral grooves closed, covered by ossicles; tube feet without suckers and not used for locomotion; pedicellariae absent. Examples: *Ophiura, Gorgonocephalus.*

Class Echinoidea (ek′i-noi′de-a; Gr. *echinos,* sea urchin, hedgehog, + *eidos,* form). Sea urchins, heart urchins, and sand dollars. More or less globular or disc-shaped, with no arms; compact skeleton, or test, with closely fitting plates; movable spines; ambulacral grooves closed and covered by ossicles; tube feet with suckers; pedicellariae present. Examples: *Arbacia, Strongylocentrotus, Lytechinus, Mellita.*

Class Holothuroidea (hol′o-thu-roi′de-a; Gr. *holothourion,* sea cucumber, + *eidos,* form). Sea cucumbers. Cucumber-shaped, with no arms; spines absent; microscopic ossicles embedded in thick muscular wall; anus present; ambulacral grooves closed; tube feet with suckers; circumoral tentacles (modified tube feet); pedicellariae absent; madreporite plate internal. Examples: *Thyone, Parastichopus, Cucumaria.*

Name_____

Date_____

Section_____

Comparative Table of Echinoderm Characteristics

Characteristic	Sea star	Brittle star	Sea urchin	Sea cucumber

Phylum Chordata
Protochordates
Subphylum Urochordata
Subphylum Cephalochordata

What Defines a Chordate?

C hordates show a remarkable diversity of form and function, ranging from the **protochordates** to humans. Most chordates are vertebrates, but the phylum also includes a few invertebrate groups. All animals that belong to phylum Chordata must have at some time in their life cycle the following characteristics:

1. **Notochord.** The notochord (Gr. *nōton,* back, + L. *chorda,* cord) is a slender rod of cartilage-like connective tissue (Figure 15.1) lying near the dorsal side and extending most of the length of the animal. It is regarded as an early endoskeleton and has the functions of such. In most vertebrates, it is found only in the embryo.

2. **Pharyngeal gill slits.** The pharyngeal gill slits (see Figure 15.3) are a series of paired slits in the pharynx that serve as passageways for water to the gills. In some vertebrates, they appear only in the embryonic stages.

3. **Dorsal tubular nerve cord.** A dorsal tubular nerve cord, with its modification, the brain, forms the central nervous system. It lies dorsal to the alimentary tract and has a fluid-filled cavity, in contrast to the invertebrate nerve cord, which is ventral and solid.

4. **Endostyle or thyroid gland.** The endostyle or its derivative, the thyroid gland, is found in all chordates, but in no other animals. The endostyle secretes mucus and traps small food particles for protochordates and lamprey larvae. Some cells

---

What Defines a Chordate?
EXERCISE 15A
Subphylum Urochordata—*Ciona,* an Ascidian
Ciona
EXERCISE 15B
Subphylum Cephalochordata—Amphioxus
Amphioxus
Classification: Phylum Chordata

---

in the endostyle are homologous with cells of the thyroid gland found in the remainder of vertebrates.

5. **Postanal tail.** A postanal tail projects beyond the anus at some stage and serves as a means of propulsion in water. It may or may not persist in the adult. Along with body muscles and the stiffened notochord, it provides motility for a free-swimming existence.

These features vary in chordates. Some ancestral chordates have all of these structures throughout life. In many chordates having more derived characteristics, the gill slits never break through from the pharynx but merely form pouches that have no function; the notochord is replaced by the vertebral column; and only the dorsal nerve cord actually persists in the adult as a diagnostic chordate character. The **protochordates** demonstrate each of the chief chordate characteristics at some point in their life cycle.

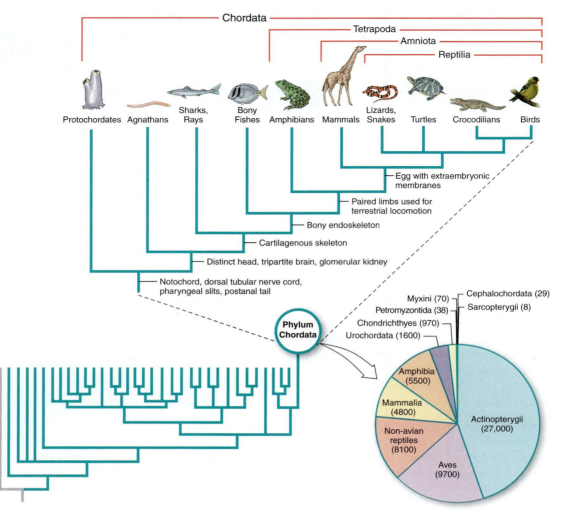

Figure 15.1
Structure of the notochord and its surrounding sheaths. Cells of the notochord proper are thick-walled, pressed together closely, and filled with semifluid. Stiffness is caused mainly by turgidity of fluid-filled cells and surrounding connective tissue sheaths. The primitive type of endoskeleton is characteristic of all chordates at some stage of the life cycle. The notochord provides longitudinal stiffening of the main body axis, a base for trunk muscles, and an axis around which the vertebral column develops.

EXERCISE 15A
Subphylum Urochordata— *Ciona*, an Ascidian

Core Study

Ciona

Phylum Chordata
  Subphylum Urochordata
    Class Ascidiacea (sea squirts)
      Order Enterogona
        Family Cionidae
          Genus *Ciona*
            Species *Ciona intestinalis*

Urochordata (Gr. *oura,* tail, + L. *chorda,* cord) are commonly called tunicates because of their leathery covering,

or tunic. They are divided into three classes: Ascidiacea, sea squirts; Thaliacea, salpians; and Appendicularia. The largest group is the ascidians (Gr. *askidion,* leather bag or bottle), which are also the most generalized. They are called sea squirts because of their habit, when handled, of squirting water from the excurrent siphon. Any of the small, translucent ascidians may be used for this exercise. Adult tunicates are all sessile, whereas the larvae undergo a brief free-swimming existence.

Where Found

Tunicates are found in all seas and at all depths. Most of them are sessile as adults, although some are pelagic (found in the open ocean). *Ciona intestinalis* (Gr. *Chionē,* demigoddess of mythology) is a cosmopolitan species commonly found in shallow water. Like most shallow-water sea squirts, it attaches to almost any available rigid surface, especially wharf pilings, anchored and submerged objects (rocks, shells, and ship bottoms), and eelgrass. It grows to 15 cm in its largest dimension. Although sea squirts are common in marine environments, they are often overlooked. One of the best places to see sea squirts is on pilings, where they may cover the surface.

External Features and Behavior

Sea squirts are fairly hardy in a marine aquarium.

☞ Examine, in a finger bowl of seawater, a living solitary tunicate such as *Ciona* or *Molgula,* or a portion of a colony of *Perophora* or some other ascidian as available.

Observe the use of the two openings, or **siphons.** When fully submerged and undisturbed, the siphons are open, and respiratory water, kept moving by ciliary action, enters the more terminal siphon (called the **incurrent** or **oral siphon**) at the mouth, circulates through a large pharynx, and leaves through the **excurrent** or **atrial siphon** on one side (the dorsal side) (Figures 15.2 and 15.3).

☞ Release a little carmine suspension near the animal to verify the flow of water through the siphons.

Figure 15.2

Ciona intestinalis, a solitary tunicate, showing its siphons in use. Throughout its life, the animal remains anchored to one spot on the seafloor. Its free-swimming larva bears all the chordate hallmarks: notochord, gill slits, dorsal nerve cord, and postanal tail.

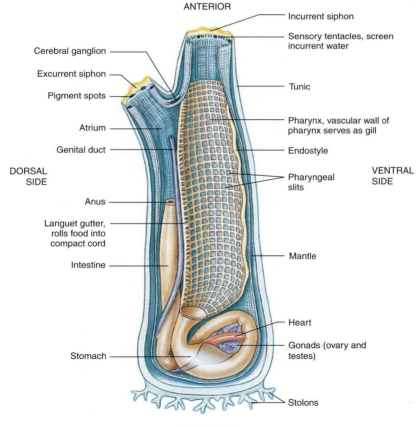

ANTERIOR

Incurrent siphon

Sensory tentacles, screen incurrent water

Cerebral ganglion

Excurrent siphon

Pigment spots

Tunic

Pharynx, vascular wall of pharynx serves as gill

Atrium

Genital duct

Endostyle

DORSAL SIDE

VENTRAL SIDE

Pharyngeal slits

Anus

Languet gutter, rolls food into compact cord

Intestine

Mantle

Heart

Gonads (ovary and testes)

Stomach

Stolons

POSTERIOR

Figure 15.3

Structure of a solitary sea squirt, *Ciona.*

The outer covering of a tunicate is called the **tunic,** or **test** (Figure 15.3). It is secreted by the **mantle,** which lies just inside it, and it contains cellulose—an uncommon substance in animals. The mantle contains the muscle fibers by which the body can contract. If the tunic is translucent enough and the light is properly adjusted, you may be able to see some of the internal structure.

Neuromuscular System. The tunicate nervous system is reduced and not well understood. A cerebral ganglion (closely associated with a subneural gland of uncertain function) is located between the siphons. The tunic probably has no sensory nerves, but pressure on the tunic may be transmitted to nerves in the mantle. Both direct and crossed reflexes have been observed in some ascidians and can be tested in a living *Ciona* or other tunicate by touching selected areas with the tip of a glass rod or dissecting needle.

Direct reflexes result from mechanical stimulation of the *outer* surface of the siphons or tunic.

☞ *Gently* stimulate various areas of the tunic and note the response. Gently touch the outer surface of one of the siphons, note the response, and then stimulate the other siphon.

What areas of the body are most sensitive? _____ **Crossed reflexes** result from mechanical stimulation of the *inner* surface of the siphons.

☞ Gently touch the inner surface of the oral siphon. What happens? _____ Try a stronger stimulus of the same siphon. Do you get the same response? _____ How do these responses differ from those occurring when the outer surface was stimulated? _____ Repeat with the atrial siphon.

Of what protective value would these reflexes be to the animal? _____

Further Study

Internal Structure

☞ Internal structure can be observed on either a living or a preserved specimen. Use fine scissors to slit the tunic longitudinally, beginning the cut at the incurrent siphon and continuing to the base of the pharynx. Be *very* careful to cut *only* the tunic and not the mantle beneath it. Slip the animal out of its tunic, and then return the animal to the bowl of seawater and study it with a dissecting microscope.

Respiratory System. The **branchial sac,** or **pharynx,** is the largest internal structure, and the space between it and the mantle is the **atrium** (Figure 15.3). The pharyngeal wall is perforated with many pharyngeal slits through which water passes into the atrium to be discharged through the excurrent siphon. The vascular wall of the pharynx serves as a gill for gas exchange.

Circulatory System. In a living specimen, with the test removed and a light properly adjusted, you should be able to see the beating of the **heart,** located near the posterior end on the right side. The tubular heart empties into two vessels, one at each end. Its peristaltic waves are of unusual interest, because they send the blood in one direction for awhile and then reverse direction and pump the blood in the opposite direction. Apparently there are two pacemakers that initiate contractions, one at each end of the heart, and they alternate in dominance over each other. The tunicate open circulatory system is an open type of system. Blood cells are numerous and colorful. There are no respiratory pigments.

Digestive System. At the junction of the **mouth** and the **pharynx** is a circlet of **tentacles** forming a grid that screens the incurrent water. By dropping a grain of sand into the incurrent siphon of a living tunicate, you may be able to observe the ejection reflex.

Inside the pharynx along the midventral wall is the **endostyle,** which is a ciliated groove that secretes a great deal of mucus. Cilia on the walls of the pharynx distribute the mucus. Food particles become tangled in the mucus and are propelled by cilia to a dorsal gutter in which the mucus with its trapped food becomes rolled into a compact cord. In *Ciona* and a few other tunicates, the gutter is lined with a row of curved, ciliated, fingerlike processes called languets (F. *languette,* small tongue) (Figure 15.3). The cord is propelled posteriorly to the esophagus and stomach.

☞ If living ascidians are available, examine one that has been submerged for some time in a suspension of carmine particles in seawater. Open the pharynx by cutting through the incurrent siphon and downward, a little to one side of the midventral line. Then cut around the base and lay the animal open in a pan of water. You should be able to see a concentration of carmine particles in the middorsal area. Cut out a small piece of the pharyngeal wall (free from the tunic and mantle) and mount it on a slide to observe the gill slits and beating cilia.

It may be difficult to differentiate the **esophagus, stomach,** and **intestine.** The **anus** empties into the atrium near the excurrent siphon.

Excretion. A ductless structure near the intestine is assumed to be a type of nephridium that is excretory in function.

Reproduction. *Ciona,* like most tunicates, is hermaphroditic and bears a single ovary and testis, each with a gonoduct that opens into the atrium.

Some tunicates are colonial. In certain colonial tunicates, zooids are separate and attached by a stolon, as in

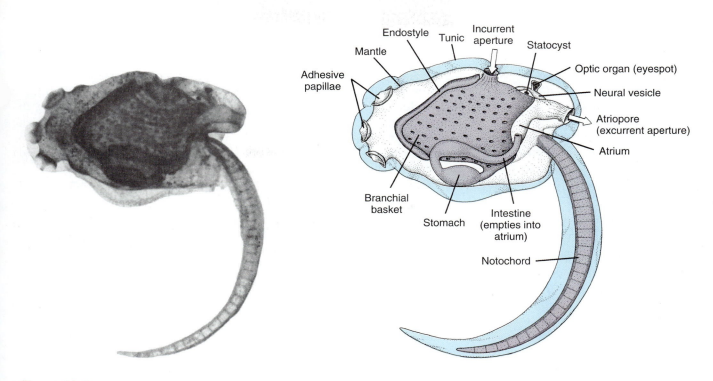

Figure 15.4
Tadpole larva of a tunicate, photographed from a stained slide specimen.

Perophora. In others, the zooids are regularly arranged and partly united at the base by a common tunic; all the incurrent siphons are at one side and the excurrent siphons at the other. Another type of colonial tunicate has a system of zooids that share a common atrial (excurrent) chamber. Colonies are formed asexually by budding.

Solitary tunicates generally shed their eggs from the excurrent siphon, and development occurs in the sea. Colonial species usually brood their eggs in the atrium, and the microscopic larvae leave by the excurrent siphon. For a very brief period, the larvae are nonfeeding and live a planktonic, free-swimming existence; then they settle down, attach to the substrate, and metamorphose.

Ascidian Larvae (Study of Stained Slides)

Ascidian larvae are free-swimming (Figure 15.4). Why are the larvae often called "tadpole" larvae? _____
They do not look like sessile adult sea squirts and are actually more characteristic of the chordates than are the adults. They possess not only gill slits, but also a notochord, a dorsal tubular nerve cord, and a tail, structures that have been lost in the adult. You may be able to identify some of these structures on a stained slide.

Adhesive papillae at the anterior end of the larva are used to attach to some object during metamorphosis, which occurs within a short time after hatching. The **notochord** can be identified in the long tail. This character, which becomes lost in the adult, gives subphylum Urochordata ("tail-cord") its name. A **nerve cord** dorsal to the notochord enlarges anteriorly into a neural vesicle. Can you identify a pigmented, photoreceptive **eyespot?** _____ A smaller pigmented area anterior to the eye is a **statocyst.** What is the function of the statocyst? _____

At metamorphosis, these portions of the nervous system degenerate, and a ganglion serves as the nerve center.

Look for the anterior **oral (incurrent) aperture** and the more posterior **atriopore (excurrent aperture).** Perhaps you can identify the **branchial basket** (pharynx) with **gill slits, stomach, intestine, atrium,** and **endostyle.**

EXERCISE 15B
Subphylum Cephalochordata— Amphioxus

Core Study

Amphioxus
Phylum Chordata
    Subphylum Cephalochordata
        Genus *Branchiostoma* (= Amphioxus)
           Species *Branchiostoma lanceolatus*

The little lancelet, *Branchiostoma* (Gr. *branchia*, gills, + *stoma*, mouth), commonly called amphioxus, illustrates basic chordate structure and is considered similar to the ancestor of the vertebrates. Besides the basic characteristics—**notochord, pharyngeal gill slits, dorsal tabular nerve cord, endostyle,** and **postanal tail**—it also possesses the beginning of a **ventral heart** and a **metameric arrangement** of muscles and nerves. There are only two genera of Cephalochordates—*Asymmetron* (Gr. *asymmetros,* ill-proportioned) and *Branchiostoma*.

Where Found

Branchiostoma is common along the southern Pacific and southern Atlantic coasts of the United States, as well as on the coasts of China and the countries bordering the Mediterranean Sea. On sandy bottoms, it dives in headfirst, and then twists upward so that the tail remains buried in sand and the anterior end is thrust upward into the water.

External Structure

☞ Place a preserved mature specimen in a watch glass and cover with water. Do not dissect or mutilate the specimen.

How long is it? _____ Why is it called a lancelet? _____ Does it have a distinctive head? _____ Observe the **dorsal fin,** which broadens in the tail region **(caudal fin)** and continues around the end of the tail to become the **ventral fin.**

The anterior tip is the **rostrum** (Figure 15.5A). With a hand lens, find the opening of the **oral hood,** which is fringed by a number of slender oral tentacles, also called buccal cirri, that strain out large particles of sand and are sensory in function.

On the flattened ventral surface are two **metapleural folds** of skin extending like sled runners to the ventral fin. Find the **atriopore,** which is anterior to the ventral fin. The atriopore is the opening of the atrium (L., entrance hall), a large cavity surrounding the pharynx. The **anus** opens slightly to the left of the posterior end of the ventral fin.

In mature specimens, little blocklike **gonads** (testes or ovaries) lie in the atrium anterior to the atriopore and just above the metapleural folds on each side. They can be seen through the thin body wall.

Study of the Whole Mount

☞ Examine with low power a stained and cleared whole mount of an immature specimen.

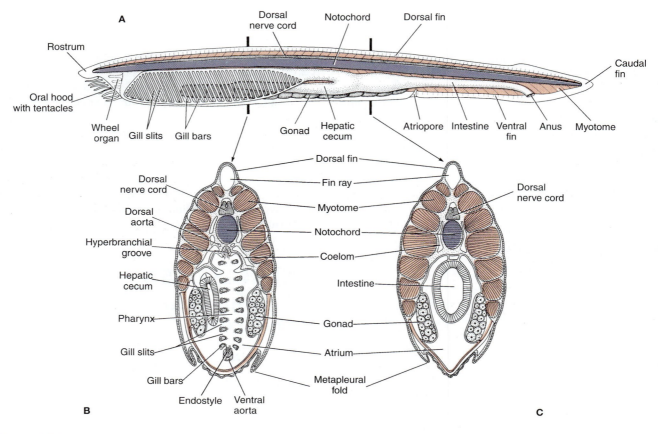

Figure 15.5
Structure of amphioxus, showing external structure and transverse sections through the pharynx and intestine.

Chevronlike **myotomes** (Gr. *mys,* muscle, + *tomos, slice*) along the sides of the animal are segmentally arranged muscles. Does the myotome of an amphioxus zigzag more or less than the myomere of a bony fish? _____ How might the differences in musculature help us predict whether the amphioxus or the fish would be the better swimmer? _____ Identify the various parts of the **fin** and note its skeletal support, the transparent **fin rays.** You may have to reduce the light to see the fin rays.

Beneath the rostrum is a large chamber called the **buccal cavity,** which is bounded laterally by fleshy, curtainlike folds and is open ventrally. The rostrum and lateral folds together make up the **oral hood** (Figure 15.5A). The roof of the oral hood bears the notochord, which may have a supporting function in spreading the hood open. Each of the oral tentacles is stiffened by a skeletal rod of fibrous connective tissue. Behind the buccal cavity is an almost perpendicular membrane, the **velum** (L., veil), pierced ventrally by a small opening, the true **mouth,** which is always open and leads into the **pharynx.** On the walls of the buccal cavity, projecting forward from the base of the velum, are several fingerlike ciliated patches that compose the **wheel organ.** The rotating effect of these cilia helps maintain a current of water flowing into the mouth. Around the mouth, projecting posteriorly from the velum, are about a dozen delicate **velar tentacles,** also ciliated. Both the oral tentacles and the velar tentacles have chemoreceptor cells for monitoring the incurrent water.

The large **pharynx** narrows into a straight **intestine** extending to the **anus** (Figure 15.5). The sidewalls of the pharynx are composed of a series of parallel, oblique **gill bars,** between which are **gill slits.** Just posterior to the pharynx is a diverticulum of the intestine called the **hepatic cecum,** or liver, which extends forward along one side of the pharynx. Surrounding the pharynx is the **atrium,** a large cavity that extends to the **atriopore.** Water entering the mouth filters through the gill slits into the atrium and then out the atriopore.

Cephalochordates such as sponges, clams, and tunicates are filter-feeders. They use a mucus-ciliary method, feeding on minute organisms. As the animal rests with its head out of the sand, the ciliated tentacles, wheel organ, and gills draw in a steady current of food-laden water, from which the cirri and velar tentacles strain out large or unwanted particles. On the floor of the pharynx is an **endostyle** (Figure 15.5) consisting of alternating rows of ciliated cells and mucus-secreting cells, and in the roof of the pharynx is a ciliated **hyperbranchial groove** (= epipharyngeal groove). Particles of food entangled in the stream of mucus secreted by the endostyle are carried upward by cilia on the inner surface of the gill bars, and then backward toward the **intestine** by cilia in the hyperbranchial groove. Digestion occurs in the intestine.

Oxygen–carbon dioxide exchange occurs in the epithelium covering the gill bars. The **notochord** just dorsal to the digestive system is transversely striated and is best seen in the head and tail regions. It provides skeletal support and a point of attachment for the muscles. Note that it extends almost to the tip of the rostrum. Above and parallel to the notochord is the **dorsal nerve cord** (Figure 15.5). The row of black spots in the nerve cord are pigmented **photoreceptor cells.** Chemoreceptors are scattered over the body but are particularly abundant on the oral and velar tentacles. Touch receptors are located over the entire body.

Further Study

Circulatory System

Amphioxus does not have a heart; peristaltic contractions of the **ventral aorta** keep the colorless blood in motion, sending it forward and then upward through **afferent branchial arteries** (Figure 15.6) to capillaries in the gill bars for gas exchange. Blood is carried from the gills by **efferent branchial arteries** up to a pair of **dorsal aortas.** These join posterior to the gills to form a **median dorsal aorta,** which gives off **segmented arteries** to the capillaries of the myotomes and to the capillaries in the wall of the intestine. From the intestinal wall, blood, now rich in digested food nutrients, is picked up by the **subintestinal vein** and carried forward to the **hepatic portal vein,** which enters the **hepatic cecum.** In the capillaries of the liver, nutrients are either removed and stored in liver tissue or processed and returned to the blood as needed. Blood returns to the ventral aorta by way of the **hepatic vein.** Blood containing waste products from the muscular walls returns by way of left and right **precardinal** and **postcardinal veins,** which empty into left and right **ducts of Cuvier** and then into the ventral aorta.

Oral Report

Compare circulation in amphioxus with that in the earthworm and the crayfish. Be able to trace a drop of blood to various parts of the body and back to the ventral aorta, and explain what the blood gains and loses in each of the capillary beds through which it passes. Locate as many of the main vessels as you can on any transverse sections you may have.

Cross Section—Stained Slide

Look at a cross section through the pharynx with the unaided eye and understand how the section is cut with reference to the whole animal. Note the **dorsal fin,** its supporting **fin ray,** and the ventral **metapleural folds** (see Figure 15.5). With the microscope, examine the **epidermis,** a single layer of columnar epithelial cells, and the

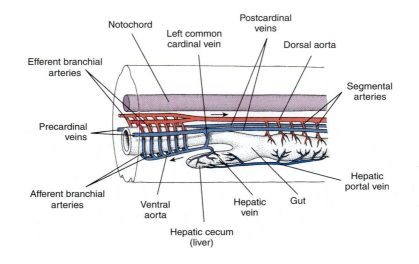

Figure 15.6

Major circulatory vessels of amphioxus. The ventral aorta pumps the blood forward and then upward to the gill capillaries via what arteries? _____ What arteries come from the dorsal aorta and feed the capillaries of the myotomes and intestinal wall? _____ Blood from the intestine, now rich in digested food nutrients, is carried forward to the hepatic cecum via what vein? _____ Blood returns to the ventral aorta by way of what vein? _____

dermis, a gelatinous connective tissue layer. The large **myotomes,** or muscles, are paired, but members of a pair are not opposite each other. The myotomes are separated by connective tissue, called **myosepta.**

The **nerve cord,** enclosed within the **neural canal,** has in its center a small **central canal,** which is prolonged dorsally into a slit. In some sections, dorsal **sensory nerves** or ventral **motor nerves** may be seen. These are given off alternately from the cord to the myotomes. The large, oval **notochord** with vacuolated cells is surrounded by the **notochordal sheath.**

The cavity of the **pharynx** is bound by a ring of triangular **gill bars** separated by **gill slits** that open into the surrounding **atrium** (see Figure 15.5B). From your study of the whole mount, why do you think the cross section shows the gill bars as a succession of cut surfaces? _____ The somewhat rigid gill bars contain blood vessels and are covered by ciliated **respiratory epithelium,** where gas exchange takes place. On the dorsal side of the pharynx, find the ciliated **hyperbranchial groove** and on the ventral side, the **endostyle.** The latter secretes mucus in which food particles are caught. What is the function of the cilia in the grooves? _____

The **gonads** (ovaries or testes) lie on each side of the atrial cavity. The reduced **coelom** consists of spaces, usually paired, on each side of the notochord and the hyperbranchial groove, as well as on each side of the gonads and below the endostyle. In favorable specimens, little **nephridial tubules** may be found in the dorsal coelomic cavities (see Figure 15.5). What is their function? _____

Ventral to the notochord are the paired **dorsal aortas.** The **ventral aorta** lies ventral to the endostyle.

Drawings

On separate paper, make drawings of sections of amphioxus as your instructor requests.

Classification: Phylum Chordata

Subphylum Urochordata (u'ro-kor-da'ta; Gr. *oura,* tail, + L. *chorda,* cord) **(Tunicata).** Tunicates. Only larval forms have all the chordate characteristics; adults sessile, without notochord and dorsal nerve cord; body enclosed in tunic. Example: *Molgula,* a sea squirt.

Subphylum Cephalochordata (sef'a-lo-kor-da'ta; Gr. *kephalē,* head, + L. *chorda,* cord). Lancelet. Notochord and nerve cord persist throughout life; lance-shaped. Example: *Branchiostoma* (amphioxus).

Subphylum Vertebrata (ver'te-bra'ta; L. *vertebratus,* backboned). Enlarged brain enclosed in cranium; nerve cord surrounded by bony or cartilaginous vertebrae; notochord in all embryonic stages and persists in adults of some fishes; typical structures include two pairs of appendages and body plan of head, trunk, and postanal tail.

Superclass Agnatha (ag'na-tha; Gr. *a,* without, + *gnathos,* jaw). **(Cyclostomata)** No jaws or ventral fins; notochord persistent.

Class Cephalaspidomorphi (sef-a-lass'pe-do-morph'e) (Petromyzontes). Lampreys.

Class Myxini (mik-sy'ny). Hagfishes.

Superclass Gnathostomata (na'tho-sto'ma-ta; Gr. *kephalē,* head, + *aspidos,* shield, + *morphē,* form). Jaws present; usually paired limbs; notochord persistent or replaced by vertebral centra.

Class Chondrichthyes (kon-drik'thee-eez; Gr. *chondros,* cartilage, + *ichthys,* a fish). Sharks, skates, rays, and chimaeras.

Classes Actinopterygii (ak'ti-nop-te-rij'ee-i) and **Sarcopterygii** (sar-cop-te-rij'ee-i), formerly grouped within the class Osteichthyes (os'te-ik'thee-eez) (Gr. *osteon,* bone, + *ichthys,* a fish). Bony fishes.

Class Amphibia (am-fib'e-a; Gr. *amphi,* both or double, + *bios,* life). Amphibians. Frogs, toads, and salamanders.

Class Reptilia (rep-til'e-a; L. *repere,* to creep). Nonavian reptiles: Snakes, lizards, turtles, crocodiles, and others.

Subclass Aves (ay'veez; L. pl. of *avis,* bird). Birds.

Class Mammalia (ma-may'lee-a) (L. *mamma,* breast). Mammals.

The Fishes
Lampreys, Sharks, and Bony Fishes

EXERCISE 16A
Class Petromyzontida— Lampreys (Adult and Ammocoete Larva)

Core Study

Lamprey

Phylum Chordata
 Subphylum Vertebrata
  Superclass Agnatha
   Class Petromyzontida
    Genus *Petromyzon*
     Species *Petromyzon marinus*

Hagfishes and lampreys are the only living descendants of the earliest known vertebrates, a group of Paleozoic jawless fishes collectively called ostracoderms. Hagfishes and lampreys are conventionally grouped together in the superclass Agnatha ("without jaws") and share certain characteristics, including the absence of jaws, internal ossification, scales, and paired fins. In other respects, however, hagfishes and lampreys are radically different from each other.

Lampreys have a worldwide distribution, and most are **anadromous** (an-ad′ruh-mus; Gr. *anadromos,* running upward), meaning that they ascend rivers and streams to spawn. The species *Petromyzon marinus* (Gr. *petros,* stone, + *myzon,* sucking, referring to its habit of holding its position in a current by grasping a stone with its mouth) is the sea lamprey that lives in the Atlantic drainages of Canada, the United States, Iceland, and Europe and is landlocked in the Great Lakes. It grows to be 1 m long and can live both in fresh water and in the sea. It is a marine species that migrates up freshwater streams to spawn.

The young larvae, known as **ammocoetes** (Gr. *ammos,* sand, + *koitē,* bed; Fr. *keisthai,* to lie, referring to the preferred larval habitat), live in sand for 3 to 5 years and then metamorphose rapidly and become parasites of fishes. Attaching themselves with their suckerlike mouth, they rasp away the fish's flesh with their horny teeth and suck out blood and body fluids. Adults grow rapidly for a year, spawn in the winter or spring, and soon die. After

invading the Great Lakes in the nineteenth century, sea lampreys devastated the important commercial fisheries there. Wounding rates of fishes are lower now, but sea lampreys remain a threat to the commercial fishing trade.

Freshwater lampreys, known as brook or river lampreys, belong to the genera *Lampetra* (L. *lambo,* to lick or lap up) and *Ichthyomyzon* (Gr. *ichthyos,* fish, + *myzon,* sucking), of which there are about 33 species. They have larval habits similar to those of the marine form, but adults in about half the species are not parasitic. Nonparasitic forms do not eat as adults and live only a month or so after emerging from the sand to spawn.

Drawings

On separate paper, draw any transverse sections required by your instructor. Label them fully. How many features of the amphioxus adult do you find repeated in the ammocoete? _____ Do you think these similarities might be interpreted as an example of homology? _____ Keep in mind the structure of these early chordate forms so that you will be able to compare them with those of the fish, amphibian, and mammal forms you will study later.

Adult Lampreys

External Structure

☞ Examine a preserved specimen of an adult lamprey.

In addition to the obvious difference in size, it will be immediately evident that the adult lamprey differs in many anatomical details from the ammocoete. During the dramatic metamorphosis from larva to adult, the body becomes rounder and shorter, the pharynx divides longitudinally, the larval hood is replaced by an oral disc with teeth, the eyes enlarge, and the nostril shifts to the top of the head. These changes are essential to the shift from a larval life habit of filter feeding to an adult existence as a parasite of fish.

Note the eel-like shape of the lamprey and its tough, scaleless skin. Among the epithelial cells are numerous gland cells that produce a protective slime. Identify the two **dorsal fins** and the **caudal fin** (Figure 16.1A). There are no paired appendages. What does the lack of

paired fins tell you about how a lamprey swims compared to how a bony fish swims? _____

Examine the hood-shaped **buccal funnel,** which is supported by a cartilaginous ring that serves as a sucking disc for attachment to the host. The opening is fringed by numerous fingerlike sensory papillae, and the interior of the funnel bears horny "teeth"; these are actually epidermal thickenings and not homologous to true vertebrate teeth, which are derived from mesoderm. Locate the **mouth** at the back of the buccal funnel and dorsal to the **tongue.** The tongue also bears sharp, horny teeth used for rasping.

A single **nostril** located middorsally on top of the head opens into an olfactory sac (the latter visible in the sagittal section, Figure 16.1B). Just behind the nostril is a small oval area marking the position of the so-called third eye, the **pineal organ.** It is not an eye in the true sense, but it does contain photoreceptors that detect changes in illumination and serve to adjust the lamprey's internal activities. The pineal organ is present in most fishes, but it is better developed in lampreys than in any other living vertebrate except certain reptiles. Note the functional, lidless **eyes** and the seven **external gill slits** just behind the eye on each side of the head.

The **lateral line system,** characteristic of nearly all fishes, consists of specialized receptors located in small patches on the head and trunk of the lamprey. They can be

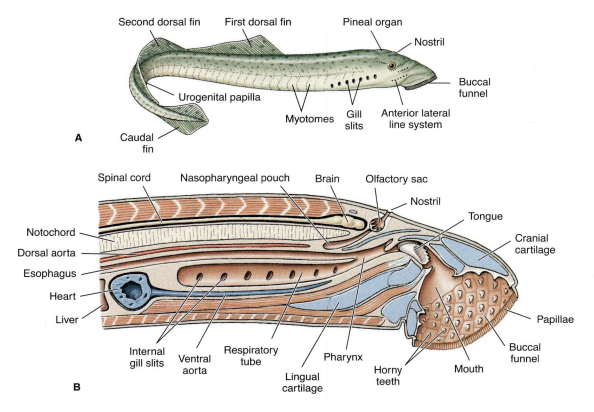

Figure 16.1
Adult lamprey. **A,** External structure. **B,** Sagittal section.

seen as groups of pores extending below and caudally from the eye on either side of the head. These receptor cells are sensitive to currents and water movement. How would they be useful to the lamprey? _____ In the lamprey, the receptors are open to the exterior and not protected within canals as they are in bony fishes.

Find the **urogenital[1] sinus,** with its projecting urogenital papilla, and the ventral juncture of the trunk and the tail. There is an **anal opening** in front of the urogenital sinus.

How do the structures of the myotomes compare to those of amphioxus? _____ _____

Ammocoete Larva

Although the ammocoete larva resembles amphioxus in appearance, life habit, and many anatomical details, it bears several characteristics that anticipate the vertebrate body plan and that are lacking in amphioxus. How is blood propelled in amphioxus? _____ _____

The ammocoete has a two-chambered heart, two median eyes each with a lens and receptor cells, a three-part brain, thyroid and pituitary glands, and a pronephric kidney. Instead of the numerous gill slits found in the amphioxus, the ammocoete larva has only seven pairs.

 Examine a preserved ammocoete larva as well as a stained whole mount of a small specimen.

Study of the Preserved Larva

 Cover the preserved larva with water in a water glass. Use a hand lens or dissecting microscope.

List two differences between the preserved specimen and a mature amphioxus: _____ _____

_____
[1]Urogenital = urinogenital. The stem **uro-** derives from the Greek *ouron,* meaning urine. Unfortunately, the stem is used in other terms to mean "tail" (Gr. *oura,* tail) as, for example, in *Urodela, urostyle,* and *uropod.*

_____ Note the **myotomes** (segmental muscles) appearing faintly on the surface (Figure 16.2). Do the myotomes have the same arrangement as those of the amphioxus? _____ Note the **oral hood** with **oral papillae** attached to the roof and sides of the hood. They are used, as in amphioxus, for filter feeding. The **lateral groove** on each side contains seven small **gill slits.** The anus, or **cloacal opening,** is just anterior to the **caudal fin.** Is the caudal fin continuous with the **dorsal fin?** _____ Note the **chromatophores** scattered over the body.

Study of the Stained Whole Mount

 With low power, examine a stained whole mount of an ammocoete larva.

On the whole mount, find the darkly stained, dorsal, hollow **nerve cord,** enlarged anteriorly to form the **brain.** Immediately below it is the lighter **notochord** (Figure 16.2).

The oral hood encloses a **buccal cavity,** to the back and sides of which are attached oral papillae. Posterior to the buccal cavity is the **velum,** a large pair of flaps that create water currents. The large **pharynx** has **internal gill slits,** which open into **gill pouches.** The gill pouches open to the outside by small **external gill slits.** How many pairs of gill slits are there in the ammocoete? _____ Using low power, focus upward onto the outer surface of the animal to see the row of small external gill slits. Between the internal gill slits are cartilaginous rods, the **gill bars,** which strengthen the pharynx walls. Note the **gill lamellae** on the pharynx walls. They are rich in capillaries, through which the blood gives up its carbon dioxide and takes up its oxygen from the water.

The ammocoete is a filter-feeder. Was amphioxus also a filter-feeder? _____ Water is kept moving through the pharynx by muscular action of both the velum and the whole branchial basket. This contrasts with the amphioxus, in which water is moved by ciliary action.

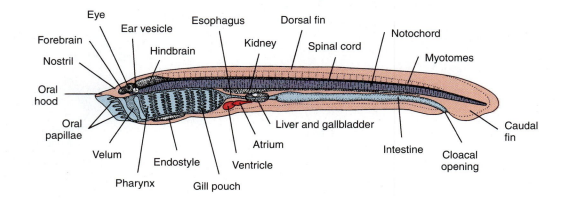

Figure 16.2
Ammocoete larva, sagittal section.

The **endostyle** (subpharyngeal gland) in the floor of the pharynx is a closed tube the length of four gill slits. It secretes mucus via a duct into the pharynx. Food particles brought in by water currents are trapped in the mucus and carried by ciliary action to the **esophagus.** During metamorphosis of the larva, a portion of the endostyle becomes a part of the thyroid gland of the adult.

The narrow esophagus widens to become the **intestine,** the posterior end of which, called the **cloaca,** also receives the kidney ducts. The **anus** opens to the outside a short distance in front of the postanal tail.

The **liver** lies under the posterior end of the esophagus, and embedded in it is the **gallbladder,** which appears as a clear, round vesicle. The two-chambered **heart** lies under the forepart of the esophagus.

Over the heart and around the esophagus is the **pronephric kidney,** consisting of a number of small tubules that empty into the cloaca through pronephric ducts (not easily distinguished on the whole mounts). Later, a mesonephric kidney will develop above the intestine, using the same ducts, and the pronephros will degenerate.

The tubular dorsal **nerve cord** enlarges anteriorly into a three-lobed **brain,** visible in most slides. How does the nerve cord differ in location from that of the flatworms? _____ How does the nerve cord differ from those of annelids and arthropods? _____ The **forebrain** contains the **olfactory lobe.** In front of the forebrain is the **nasohypophyseal canal,** opening dorsally to the outside by a median **nostril.** The darkly pigmented **eyes** connect with each side of the **midbrain.** At this stage, the eyes are covered with skin and muscle and have little sensitivity

to light; however, the ammocoete larva does have photoreceptors in its tail. Find the **hindbrain** and, with careful focusing, try to see one of the clear oval **ear vesicles** that flank each side of it. Which lobe of the brain is chiefly concerned with the sense of smell? _____ Of sight? _____ Of hearing and equilibrium? _____ How is the nervous system of this agnathan advanced over that of the cephalochordate amphioxus? _____

Transverse Sections of Ammocoetes

☞ Your slide may contain four typical sections—one each through the brain, the pharynx, the intestine, and the postanal tail. Or it may contain 15 to 20 sections through the body, arranged in sequence. As you study each section, refer to the whole mount again to interpret relationships. Use the low power of the microscope.

Sections Anterior to the Pharynx. Sections through the **forebrain** may include the **oral papilla** and the **oral hood.** Sections through the **midbrain** may include the **eyes,** the **buccal chamber,** and portions of the **velar flaps.** Sections through the **hindbrain** may include the **ear (otic) vesicles** and the buccal chamber (Figure 16.3A) or the forepart of the pharynx. Compare the size of the brain with that of the spinal cord in more posterior sections. Do you find a **notochord** lying just below the brain in any of the sections? _____ Why is the notochord considered a primitive type of endoskeleton? _____

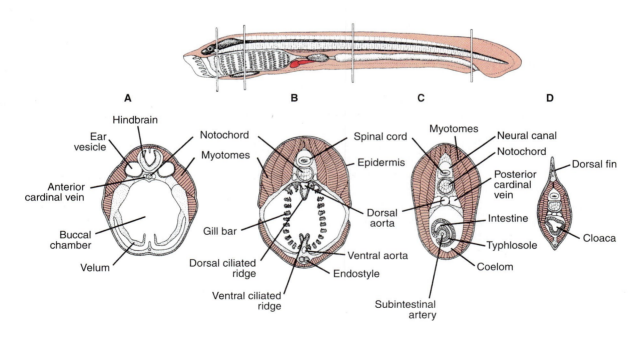

Figure 16.3
Transverse sections of an ammocoete larva. **A,** Posterior part of buccal chamber. **B,** Pharynx. **C,** Intestine. **D,** Cloaca.

Sections Posterior to the Pharynx. Choose a section through the trunk posterior to the pharynx, and identify the following (Figure 16.3B, C): **epidermis; myotomes** (lateral masses of muscles); **nerve cord** surrounded by the **neural canal** and containing a cavity, the **neurocoel; notochord** with large vacuolated cells; **dorsal aorta** (probably contains blood cells); and **posterior cardinal veins,** one on each side of the aorta (blood cells are usually present). You may find the cardinals joining to form the duct of Cuvier. The **coelom,** or **coelomic cavity,** lined with peritoneum, contains the visceral organs. What distinguishes a true coloem from other cavities in the body? _____

_____

The visceral contents of the coelomic cavity will vary according to the location of the section.

1. Just behind the pharynx you will find the **esophagus,** composed of columnar epithelium; the paired **pronephric kidneys,** appearing as sections of small tubules; and chambers of the **heart.**

2. Sections cut posterior to the heart will show the dark **liver** and possibly the hollow **gallbladder** ventral, or lateral, to the esophagus, with sections of pronephric kidneys or their ducts located under the posterior cardinal veins.

3. Sections through the intestine will reveal it as a large tube of columnar epithelium with a conspicuous infolding, the **typhlosole,** carrying the **subintestinal artery** (Figure 16.3C). Above the intestine, you may find the **mesonephric kidneys,** with their tubules, and a small **gonad** between the kidneys. The **cloaca** is located farther back (Figure 16.3D).

4. In sections posterior to the anus (postanal tail), identify the caudal fins. What other structures can you identify? _____

Sections Through the Pharynx. The appearances of the body wall, nerve cord, and notochord will be similar to the preceding sections. The central part of the section is taken up by the large pharynx, on whose walls are the **gills,** with their platelike **gill lamellae,** extending into the pharynx or into the **gill pouch,** depending on how the section has been cut. Each lamella has lateral ridges. Some sections, such as that in Figure 16.3B, may show only the gill bars without the feathery lamellae. The gills are liberally supplied with blood vessels. Outside the gill chambers, you will find sections through cartilage rods that support the branchial basket.

In the middorsal and midventral regions of the pharynx are **ciliated ridges** bearing grooves. The cilia aid the movement toward the esophagus of mucous strands in which food particles are caught. Below the pharynx in certain sections is the bilobed **subpharyngeal gland,** whose function is probably the secretion of mucus. Later, certain portions of this gland are incorporated into the adult thyroid gland. Between the ventral ciliated ridge and the

subpharyngeal gland is the single or paired **ventral aorta.** Note the **gill pouches** lateral to the pharynx, the **lateral groove,** and in some sections the **external gill slits** to the outside. Locate the **anterior cardinal veins** on each side of the notochord and the **dorsal aorta** beneath it.

EXERCISE 16B
Class Chondrichthyes— Cartilaginous Fishes

Core Study

Squalus, Dogfish Sharks
Subphylum Vertebrata
   Superclass Gnathostomata
     Class Chondrichthyes
       Subclass Elasmobranchii
         Order Squaliformes
           Genus *Squalus*

The cartilaginous fishes are a compact, ancient assemblage of about 970 species that are characterized by their cartilaginous skeletons, powerful jaws, and well-developed sense organs. They include the sharks, skates, rays, and chimaeras. The subclass Elasmobranchii (e-laz'-mo-bran'kee-i; Gr. *elasmos,* metal plate, + *branchia,* gills) embraces the sharks, skates, and rays—cartilaginous fishes with exposed gill slits opening separately to the outside. Most are carnivores, and many are top predators. The dogfish shark is an excellent example of the generalized body plan of early jawed vertebrates.

Where Found

Dogfish sharks are small marine sharks that grow to about 1 m in length (females are slightly larger than males). Two species commonly studied belong to the genus *Squalus* (L. *squalus,* a kind of sea fish), the spiny dogfishes: *Squalus acanthias* of the North Atlantic and *Squalus suckleyi* of the Pacific coast. The two species are morphologically similar. Spiny dogfishes are distinguished by a spine on the anterior edge of both dorsal fins. Spiny dogfishes gather in huge schools of up to 1000 individuals of both sexes when immature, but only of one sex when they are mature. In detecting and capturing their main foods, bottom-dwelling fish and crabs, dogfishes are assisted by their ability to sense weak electrical fields that surround all living animals, using specialized sense organs on the head, the ampullae of Lorenzini. Spiny dogfishes are ovoviviparous (L. *ovum,* egg, + *vivus,* living, + *parere,* to bring forth); that is, they give birth to living young without dependence on placental nourishment. The embryos develop in an egg capsule in the oviduct until they hatch in the mother just before birth.

Heavy fishing pressures on dogfish sharks appear to have impacted populations. These fish are consumed in some parts of the world and body parts like their fins and tails are also marketed. They are currently considered "vulnerable" on the IUCN Red List (see also Lack, M. 2006. Conservation of Spiny Dogfish *Squalus acanthias:* a role for CITES. TRAFFIC International).

External Structure

☞ Examine an intact, preserved dogfish to identify the following features.

The body is divided into a **head** (to the first gill slit), a **trunk** (to the cloacal opening), and a **tail.** The **fins** include a pair of **pectoral fins** (anterior), which control changes in direction during swimming; a pair of **pelvic fins,** which serve as stabilizers and, in the male, are modified to form claspers used in copulation; two medium **dorsal fins,** which also serve as stabilizers; and a **caudal fin** that is **heterocercal** (asymmetric dorsoventrally) (Figure 16.4).

Identify the **mouth** with its rows of **teeth** that are adapted for cutting and shearing; two ventral **nostrils,** which lead to olfactory sacs and are equipped with folds of skin that allow continual in-and-out movement of water; and the lateral **eyes,** which lack movable eyelids but have folds of skin that can cover the eyeballs. The part of the head anterior to the eyes is called the **rostrum** (snout).

A pair of dorsal **spiracles** posterior to the eyes are modified gill slits that open into the pharynx. They can be closed by folds of skin during part of the respiratory cycle to prevent the escape of water. The spiracles serve for water intake when the shark is feeding. Five pairs of **gill slits** are the external openings of the gill chambers.

The **pharynx** is the region in back of the mouth into which the gill slits and spiracles open. A **lateral line,** appearing as a white line on each side of the trunk, represents a row of minute, mucus-filled sensory pores used to detect differences in the velocity of surrounding water currents, and thus to detect the presence of other animals, even in the dark. Note the **cloacal opening** between the pelvic fins.

The leathery skin consists of an outer layer of epidermis covering a much thicker layer of dermis densely packed with fibrous connective tissue. Draw your finger lightly over the shark's skin to feel the spines of the **placoid scales** (Gr. *placos,* tablet, plate) (Figure 16.5). Each scale is anchored in the dermis and is built much like a tooth (the shark's teeth are, in fact, modified placoid scales). The scale contains a pulp cavity and a thick layer of dentine, both derived from the dermis, and is covered with hard enamel, derived from the epidermis.

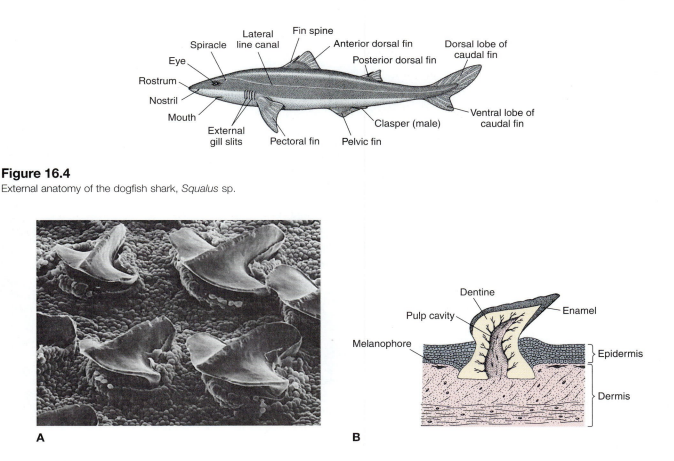

Figure 16.4
External anatomy of the dogfish shark, *Squalus* sp.

Figure 16.5
A, Surface view of shark skin showing placoid scales, or dermal denticles (SEM, ×195). **B,** Lateral section of a single placoid scale.

The spiny scales help to reduce friction-producing turbulence as the shark swims, thus lessening drag.

With its dark dorsal and light ventral surfaces **(countershading)**, the animal's coloration is protective, whether viewed from above or below.

Internal Structure

If you will not be doing a dissection, examine longitudinal and transverse sections of the shark. Note the cartilaginous skull and the vertebral column. With the help of Figure 16.6, identify as many of the internal structures as possible.

Dissection of the Shark

Open the coelomic cavity by extending a midventral incision posteriorly from the pectoral girdle through the pelvic girdle and then around one side of the cloacal opening to a point just posterior to it. On each side, make a short transverse cut just posterior to the pectoral fins and another one just anterior to the pelvic fins. Rinse out the body cavity.

The shiny membrane lining the body cavity and its organs is the **peritoneum** (Gr. *peritonaios,* stretched around). The peritoneum lining the inner surface of the body wall is called **parietal peritoneum,** and that covering the visceral organs and forming the double-membraned **mesenteries** that suspend the digestive organs is called **visceral peritoneum.**

Digestive System

Identify the large **liver,** which has two large lobes and a small median lobe (Figure 16.6). Lying along the right margin of the median lobe is a thin, tubular sac, the **gallbladder.** Bile from the liver is concentrated in the gallbladder and then discharged during meals by way of the common bile duct into the intestine. Dorsal to the

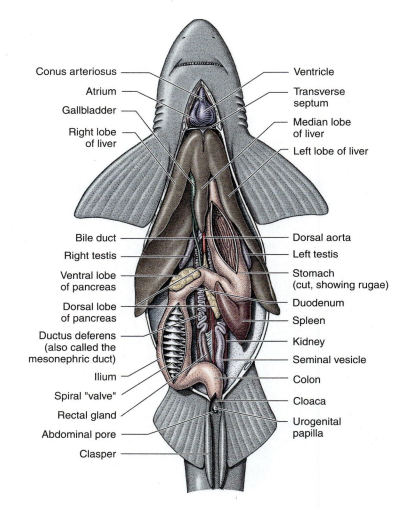

Conus arteriosus
Atrium
Gallbladder
Right lobe of liver

Ventricle
Transverse septum
Median lobe of liver
Left lobe of liver

Bile duct
Right testis
Ventral lobe of pancreas
Dorsal lobe of pancreas
Ductus deferens (also called the mesonephric duct)
Ilium
Spiral "valve"
Rectal gland
Abdominal pore
Clasper

Dorsal aorta
Left testis
Stomach (cut, showing rugae)
Duodenum
Spleen
Kidney
Seminal vesicle
Colon
Cloaca
Urogenital papilla

Figure 16.6

Internal anatomy of the dogfish shark, ventral view. What large organ likely assists in keeping the shark buoyant? _____ What structure slows the passage of food through the gut? _____ What organ is formed around the posterior end of the J-shaped stomach? _____ What gland is specific to the cartilaginous fishes and helps regulate the fishes' salt imbalance? _____ What organ functions as the primary source of red blood cell production? _____

liver is the large **esophagus,** which leads from the pharynx to the J-shaped **stomach.** The digestive tract then turns caudally and gives rise to the **duodenum** (L. *duodeni,* twelve each, so called because in humans this first part of the intestine is approximately 12 fingerwidths long). Between the stomach and duodenum is a muscular constriction, the **pyloric valve,** that regulates the entrance of food into the intestine. The **pancreas** lies close to the ventral side of the duodenum, with a slender dorsal portion extending posteriorly to the large **spleen** (not a part of the digestive system). The **valvular intestine** (ileum) is short and wide and contains a **spiral valve,** or **ridge.** Extending from the dorsal wall of the short, narrow **rectum** is a **rectal gland,** which regulates ion balance. The cloaca receives the rectum and urogenital ducts.

👉 Make a longitudinal incision in the ventral wall of the esophagus and stomach. Remove and save the contents of the stomach, if any. Extend the longitudinal incision along the wall of the ileum (taking care not to destroy the blood vessels) to expose the spiral valve. Rinse out the exposed digestive tract.

Examine the contents taken from the stomach and compare them with those of other specimens being dissected. What do you infer about the shark's eating habits? _____ Examine the inner surface of the digestive tract. The walls of the esophagus bear large papillae, whereas the walls of the stomach are thrown into longitudinal folds called **rugae** (roog′ee; L. *ruga,* wrinkle). Note the structure of the pyloric valve. Observe the cone-shaped folds of the spiral valve and see if you can determine how materials pass through. The spiral valve—not really a "valve" but a fold of tissue that spirals down the ileum much like a spiral staircase—slows the passage of food through the gut. Why might slowing the passage of food be beneficial for the digestive process?

_____

_____

👉 If the intestine has been everted into the cloacal region, carefully pull it back into the body cavity.

Urogenital System

Although the excretory and reproductive systems have quite different functions, they are closely associated structurally, and so are studied together as the urogenital system (Figure 16.7).

Male. Soft, elongated **testes** lie along the dorsal body wall, one on each side of the esophagus (Figure 16.7, *left*).

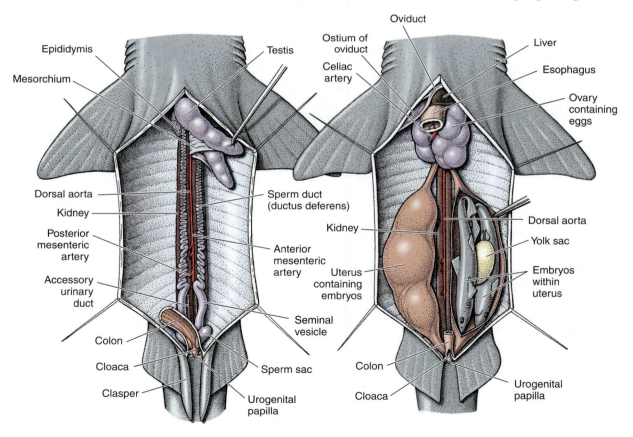

Figure 16.7
Urogenital system of the dogfish shark. **Left,** Male. **Right,** Female.

They are held in place by a mesentery called the **mesor-chium** (Gr. *mesos,* middle, + *orchis,* testicle). A number of very fine tubules (vasa efferentia) in the mesentery run from each testis to a much-convoluted **sperm duct** (also called the Wolffian duct or ductus deferens). The **kidneys** (also called opisthonephroi) are long and narrow and lie behind the peritoneum on each side of the dorsal aorta. They extend from the pectoral girdle to the **cloaca.** The sperm ducts, which also serve as urinary ducts, take a twisting course along the length of the kidneys to the cloaca and collect wastes from the kidneys by means of many fine tubules. The sperm ducts widen posteriorly into **seminal vesicles,** which dilate terminally into **sperm sacs** before entering the cloaca. The cloaca is a common vestibule into which both the rectum and the urogenital ducts empty. In the center of the cloaca, dorsal and posterior to the rectum, is a projection called the **urogenital papilla,** which is larger in males than in females. Seminal vesicles empty into the urogenital papilla, which empties into the cloaca.

☞ Slit open the cloaca to see the urogenital papilla.

A groove along the inner edge of each **clasper** is used in conducting spermatozoa to the female at copulation.

Female. A pair of **ovaries** lies against the dorsal body wall, one on each side of the esophagus (Figure 16.7, *right*). Enlarged ova may form several rounded projections on the surface of the ovaries. A pair of **oviducts** (Müllerian ducts) run along the dorsal abdominal mesentery. The anterior ends join to form a common opening into the abdominal cavity, called the **ostium tubae.** The oviducts are anteroventral to the liver but may be difficult to find except in large females. Ripened ova leave the ruptured wall of the ovary and enter the abdominal cavity. They are drawn through the ostium into the oviducts, where fertilization may occur. An expanded area of each oviduct dorsal to the ovary is a **shell gland (oviducal gland),** which in *Squalus* secretes a thin membrane around several eggs at a time. The posterior end of each oviduct enlarges into a **uterus,** the caudal end of which opens into the **cloaca.** In immature dogfish, neither the shell gland nor the uterus may be apparent.

One to six or seven eggs, depending on the species, may develop in each uterus. Vascularized villi on the wall of the uterus come in contact with the yolk sac of the embryo in a placenta-like manner. As mentioned earlier, spiny dogfishes are ovoviviparous because the embryo does not depend on placental nourishment. However, some other sharks are dependent on the mother for nourishment through the placental connection (**viviparous;** L. *vivus,* living, + *parere,* to bring forth), and some primitive sharks lay shelled eggs containing a large amount of yolk (**oviparous;** L. *ovum,* egg, + *parere,* to bring forth). Gestation periods vary from 16 to 24 months, and the young at birth range from 12 to 30 cm in length.

The slender **kidneys** extend the length of the dorsal abdominal wall, dorsal to the peritoneum. A very slender **Wolffian duct** embedded on the ventral surface of each kidney empties into the cloaca through a **urogenital papilla.**

☞ Slit open the cloaca and identify the urogenital papilla, the entrance of the rectum, and on the dorsal side, the openings from the uteri.

Further Study

Circulatory System

Because sharks do not have bone marrow, the spleen is the site of red blood cell production. The spleen also filters blood, much like the lymph system of other animals. The basic circulation plan in the shark is similar to that in the ammocoete larva.

☞ Spread the ventral body wall to reveal the cavity containing the heart **(pericardial cavity).** Lift up the **heart** to see the thin-walled, triangular **sinus venosus.**

Blood passes from the sinus venosus to the **atrium,** which surrounds the dorsal side of the muscular **ventricle;** it then flows into the ventricle, which pumps it forward into the **conus arteriosus** (Figure 16.8). **Valves** prevent backflow between compartments.

Venous System.

☞ Slit open the sinus venosus transversely, extending the cut somewhat to the left; wash out its contents.

Look for openings into the sinus venosus of one of each of the following paired veins: (1) **common cardinal** (L. *cardinalis,* chief, principal) **veins (ducts of Cuvier)** which extend laterally and into which empty the large **anterior cardinal sinuses, posterior cardinal sinuses,** (Figure 16.8), and **subclavian veins** (L. *sub,* under, + *clavus,* key); (2) **inferior jugular veins** (L. *jugulum,* collarbone) from the floor of the mouth and gill cavities (this vein is not shown in Figure 16.8); and (3) **hepatic veins,** which empty near the middle of the posterior wall of the sinus venosus and bring blood from the liver.

The **hepatic portal vein** (Figure 16.8) gathers blood chiefly from the digestive system through a system of gastric, pancreatic, and intestinal veins. The hepatic portal vein enters the right lobe of the liver and divides into several small **portal veins** (trace some of these subdivisions); it then divides into a system of capillaries, from which some of the carbohydrates brought from the intestine may be stored in liver cells as glycogen (animal starch) until needed. Blood from capillaries flows into the **hepatic veins** and from there into the sinus venosus.

Renal portal veins arise from the **caudal vein** in the tail and carry blood to the capillaries of the kidneys. Many

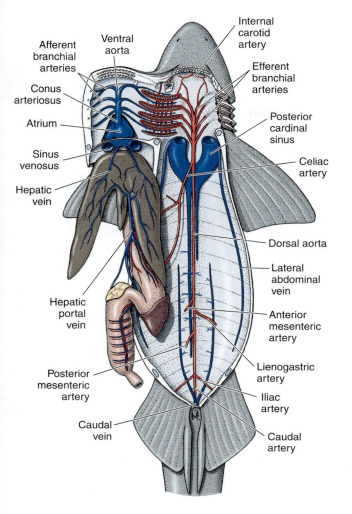

venosus, and turn the lower jaw to one side to expose the gill slits and the roof of the pharynx. Locate the spiracle internally. It represents the degenerated first gill slit. Dissect the mucous membrane lining from the roof of the mouth and pharynx to expose the four pairs of **efferent branchial arteries,** which carry oxygen-rich blood from the gill filaments and unite to form the **dorsal aorta.** By cutting the cartilages under which they pass, trace these arteries back to the gills.

The dorsal aorta extends posteriorly along the length of the body ventral to the vertebral column. It gives rise to **subclavian arteries,** which connect to the pectoral region; a **celiac artery** (Gr. *koilia,* belly) (Figure 16.8), which gives off branches to the intestinal tract and gonads; numerous **parietal arteries** to the body walls; **mesenteric arteries** to the intestine and the rectum; a **lienogastric artery** to the spleen and cardiac stomach; **renal arteries** to the kidneys; and **iliac arteries** (L. *ilia,* flanks) to the pelvic fins. The aorta continues to the tip of the tail as the **caudal artery.**

Respiratory System

In sharks, water taken in through both the mouth and the spiracles is forced laterally through five pairs of gills and leaves through five pairs of external gill slits (some elasmobranchs have a different number of gills).

 On the shark's right (intact) side, separate the gill units by cutting dorsally and ventrally from the corners of each gill slit. Now you can examine the structure of the intact gills on this side and observe the gills in cross section on the other side.

The area between the **external gill slits** and the **internal gill slits** comprises the **gill chambers (gill pouches** and **branchial chambers).** The incomplete rings of heavy cartilage supporting the gills and protecting the afferent and efferent branchial arteries are called **gill arches.** Short, spikelike projections extending medially from the gill arches are the **gill rakers.** What might be the function of the gill rakers? _____ Cartilaginous **gill rays** fan out laterally from the gill arches to support the gill tissues.

Remove an intact half of a gill arch along with its gill tissue. Examine with a hand lens. Float a small piece of gill in water and examine with a dissecting microscope.

Primary lamellae (gill filaments) are small, platelike sheets of epithelial folds arranged in rows along the lateral face of each gill. Viewing with a microscope reveals that the primary lamellae are made up of rows of tiny plates, called **secondary lamellae,** which are the actual sites of gas exchange. Blood capillaries in the secondary lamellae are arranged to carry blood inward,

Figure 16.8
Circulatory system of the dogfish shark.

small renal veins carry blood from the kidneys to the posterior cardinal sinuses and from there to the sinus venosus.

Arterial System. The arterial system includes (1) the afferent and efferent branchial arterial system, and (2) the dorsal aorta and its branches.

From the conus arteriosus, trace the **ventral aorta** forward, removing most of the muscular tissue to the lower jaw. The ventral aorta gives off three paired branches, which give rise to **five pairs of afferent branchial arteries** (Figure 16.8). In injected specimens, you can follow these arteries into the interbranchial septa, where each gives off tiny arteries to the gill lamellae.

The **efferent branchial arteries** are more difficult to dissect. With scissors, cut through the left corner of the shark's mouth and backward through the centers of the left gill slits, continuing as far as the transverse cut you made earlier at the base of the pectoral fins. Now cut transversely across the floor of the pharynx straight through the sinus or in the opposite direction from the seawater, which is flowing outward. This countercurrent flow encourages gas exchange between the blood and the water. Gill lamellae are arranged in half-gills, or **demibranchs,** on each side of the branchial arch. The two demibranchs together form the gill unit, or **holobranch.** The spiracles are believed to be remnants of the gill openings found in more primitive chordates. They are usually larger in the slow-moving, bottom-dwelling sharks than in the fast-swimming sharks, in which, because of their motion, the flow of water through the mouth is more massive.

Oral Report

Be able to identify both the external and internal features of the shark and give the functions of each organ or structure.

Be able to trace the flow of blood from the heart to any part of the body (such as the pectoral region, the kidneys, the tail, and so on) and back to the heart.

Demonstrations

1. Dogfish uterus with developing pups. Or dogfish embryos with attached yolk sac.
2. Various sharks, skates, and rays.
3. Preparation of the skull and/or skeleton of a shark.
4. Corrosion preparation of the arterial system.
5. Shark teeth.
6. Microslides of shark skin.

EXERCISE 16C
Class Actinopterygii—Bony Fishes

Core Study

Perca, Yellow Perch

Subphylum Vertebrata
  Class Actinopterygii
    Superorder Teleostei
      Order Perciformes
        Genus *Perca*
          Species *Perca flavescens*

Where Found

The yellow perch is a common freshwater fish widely distributed through the lakes of the American Midwest and parts of Canada. A closely related species is found in Europe and Asia.

Characteristics

Actinopterygii (bony fishes) represents the largest group of vertebrates both in number of species (more than 23,600) and in number of individuals. By adaptive radiation, they have developed an amazing variety of forms and structures. They flourish in fresh water or seawater, and in both deep and shallow water. Their usual chief characteristics are **dermal scales,** an **operculum over the gill chamber** of each side, a **bony skeleton,** a **terminal mouth,** a **swim bladder,** a **homocercal tail,** and both **median** and **paired fins.**

External Structure

☞ Obtain a preserved fish, and after studying its external anatomy, compare it with living fishes in the aquarium. What can their structure tell you about their living habits? _____

The body of the perch is fusiform, or torpedo-shaped. Is it compressed in any of its planes? _____ Identify the **head,** which extends to the posterior edge of the **operculum;** the **trunk,** which extends to the anus; and the **tail.** Identify the **pectoral, pelvic, anal, dorsal, and caudal fins** (Figure 16.9). How many of each are there? _____ Which of the fins are paired? _____ Note the **fin rays** that support the thin membrane of each fin. Some of the rays are soft, and others are spiny. How would you expect these spiny rays to look when the fish is threatened by a predator? _____ The caudal fin is **homocercal** (Gr. *homos,* same, + *kerkos,* tail), meaning that the upper and lower halves are equal.

The terminal **mouth** is adapted for overtaking prey while swimming. Fishes with superior mouths (those facing upward) are usually surface feeders, whereas those with inferior mouths (facing downward) are usually bottom feeders. Do the **eyes** have lids? _____ Could the perch have binocular vision? Why? _____ On each side in front of the eye, a pair of **nostrils** open into an olfactory sac. Water enters the sac through the anterior aperture, which is provided with a flaplike valve, and leaves through the posterior aperture. The **ears** are located behind the eyes, but they are not visible externally. The **lateral line** along the side of the body is a row of small pores or tubules connecting with a long tubular canal bearing sensory organs. These are sensitive to pressure and temperature changes and thus are responsive to water currents. Many microscopic sense organs are found in the skin.

Lift a gill cover, or **operculum** (L., cover), and study its structure. Along the ventral margin of the operculum, find a membrane supported by bony rays. This membrane fits snugly against the body to close the branchial cavity during certain respiratory movements. With your probe, examine the **gills** beneath the operculum.

Find the **anus** near the base of the anal fin and the small, slitlike **urogenital opening** just posterior to the anus.

Note the arrangement of the **scales.**

 Remove a scale from the lateral line region, mount it in water on a slide, and examine with low power.

The anterior, or embedded, side of the perch scale has radiating grooves. The posterior, or free, edge has very fine teeth. These are **ctenoid scales** (ten′oid; Gr., *kteis, ktenos,* comb). Note the fine concentric lines of growth. The scales are covered with a very thin epidermis that secretes mucus over the scales. This reduces friction in swimming and makes capture by a predator more difficult. Ctenoid scales are usually found on fishes with spiny rays in the fins, whereas the soft-rayed fishes usually have cycloid scales, which lack marginal teeth.

Skeletal System

 Examine the mounted specimen of a perch.

The bony skeleton of the perch consists of an **axial skeleton** (which includes the bones of the skull, vertebral column, ribs, and medial fins) and an **appendicular skeleton** (which includes the pectoral girdle and fins and the pelvic girdle and fins; Figure 16.10). Examine the mounted perch skeletons on display, and

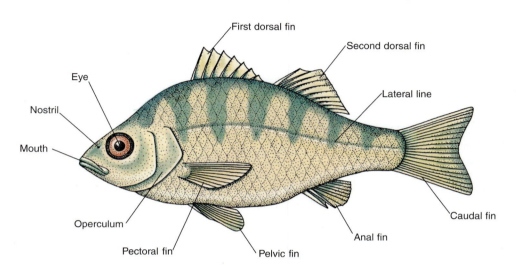

Figure 16.9
Yellow perch, external features.

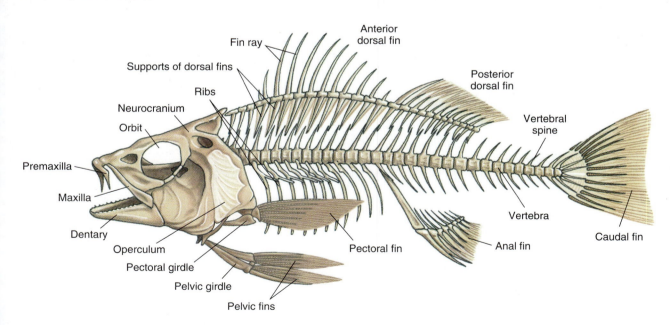

Figure 16.10
Skeleton of a perch.

compare them with skeletons of other fishes and of amphibians, birds, and mammals. Do you see any basic similarity?

Muscular System

Although the muscles of the perch are less complex than those of land vertebrates, they make up a much larger mass in relation to body size. Tetrapod locomotion results largely from the direct action of muscles on the bones of the limbs, but fish locomotion results from the indirect action of the segmental muscles—**myomeres**—on the vertebral column, a method by which a large muscle mass produces a relatively small amount of action. This type of movement is efficient in a water medium, but it would be less effective on land. The myomeres (derived from the embryonic myotomes) consist of blocks of longitudinal muscle fibers placed on each side of a central axis, the vertebral column. Their contraction, therefore, bends the body, and the action passes in waves down the body, alternating on each side.

☞ After cutting off the sharp dorsal and ventral spines, skin one side of the body and note the shape of the myomeres.

Myomeres resemble W's that are turned on their sides and stacked together. A horizontal septum of connective tissue divides the muscles into dorsal **epaxial muscles** (Gr. *epi,* upon, + *axis,* axle, meaning above the axis, or vertebral column) and ventral **hypaxial muscles** (Gr. *hypo,* under, + *axis,* axle, meaning below the vertebral column) (Figure 16.11). Posteriorly, both epaxial and hypaxial muscles are active in locomotion, but anteriorly the hypaxial muscles serve more for support of body viscera than for locomotion. Try to separate some of the myomeres. Observe the direction of the muscle fibers. Do they run zigzag as the myomeres seem to? _____ Or are they all directed horizontally—or vertically? _____

☞ Now watch the swimming motions of fishes in the aquarium and try to visualize the use of the body muscles in locomotion. What part do the fins play in locomotion? _____

Dissecting individual muscles of a fish is difficult and will not be attempted here. Muscles operating the jaws, opercula, and fins are often named according to their function; as in other vertebrates, they include adductors, abductors, dilators, levators, and so on (see Exercise 20).

Mouth Cavity, Pharynx, and Respiratory System

Before starting the dissection, cut off the sharp dorsal and ventral fins to protect your hands.

☞ Cut away the operculum from the left side, exposing the gill-bearing bars, or arches. How many arches are there? _____
Cut one gill arch, place it in water, and examine with the hand lens or dissecting microscope.

If injected, the branchial arteries will be colored. Note the **gill filaments** borne on the posterior, or aboral, side of the arch (Figure 16.12A). Are the filaments arranged in a single or double row? _____ In these filaments, which contain capillaries from the branchial arteries, exchange of gases takes place. **Gill rakers** on the

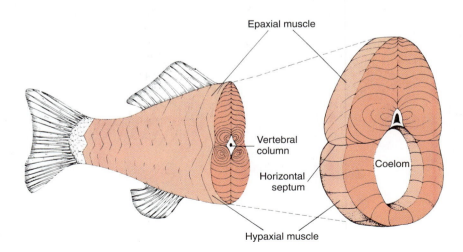

Epaxial muscle
Vertebral column
Horizontal septum
Coelom
Hypaxial muscle

Figure 16.11
Diagram of the skeletal musculature of a teleost fish.

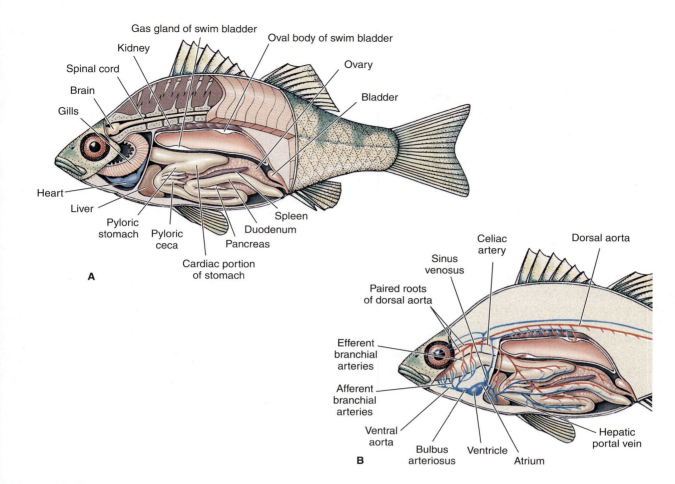

Figure 16.12
Yellow perch. **A,** Internal anatomy. **B,** Principal vessels of the circulatory system.

oral surface of each gill bar strain out food organisms and offer some protection to the gill filaments from food passing through the pharynx.

☞ **Cut through the angle of the left jaw, continuing the cut through the middle of the left gill arches to expose the mouth cavity and pharynx.**

Open the mouth wide and note the **gill slits** in the pharynx. In the mouth, locate the fine **teeth.** What would be the main function of these teeth? _____ Just behind the teeth, across the front of both the upper and lower jaw, find the **oral valves.** These are transverse membranes that prevent the outflow of water during respiration. An inflexible **tongue** is supported by the hyoid bone. Explore the spacious **pharynx,** noting the size and arrangement of the gill bars and gill slits. The gills separate the **oral cavity** from the **opercular cavity.**

Water movement across the gills involves the combined pumping action of both the oral and opercular cavities—a "double pump" system. The volume of the **oral pump** (mouth cavity) can be changed by raising and lowering the jaw and floor of the mouth. The volume of the **opercular pump** (opercular cavity) can be enlarged and decreased by muscles that swing the operculum in and out. Valves guard the opercular clefts, preventing the backflow of water. The action of the two pumps creates a pressure differential that maintains a smooth flow of water across the gills throughout nearly the entire breathing cycle.

☞ **Now watch fishes in the aquarium, observing their respiratory movements until you understand the sequence.**

Water movement across the gills is actually much smoother and less pulsatile than it appears from watching a fish respire. The reason is that the pressure in the opercular cavity is maintained *lower* than the pressure in the mouth cavity for about 90% of the respiratory cycle, and this provides the pressure that drives the water across the gills.

Further Study

Abdominal Cavity

☞ Starting near the anus and being careful not to injure the internal organs, cut anteriorly on the midventral line to a region anterior to the pelvic fins. Now, on the animal's left body wall, make a transverse cut extending dorsally from the anal region; make another cut dorsally between the pectoral and pelvic fins; then remove the left body wall by cutting between these two incisions. On the right side, make similar transverse cuts so that the right wall can be laid back, but do not remove it.

You have now exposed the abdominal cavity. This, together with the **pericardial cavity,** which contains the heart, makes up the **coelomic cavity.** What is the shiny lining of the coelom called? _____

Probably the first organ you see will be the **intestine** encased in yellow fat. Carefully remove enough of the fat to trace the digestive tract anteriorly. Find the **stomach,** lying dorsal and somewhat to the left of the intestine. Anterior to the stomach is the **liver,** dark red in life but bleached to a cream color by preservative. The **spleen** (Figure 16.12A) is a dark, slender organ lying between the stomach and the intestine. The **gonads** are in the dorsoposterior part of the cavity. The **swim bladder** lies dorsal to these organs and to the peritoneal cavity. It is long and thin-walled. Do not injure its walls or any of the blood vessels lying among the viscera. **Kidneys** are located dorsal to the swim bladder and will be seen later.

Digestive System. Run your probe through the mouth and into the opening of the **esophagus** at the end of the pharynx. Now lift up the liver and trace the esophagus from the pharynx to the large **cardiac portion** of the stomach. This ends posteriorly as a blind pouch. The short **pyloric portion** of the stomach, opening off the side of the cardiac pouch, empties by way of a **pyloric valve** into the **duodenum,** the S-shaped proximal part of the intestine. Three intestinal diverticula, the **pyloric ceca,** open off the proximal end of the duodenum near the pyloric valve (Figure 16.12A). Follow the intestine to the **anus.** Note the supply of blood vessels in the **mesentery.** The **pancreas,** a rather indistinct organ, lies in the fold of the duodenum. The **liver** is large and lobed, with the **gallbladder** located under the right lobe. Bile from the liver is drained by tubules into the gallbladder, which in turn opens by several ducts into the duodenum posterior to the pyloric ceca.

Cut open the stomach and place its contents in a glass dish with water. Compare with your neighbor's findings. Examine the stomach lining. How is its surface increased? _____ Cut open and examine the pyloric valve. Open a piece of the intestine, wash, and examine the lining with a dissecting microscope. What type of muscle would you expect to find in the intestine? _____

Swim Bladder. The swim bladder is a long, shiny, thin, but tough-walled sac that fills most of the body cavity dorsal to the visceral organs. In some fishes (not the perch), it connects with the alimentary canal. Cut a slit in it and observe its internal structure. In its anterior ventral wall, look for the **gas gland,** which contains a network of capillaries, and the **rete mirabile** (rē′tē muh-rab′uh-lē; L., wonderful net), which assists in the secretion of gases, especially oxygen, into the bladder at high pressure. Another capillary bed, the **oval,** lies in the dorsal body wall. Gases are resorbed from the swim bladder in this area. The swim bladder is a "buoyancy tank," or hydrostatic organ, that adjusts the specific gravity of a fish to varying depths of water so that the fish is always neutrally buoyant. Can you think of disadvantages to this type of buoyancy system? What will happen to the volume of the bladder and to the perch's specific gravity if the perch swims upward from some depth toward the surface? _____
When you dissected the dogfish shark, did you find a swim bladder? _____
What does a shark use for buoyancy? _____

EXPERIMENTING IN ZOOLOGY
Agonistic Behavior in Paradise Fish, *Macropodus opercularis*

Many organisms compete for resources that are limited in the environment. These resources can be food, shelter, or mates. The males of many species of animals compete for females. Competition for females often leads to aggressive encounters between conspecific males, and these encounters can escalate into fights that involve physical contact. Thus, natural selection has favored aggressive behavior in the males of some species so much that their aggression is innate (instinctive). Males in these species interact aggressively even when females are not present. Scientists who have observed aggressive interactions between males have suggested that the intensity of the encounter can depend on how closely matched the males are in size and strength. Small males can become injured when fighting larger males and may choose not to fight. However, when males are closely matched in size and strength, aggressive encounters tend to be intense as each male attempts to establish its dominance and ultimately its access to the resource (females, in this case).

Paradise fish, *Macropodus opercularis* (Figure 16.13), are tropical fish from Southeast Asia. They have special organs in their mouth that allow them to breathe air in addition to using their gills. Males are more colorful than females and have red and blue bars that run vertically down their body. Several paradise fish can be kept together in a single tank; however, when two males are isolated from the other fish, they begin to display aggressive interactions.

Getting Ready

Work with a partner for this exercise. Try to work in a room that can be darkened. You should have access to a fish tank (approximately 3 to 10 gallons) that has an overhead tank light. If an overhead light is not available, illuminate the tank from the side with a small desk lamp. Turning off the room light will minimize the fishes' ability to see you as well as minimizing their distractions. Have a stopwatch available for timing fish behavior during your experiment.

How to Proceed

Select two male paradise fish from a communal tank. Try to select two males that appear very similar in size and other physical traits. Make sure to add the two fish to the experimental tank at exactly the same time. For the next 15 minutes, quantify the interactions between the fish. The easiest way to measure aggressive interactions is to record the amount of time the two males stay very close to each other. You might elect to record all of the time the fish spend within one body length of one another. If the two males display aggression to each other, you will notice several different types of behaviors. For example, males might chase each other, raise their dorsal fins, swim parallel very closely together, or even come side-to-side and vibrate, thereby sending mechanical waves toward each other. If you decide to use categories of behaviors as well as total

Figure 16.13
Two male paradise fish interacting.

time spent near each other, you might consider observing two "practice" males before starting your experiment so that you can become familiar enough with the various behaviors to be able to count them during the experiment.

After recording your observations and data from the two males that are closely matched in size, report the trial by selecting two new males. This time, however, choose two males that differ conspicuously in size (although make sure you have two males and are not placing a male and female together). Repeat the experiment, keeping track of the time the males spend together and the number of aggressive behaviors displayed.

Compile the data from the entire class and calculate the overall mean times spent together in closely matched contests versus contests in which males differed in size.

Do males that are closely matched in size interact more than males that differ in size? _____ Do the closely matched males have encounters that escalate into more aggressive behaviors (e.g., biting, nipping) than males that differ in size? _____

Questions for Independent Investigations

1. Keeping the fish matched in size, how might light levels affect aggressive interactions?

2. Scientists who study fish aggression have found that tank residents are more aggressive than fish recently added to a tank. How could you test this with paradise fish?

3. Many kinds of fish release chemical alarm pheromones when they are injured. Gently remove a couple of scales from a paradise fish and rinse his body with filtered water. This rinse water may now contain alarm pheromones. Add these chemicals to a tank containing two interacting males. How might these chemicals affect the aggression levels of the males? Why?

4. What might happen to aggressive interactions if paradise fish were exposed to a potential predator?

References

Csanyi, V., J. Haller, and A. Miklosi. 1995. The influence of opponent-related and outcome-related memory on repeated aggressive encounters in paradise fish. Bio. Bull. **188:**83–86.

Francis, R. C. 1983. Experiential effects on agonistic behavior in the paradise fish, *Macropodus opercularis*. Behaviour **85:**292–313.

Gerlai, R. 1993. Can paradise fish recognize a natural predator? An ethological analysis. Ethology **94:**127–136.

Jakobsson, S., O. Brick, and C. Kullberg. 1995. Escalated fighting behaviour incurs increased predation risk. Animal Behav. **49:**235–249.

EXPERIMENTING IN ZOOLOGY
Analysis of the Multiple Hemoglobin System in *Carassius auratus,* the Common Goldfish

Carassius auratus, the Common Goldfish

Subphylum Vertebrata
  Class Actinopterygii
    Order Cypriniformes
      Family Cyprinidae (carps and minnows)
        Genus *Carassius*
          Species *Carassius auratus*

Cells require oxygen to utilize aerobic pathways in the production of ATP. For large multicellular animals, diffusion of oxygen from the environment into the animal is insufficient to supply their metabolic needs. This problem led to the development of complex oxygen exchange surfaces and organs in animals and also to the development of oxygen transport molecules within circulatory systems. In vertebrates, most oxygen is transported in the blood bound to hemoglobin within red blood cells. Oxygen is picked up by the hemoglobin in the gills or lungs at the exchange surfaces and delivered to tissues by the circulating blood. The hemoglobin molecule in vertebrates consists of four noncovalently linked subunits that form a tetrameric molecule. Typically, this tetramer is assembled from two α (alpha) polypeptide chains and two β (beta) polypeptide chains. Each polypeptide subunit of hemoglobin binds a heme group containing a ferrous iron responsible for binding the oxygen molecule (Figure 16.14).

Oxygen availability is relatively constant in the terrestrial environment; however, in the aquatic environment, oxygen availability is highly variable because oxygen solubility in water is dependent upon temperature, ionic concentration, pH, and the amount of biomatter consuming or producing oxygen. Cyprinid fishes, which include

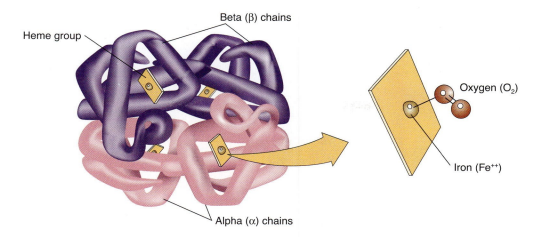

Beta (β) chains

Heme group

Oxygen (O$_2$)

Iron (Fe^{++})

Alpha (α) chains

Figure 16.14

Human hemoglobin molecule showing the four polypeptide chains (two α and two β) each associated with a heme group to which an oxygen molecule will bind.

carp, minnows, and goldfish, often inhabit shallow ponds in which temperature may vary by as much as 10° or 20°C each day. Typically, as the temperature of water increases, the oxygen solubility decreases. A change in water temperature from 0°C to 40°C causes available oxygen to drop by about one-half. An additional confounding factor is that animal tissue metabolism increases as temperature increases.

The decrease in oxygen availability combined with increased oxygen demand as water temperature increases poses a problem for maintaining aerobic respiration in fishes. One adaptation that fish have evolved to deal with variable oxygen availability is to produce multiple forms of hemoglobin. These multiple forms of hemoglobin in fishes may be distinguished electrophoretically (Riggs, 1970; di Prisco and Tamburrini, 1992) and may differ in physiological properties (Binotti et al., 1971; di Prisco and Tamburrini, 1992). The differences in physiological properties may allow for enhanced oxygen transport when oxygen availability is low in the environment. Trout are known to have up to nine distinctly different hemoglobin forms (I–IX), while goldfish exhibit three different hemoglobin forms, called G$_1$, G$_2$, and G$_3$. These three forms in goldfish vary in concentration as temperature varies, presumably brought about by rapid changes in the aggregation of different subunits (Houston and Cyr, 1974; Houston et al., 1976; Houston and Rupert, 1976).

How to Proceed

We can visualize the multiple hemoglobin systems found in common goldfish using simple native polyacrylamide gel electrophoresis. In this procedure, we will remove red blood cells from goldfish that have

been maintained in the laboratory at about 20° to 25°C using a heparinized syringe. (Heparin is a compound that prevents coagulation of blood removed from the circulatory system.) The cells will be washed in isotonic saline (0.9% NaCl) to remove serum proteins and lysed in distilled water to release the hemoglobins. (Distilled water is hypotonic to the red blood cells; therefore, the cells will swell and burst, releasing hemoglobin and other cellular components.) The hemoglobin lysate will be mixed 1:1 with a loading buffer, and a 20 μl sample will be applied to a well on an 8% native polyacrylamide gel (Figure 16.15). A sample of human hemoglobin prepared in the same way as the goldfish hemoglobin will serve as a marker on the gel. Electrophoretic separation will proceed for 30 to 45 minutes at 80 to 90 volts. Following electrophoresis, you should be able to see three faint reddish-orange bands in the goldfish lane, indicating the presence of the three isoforms of hemoglobin. Typically, the G$_2$ band (middle band) predominates, while the G$_1$ (upper) and G$_3$ (lower) bands are fainter. Only one band will appear in the human hemoglobin lane. To assist visualization of the hemoglobin bands, the gel will be removed from its case, stained with Coomassie Blue dye for a few minutes, and then destained to remove dye not bound to protein. A picture of the gel may be taken if appropriate equipment is available.

Questions for Thought

1. How might possession of multiple hemoglobin isoforms allow for adaptation to different environments?
2. Why do you think trout have so many different forms of hemoglobin?
3. Can you suggest any other species of fish that might show multiple forms of hemoglobin?

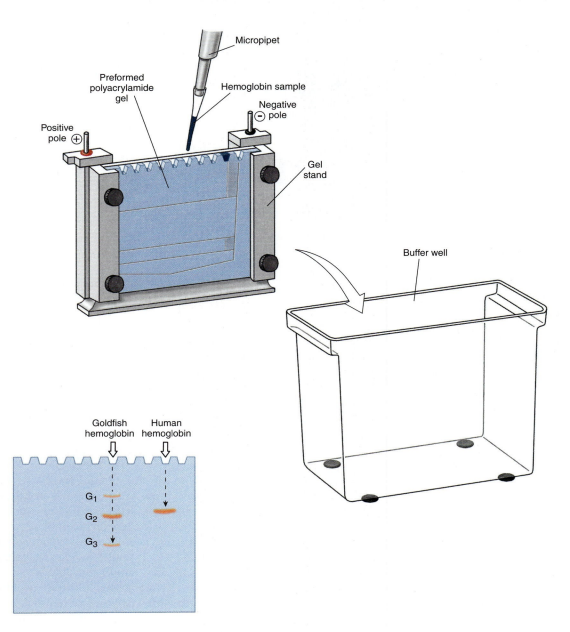

Figure 16.15

Diagram outlining the preparation, loading, and electrophoresis of a vertical polyacrylamide gel to separate the isoforms (G_1, G_2, and G_3) of hemoglobin from goldfish.

References

di Prisco, G., and M. Tamburrini. 1992. The hemoglobins of marine and freshwater fish: the search for correlations with physiological adaptation. Comp. Biochem. Physiol. **102B:**661–671.

de Souza, P. C., and G. O. Bouilla-Rodriguez 2007. Fish hemoglobins. Brazilian J. Med. Biol. Research **40:**769–778.

Houston, A. H., and D. Cyr. 1974. Thermoacclimatory variation in the haemoglobin systems of goldfish *(Carassius auratus)* and rainbow trout *(Salmo gairdneri).* J. Exp. Biol. **61:**455–461.

Houston, A. H., and R. Rupert. 1976. Immediate response of the hemoglobin system of the goldfish, *Carassius auratus,* to temperature change. Can. J. Zool. **54:**1737–1741.

Houston, A. H., K. M. Mearow, and J. S. Smeda. 1976. Further observations upon the hemoglobin systems of thermally acclimated freshwater teleosts: pumpkinseed *(Lepomis gibbosus),* white sucker *(Catostomus commersoni),* carp *(Cyprinus carpio),* goldfish *(Carassius auratus)* and carp-goldfish hybrids. Comp. Biochem. Physiol. **54A:**267–273.

Riggs, A. 1970. Properties of fish hemoglobins. In Hoar, W. S., and D. J. Randall (eds.). Fish physiology. Vol. 4. New York, Academic Press.

EXERCISE 17

Class Amphibia
Frogs

The amphibians are a transition group between aquatic and strictly land vertebrates. They still have the soft, moist epidermis of aquatic forms and therefore cannot stray too far from water or moist surroundings. Amphibian eggs lack the tough protective shell and specialized extraembryonic membranes characteristic of the eggs of terrestrial vertebrates and so remain adapted to an aquatic habitat. Many amphibians have developed lungs for breathing air but still have aquatic larvae with external gills. The moist skin is also a respiratory organ. The evolution of the lung has brought a change in circulation that includes a pulmonary circuit as well as a systemic circuit. Amphibians have a three-chambered heart (two atria and one ventricle). They are usually four-limbed for walking or jumping but often have webbed feet for swimming.

Frogs and toads belong to the order Anura (Gr. *an,* without, + *oura,* tail), the largest and most diverse group of living amphibians. The anurans differ from the salamanders (order Caudata) and the tropical caecilians (order Gymnophiona) in several distinctive ways that are associated with a specialized jumping mode of locomotion. A frog's body is extremely shortened and virtually fused with the head, an adaptation that provides rigidity to the skeletal framework. A frog lacks true ribs. The caudal vertebrae (which would normally form the tail) are fused into a single pillarlike bone, the urostyle. The hindlegs are much larger and more powerful than the forelegs. Anurans are the most diverse of the amphibians and are commonly used in the general zoology laboratory because of their ready availability. The large bullfrog *Rana catesbeiana* and the small leopard frog *Rana pipiens* are commonly used for dissection.

Frogs

Phylum Chordata
　　Subphylum Vertebrata
　　　　Class Amphibia
　　　　　　Order Anura (= Salientia)
　　　　　　　　Family Ranidae
　　　　　　　　　　Genus *Rana*

EXERCISE 17A
Behavior and Adaptations
　Frogs

EXERCISE 17B
Skeleton
　A Body for Life on Land

EXERCISE 17C
Skeletal Muscles
　Directions for Study of Frog Muscles

EXERCISE 17D
Digestive, Respiratory, and Urogenital Systems
　Mouthparts
　Dissection of a Frog

EXERCISE 17E
Circulatory System
　Venous System
　Arterial System
　Heart

Where Found

Ranid frogs (family Ranidae) are almost worldwide in distribution. Their favorite habitats are swamps, low meadows, brooks, and ponds, where they feed on flies and other insects. Their young, the tadpoles, develop in water and are herbivorous. It should be noted that ranids, like many species of frogs, appear to be declining in numbers around the world. It appears that there are several possible causes for the decline and many scientists are now investigating the conservation biology of amphibians.

EXERCISE 17A
Behavior and Adaptations

Core Study

☞ Place a live frog on a piece of wet paper toweling in a jar large enough not to cramp it. Do not let its skin become dry. Do not excite it unnecessarily.

Observe its adaptations. Make notes of your observations. Have a mounted frog skeleton at your table for comparison.

Is the skin smooth or rough? _____ Moist or dry? _____ Examine the frog's feet. How are they adapted for jumping? _____ For swimming? _____ For landing on a slippery rock or log? _____ Note the sitting position and compare with the mounted skeleton. In what ways is the body form of the frog adapted as a lever system that can catapult the animal into the air? _____

Compare the color of the dorsal and ventral sides. Imagine a predator approaching a swimming frog from above or from below. What is the protective advantage of this dorsal-ventral difference in coloration? _____

Knowing that a frog captures prey by striking at it with its protrusible tongue, how is the position of the eyes advantageous to the animal? _____ How do the eyes close? _____ Examine the skeleton and see how this is possible. Note the transparent nictitating membrane. How is it used? _____

Feeding Reactions

Place some live fruit flies in the jar with the frog. If the frog is hungry and not excited, it may feed. Describe how it seizes prey.

Breathing

Observe movement of the throat and nostrils. Air is drawn into the mouth; then the nostrils close, and the throat muscles contract to force air into the lungs. This form of breathing is called positive pressure breathing. The nostrils (nares) can be opened or closed at will. Record the number of movements per minute of the throat and then of the nostrils. _____ Do their rates coincide? _____ Do the sides of the body move during breathing? _____ Excite the animal by prodding it; then, as soon as it becomes quiescent, count the rate of breathing again.

Righting Reaction

Place the frog on its back, release it, and note how it rights itself. Repeat this experiment, but hold the frog down gently with your hand until it ceases to struggle. When you release it, the frog may remain in this so-called hypnotic state for some time.

Locomotion

Observe a frog in an aquarium. What is the floating position? _____ How does it use its limbs in swimming? _____ How does it dive from a floating position? _____ Note how it jumps.

Written Report

Record your observations in the lab report on p. 223.

External Structure

Study a preserved frog and compare it with a live frog and a mounted skeleton.

Note the **head** and **trunk.** Note the **sacral hump** produced by the protrusion of the pelvic girdle. Find this hump on the mounted skeleton. The **cloacal opening** is at the posterior end of the body.

On the **forelimbs,** identify the arm, forearm, wrist, hand, and digits. On the **hindlimbs,** find the thigh, shank, ankle, foot, and digits. How does the number of digits compare with your own? _____ Find the rudimentary thumb (prepollux) and the rudimentary sixth toe (prehallux). During the breeding season, the inner (thumb) digit of the male is enlarged into a nuptial pad for clasping the female. Observe the **webbed toes** of the hindfoot. How is the long ankle advantageous to the frog? _____

The **eyes** are protected by **eyelids** and by a transparent **nictitating membrane** (L. *nictare,* to wink). Look in the corner of your neighbor's eye to see a vestige of this membrane, the semilunar fold.

The **tympanic membrane** (eardrum) is a circular region of tightly drawn skin just behind the eye. The frog has no external ear—only the middle and internal ears.

Name ______________________

Date ______________________

Section ______________________

LAB REPORT

Observations of Frog Behavior

Feeding reaction ______________________

____

____

____

Breathing ______________________

____

____

____

Righting reaction ______________________

____

____

____

Locomotion ______________________

____

____

____

Other observations ______________________

____

____

____

____

____

EXERCISE 17B
Skeleton

Core Study

A Body for Life on Land

In amphibians, as in fishes, the well-developed skeleton provides a framework for the muscles used in movement and for protection of the viscera and nervous systems. Living on land and the necessity of transforming paddlelike fins into tetrapod legs capable of supporting the body's weight introduced a new set of stress and leverage problems. These changes are most noticeable in frogs and toads, whose entire musculoskeletal system is specialized for jumping and swimming by simultaneous extensor thrusts of the hindlimbs. Frogs no longer move in the typical sinuous or fishlike motion of their aquatic ancestors. Consequently, the vertebral column has lost its flexibility and, together with the enlarged pelvic girdle, has become a rigid frame for transmitting force from the hindlimbs to the body.

☞ Prepared skeletons are brittle and delicate. Handle them with care and do not deface the bones in any way. Use a probe or dissecting needle, not a pencil, for pointing.

The **axial** (L., axis) **skeleton** includes the skull, the vertebral column, and the sternum. The **appendicular** (L. *ad,* to, + *pendare,* to hang) **skeleton** includes the pectoral and pelvic girdles and the forelimbs and hindlimbs.

For a discussion of the structure and growth of bones and their articulations, see Exercise 20A.

Axial Skeleton

Skull. The skull and jaws of a frog serve to protect the brain and special sense organs. As with the rest of the frog skeleton, the skull is vastly altered compared to that of early amphibian ancestors. It is much lighter in weight, is more flattened in profile, and contains fewer bones and less ossification. The front part of the skull, which houses the eyes, nose, and brain, is better developed, whereas the back of the skull, which in fishes contains the gill apparatus, is much reduced. Lightening of the skull was essential to mobility on land, and the other changes fitted the frog for its improved senses and means of feeding and breathing.

The skull includes the **cranium,** or braincase, the **visceral skeleton,** made up of the bones and cartilage of the jaws, the hyoid apparatus, and the little bones of the ears. All these elements of the visceral skeleton[1] are derived from the jaws and gill apparatus of fish ancestors. The **orbital fossae** and the **nasal fossae** are the dorsal openings where the eyes and external nares are located.

Vertebral Column. The backbone of the frog consists of only nine vertebrae and a **urostyle** (Gr. *oura,* tail, + *stylos,* pillar), the latter representing the fusion of several caudal vertebrae. The first vertebra, the **atlas** (Gr. *Atlas,* a Titan of Greek mythology, who bore the heavens on his shoulders), articulates with the skull. The ninth, or **sacral**[2] (L. *sacer,* sacred), vertebra has transverse processes for articulation with the ilia of the pelvic girdle.

No frogs of the large family Ranidae have ribs, either as larvae or as adults; ribs do appear, however, in two other frog families.

The sternum (L., breastbone) provides ventral protection for the heart and lungs and a center for muscular attachment. Its four parts, beginning at the anterior end, are the **episternum** (Gr. *epi,* upon), a cartilaginous rounded end often not seen in prepared skeletons; the **omosternum** (Gr. *omos,* shoulder); the **mesosternum** (Gr. *mesos,* middle), the section located posterior to the coracoid bone of the pectoral girdle; and the **xiphisternum** (Gr. *xiphos,* sword), a cartilaginous, heart-shaped end often not present on prepared skeletons. The episternum and omosternum are not visible in Figure 17.1.

Appendicular Skeleton

Pectoral Girdle and Forelimbs. The pectoral girdle supports the forelimbs, which frogs use mainly to absorb the shock of landing after a jump. The pectoral girdle articulates with the sternum ventrally. Each half of the girdle includes a suprascapula and a scapula (L., shoulder blade) and a clavicle (collar bone) (L., *clavicula,* small key) lying anterior to the coracoid (Gr. *korax,* crow, + *eidos,* form). Consult Figure 17.1 to see the bones of the forelimb.

Pelvic Girdle and Hindlimbs. The pelvic girdle supports the hindlimbs. Each half is made up of the long **ilium** (L., flank), the anterior **pubis** (L. *pubes,* mature), and the posterior **ischium** (Gr. *ischion,* hip). Consult Figure 17.1 to see the bones of the hindlimb.

Oral Report

💬 Be familiar with the parts of the skeleton and the purpose served by each part.

---

[1]The term "visceral" (from the Latin *viscera,* bowels) refers to elements and structures of the body associated with the gut tube. The visceral skeleton represents specialized parts of the primitive gut of jawed vertebrates.

[2]The origin of the term sacrum, or "holy bone," is unknown. One hypothesis is that the curved appearance of the human sacrum suggested to Renaissance anatomists a resemblance to an obsidian knife used in ancient sacrifice.

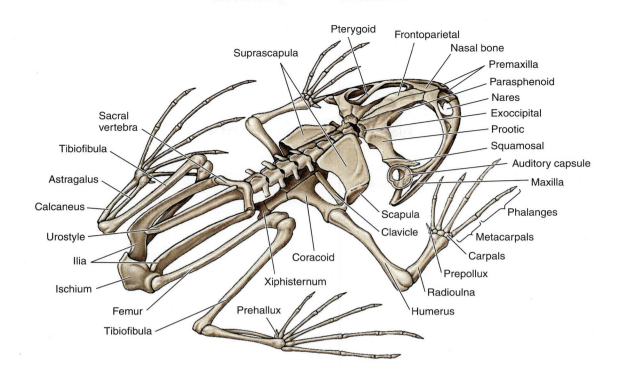

Figure 17.1
Skeleton of a frog.

EXERCISE 17C
Skeletal Muscles

Core Study

A frog has hundreds of muscles, but this exercise describes only the most important ones (more correctly, the most conspicuous ones, since they are all important to a frog).

Learning a few of the Greek and Latin roots of some of the muscle names will give you clues to their orientation or action. For example, **rectus** means "straight," so the fibers of rectus muscles generally run along the long axis of the body. **Gracilis** means "slender." **Triceps** means "three heads," and describes a muscle that has three tendons of origin; the **biceps** has two heads of origin. Long muscles are called **longus,** and short muscles are called **brevis;** large muscles are termed **magnus** or **major. Anticus** means "anterior." Other muscles are named for specific movements. The **sartorius** is derived from the Latin word for "tailor" and is homologous to a human muscle of the same name that is active in crossing the legs (tailors, before the days of sewing machines, sat on the floor with crossed legs). The **gastrocnemius** (Gr. *gaster,* stomach, + *kneme,* tibia) is named for its fat "belly."

Note: You will also need to be familiar with the terminology relating to muscle connections and their actions. Refer to the general discussion of skeletal muscles in Exercise 20B.

Directions for Study of Frog Muscles

☞ To skin the frog, slit the skin midventrally from the anal region to the chin, keeping the scissors point up to keep from injuring the underlying muscles. Make a transverse cut completely around the body just above the hindlegs and another one anterior to the forelegs. A middorsal cut the length of the back will divide the skin into portions that can be peeled off easily. Loosen the skin with a blunt instrument, and carefully pull off the skin over a leg wrong side out. Be careful not to tear thin muscle attached to the skin. The skin can be pulled over the head and eyes in the same way.

The large spaces between skin and muscle where the skin is not attached are **subcutaneous lymph sacs.**

☞ In separating the muscles from each other, first observe the direction of the muscle fibers and the extent of the muscle; then use your fingers, a blunt probe, or the handle of a scalpel to loosen the tissues. *Never use scissors, scalpel blade, or needle for dissecting muscles.* Never cut a muscle unless instructed to do so. If it is necessary to cut superficial muscles to find deep muscles, cut squarely across the belly (middle fleshy portion) of the muscle, leaving the origin and insertion in place.

TABLE 17.1

Major Trunk and Leg Muscles of Frogs

Muscle	Origin	Insertion
Ventral Trunk Muscles		
Pectoralis	Sternum and fascia of body wall	Humerus
Rectus abdominis	Pubic border	Sternum
External oblique	Dorsal fascia of vertebrae, also ilium	Sternum, linea alba
Transversus	Ilium and vertebrae	Linea alba
Ventral Thigh Muscles		
Sartorius	Pubis	Tibiofibula
Adductor magnus	Pubic and ischial symphysis	Distal end of femur
Gracilis major	Ischium	Tibiofibula
Gracilis minor	Ischium	Tibiofibula
Adductor longus	Ilium	Femur
Dorsal Thigh Muscles		
Triceps femoris	Three divisions: one head on acetabulum, two on ilium	Tibiofibula
Biceps femoris	Ilium	Tibiofibula
Semimembranosus	Ischium	Tibiofibula
Gluteus	Ilium	Femur
Shank Muscles		
Gastrocnemius	Two heads: distal end of femur and tendon from triceps femoris	By Achilles tendon to sole of foot
Peroneus	Femur	Distal end of tibiofibula; head of calcaneus
Tibialis anterior longus	Femur; divides into two bellies	By two tendons on ankle bones
Extensor cruris	Femur	Ventral surface of tibiofibula
Tibialis posterior	Side of tibiofibula	Ankle

Trunk Muscles

The trunk muscles of the frog, no longer required to produce the lateral flexion movements of the amphibians' swimming ancestors, have been modified to brace the back and to support the viscera in air. The ventral trunk muscles (Table 17.1) are arranged in layers that run in different directions. The **rectus abdominis** runs longitudinally, forming a sling from pubis to sternum (Figure 17.2). In the midventral line is a thin but tough band of connective tissue, the **linea alba** (literally, "white line"). Inserting on the linea alba are two oblique muscle bands, the **external oblique** and the **transversus,** the latter lying beneath the external oblique and the rectus abdominis. (The transversus is not shown in Figures 17.2 or 17.3.) These assist the rectus abdominis in supporting the viscera.

Thigh Muscles

A frog's limb muscles are derived from muscles that raised and lowered the fins of fishes, now much modified to brace and move the limbs for walking and thrust-swimming. Not surprisingly, the limb muscles are complex because they perform several actions. Nevertheless, we can recognize two major groups of muscles on any limb: (1) an anterior and ventral group that pulls the limb forward (protraction) and toward the midline (adduction) and (2) a set of posterior and dorsal muscles that draws the limb back (retraction) and away from the body (abduction). Only the more conspicuous thigh muscles are included in this discussion (Table 17.1).

In the first group (protractors and adductors) are the **sartorius, adductor magnus, gracilis major,** and **adductor longus** (visible on the ventral surface, Figure 17.2) and the **biceps femoris** and **gracilis minor** (visible on the dorsal surface, Figure 17.3). All of these act to draw the limb forward and toward the midline, or to flex more distal parts of the limb. In the second group (retractors and abductors) is the **triceps femoris,** a large muscle having three divisions, all of which serve to abduct the thigh and extend the shank (Figure 17.3).

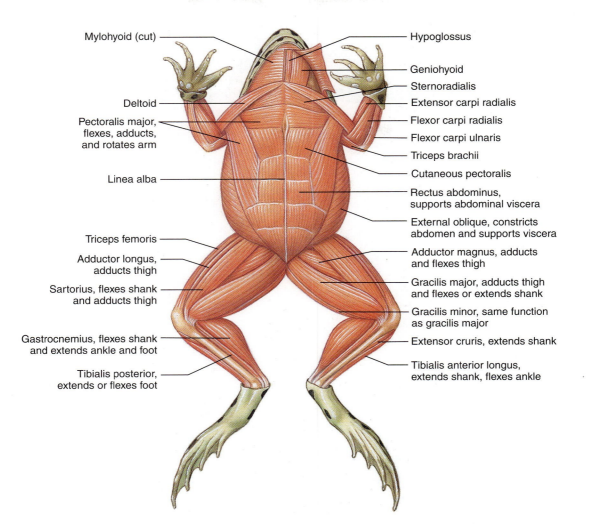

Figure 17.2
Muscles of a frog, ventral view.

The following labels appear in the figure:

- Mylohyoid (cut)
- Deltoid
- Pectoralis major, flexes, adducts, and rotates arm
- Linea alba
- Triceps femoris
- Adductor longus, adducts thigh
- Sartorius, flexes shank and adducts thigh
- Gastrocnemius, flexes shank and extends ankle and foot
- Tibialis posterior, extends or flexes foot
- Hypoglossus
- Geniohyoid
- Sternoradialis
- Extensor carpi radialis
- Flexor carpi radialis
- Flexor carpi ulnaris
- Triceps brachii
- Cutaneous pectoralis
- Rectus abdominus, supports abdominal viscera
- External oblique, constricts abdomen and supports viscera
- Adductor magnus, adducts and flexes thigh
- Gracilis major, adducts thigh and flexes or extends shank
- Gracilis minor, same function as gracilis major
- Extensor cruris, extends shank
- Tibialis anterior longus, extends shank, flexes ankle

Shank Muscles

The most conspicuous shank muscle is the **gastrocnemius,** which extends from the femur to the foot and inserts by the Achilles tendon; it extends the ankle during jumping and swimming. Because the gastrocnemius is easily dissected out together with the sciatic nerve that innervates it, it is commonly used in physiological studies of skeletal muscle. The ankle is flexed by the **tibialis anterior longus** and other muscles not shown in Figure 17.3. The major action of the **peroneus** is to flex the ankle joint and extend the shank.

If the frog you are dissecting is in the *Rana* group, it likely has well-developed leg muscles for swimming and jumping. Toads, another large group of frogs, are toxic and have less-developed hindlimb muscles. Why might a toxic animal have less-developed leg muscles? _____
_____

Other Muscles

With the help of Figures 17.2 and 17.3, you can identify many of the muscles of the back, shoulder, head, and arm.

How the Muscles Act

A muscle has only one function—to contract. For effective action, muscles must be arranged in antagonistic pairs. The gastrocnemius and the tibialis anterior longus represent such an antagonistic pair. Loosen the body of each of these muscles, pull on the gastrocnemius, and see what happens. Now pull on the tibialis anterior longus. Which of these muscles would be used in jumping or diving? _____ Which in sitting? _____
See whether you can locate other antagonistic pairs.

For most movements, groups of muscles rather than single muscles are required. By varying the combination

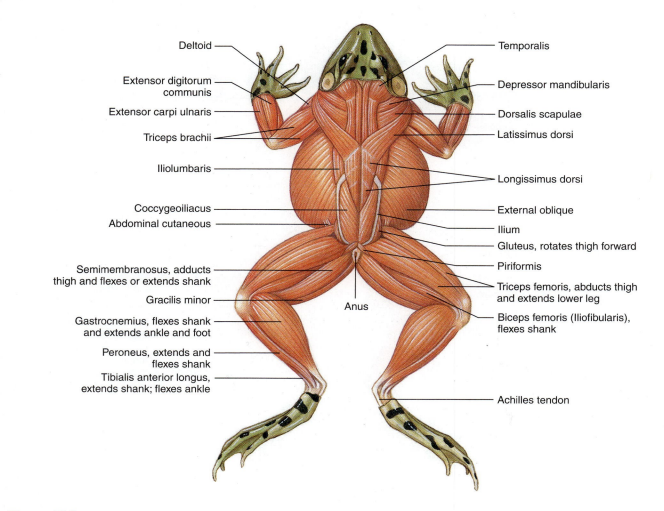

Figure 17.3
Muscles of a frog, dorsal view.

of these groups, many complicated movements are possible.

Oral Report

Be able to demonstrate a careful dissection of the muscles and to name the muscles and their actions.

Written Report

In the lab report on p. 229, name (1) the principal muscles of the hindleg that are involved when a frog leaps, and (2) the principal muscles that contract when the frog resumes a sitting position.

Principal Hindleg Muscles Involved in Leaping

Principal Hindleg Muscles Involved in Sitting

EXERCISE 17D
Digestive, Respiratory, and Urogenital Systems

Core Study

Mouthparts

☞ Pry open the mouth, cutting the angle of the jaw if necessary, and wash the mouth cavity in running water.

The posterior portion of the mouth cavity is the **pharynx,** which connects with the **esophagus.** Feel the **maxillary teeth** along the upper jaw and the **vomerine teeth** in the roof of the mouth (Figure 17.4). Are these better adapted for biting and chewing or for holding the prey to prevent its escape? _____ Find the **internal nares** (sing., **naris**) in the roof of the mouth, and note how they connect with the external nares. Note how the ridge on the lower jaw fits into a groove in the upper jaw to make the mouth closure airtight. This is important in the frog's respiratory movements.

Eustachian tubes, which connect with and equalize the air pressure in the middle ear, open near the angle of the jaws. In male frogs, openings on the floor of the mouth slightly anterior to the eustachian tubes lead to **vocal sacs,** which, when inflated, serve as resonators to intensify the mating call. Examine the **tongue** and note where it is attached. Which end of the tongue is flipped out to catch insects? _____ Feel the **sensory papillae** on the tongue surface. The free end of the tongue is highly glandular and produces a sticky secretion that adheres to the prey. Behind the tongue is a slight elevation in the floor of the mouth containing the **glottis,** a slitlike opening into the **larynx.**

Dissection of a Frog

☞ Make an incision through the abdominal wall from the junction of the hindlegs to the lower jaw, cutting through the bones of the pectoral girdle as you go. Make transverse cuts anterior to the hindlegs and posterior to the forelegs, and pin back the flaps of muscular tissue.

Note the three layers of the body wall: **skin, muscles** (with enclosed skeleton in some places), and **peritoneum,** which lines the large **coelom.**

In a mature female, the ovaries with their dark masses of eggs may fill much of the coelomic cavity. In this case, remove the left ovary and its white, convoluted oviduct.

Note the **heart** enclosed in its **pericardial sac** and surrounded by lobes of the **liver.** Lift up the heart to find the **lungs.**

Digestive System

The digestive tract is relatively short in adult amphibians, a characteristic of most carnivores. However, the larval (tadpole) stages of the frog are herbivorous, feeding on pond algae and other vegetation; they have a relatively long digestive tract, since their bulky food must be submitted to time-consuming fermentation before useful products can be absorbed.

You have seen the **mouth** and the **pharynx.** Lift the heart, liver, and lungs to see where the **esophagus** empties into the stomach (Figure 17.5). A **pyloric valve** controls movements of food into the **small intestine.** Note the blood vessels in the mesentery, which holds the stomach and small intestine in place. Why must the digestive tract be so well supplied with blood? _____

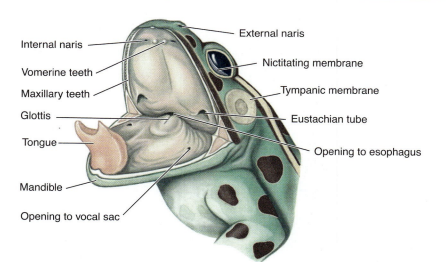

Internal naris
Vomerine teeth
Maxillary teeth
Glottis
Tongue
Mandible
Opening to vocal sac
External naris
Nictitating membrane
Tympanic membrane
Eustachian tube
Opening to esophagus

Figure 17.4
Mouthparts of a frog.

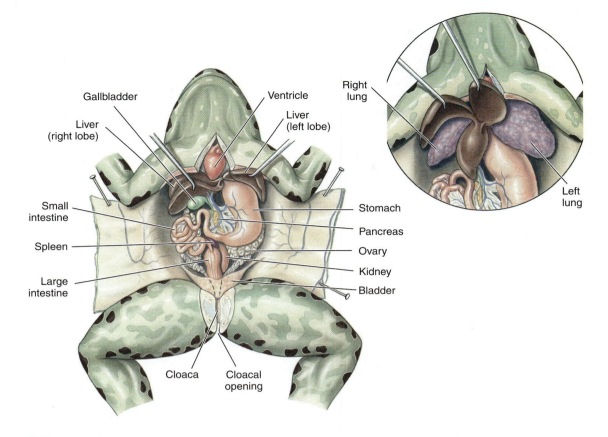

Figure 17.5
Abdominal cavity of a frog, ventral view. **Inset,** The liver has been lifted up and turned back to expose the lungs.

The **liver,** the largest gland in the body, secretes bile, which is carried by a small duct to the **gallbladder** for storage. Find the gallbladder between the right and median lobes of the liver. The **pancreas** is thin and inconspicuous, lying in the mesentery between the stomach and duodenum.

The **large intestine** narrows down in the pelvic region to form the **cloaca,** which also receives urine from the kidneys and products from the reproductive organs. It empties through the **cloacal opening.** Not all vertebrate animals have cloacas; mammals do not.

Respiratory System

Amphibians breathe through their skin **(cutaneous respiration)** as well as with their lungs (although some adult amphibians do not have lungs and rely primarily on gas exchange through the skin). Some respiration also occurs through the lining of the mouth, which is highly vascular. Cutaneous respiration is very important for the frog, especially in winter, when it burrows into the bottom mud of ponds and ceases all lung breathing.

The frog has no diaphragm. Thus, it draws air into its mouth cavity through the nares by closing the **glottis** (the opening into the windpipe) and depressing the floor of its mouth. Then, by closing the nares and raising the floor of its mouth cavity, the frog forces air from the mouth through the glottis into the lungs (positive pressure breathing). Air is expelled from the lungs by the contraction of the muscles of the body wall and the elastic recoil of the stretched lung.

☞ Probe through the **glottis** into the **larynx.** Find the short **bronchus,** connecting each **lung** to the larynx. Slit open a lung and observe its internal structure.

Note the little pockets, or **alveoli** (sing., **alveolus**), in the lining. What is the purpose of this arrangement? _____

Did you find any parasites in the lungs of your specimen? _____

Further Study

Urogenital System

Functionally, the urogenital system is two systems: the **urinary** (or **excretory**) **system** and the **reproductive system.** However, because some structures function in both systems, they are usually considered together.

Be careful not to injure the blood vessels as you study this system.

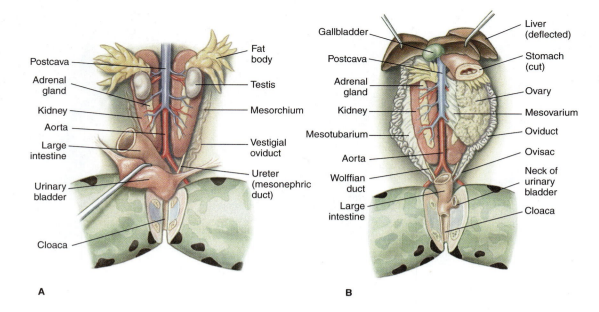

Figure 17.6
Urogenital system of a frog, ventral views. **A,** Male. **B,** Female.

Excretory System. The **mesonephric kidneys** separate the urine from the blood. They rid the body of metabolic wastes (aided by the lungs and skin), and they maintain a proper water balance in the body and a general constancy of content in the blood.

The kidneys (Figure 17.6) lie close to the dorsal body wall, separated from the coelom by a thin peritoneum. The **urinary bladder,** when collapsed, appears as a soft mass of thin tissue just ventral to the large intestine. It is bilobed and empties into the cloaca. The **ureters** connect the kidneys with the cloaca.

☞ To expose the cloaca, cut through the ischiopubic symphysis with a scalpel and push the pelvic bones aside. If you wish, you may open the cloaca just left of the midventral line, separate the cut edges, and find the bladder opening on the ventral wall of the cloaca and the openings of the ureters on the dorsal wall.

The **adrenal glands,** a light stripe on the ventral surface of each kidney, are endocrine glands, not urogenital organs. **Fat bodies** attached to the kidneys, but lying in the coelom, are for fat storage. They may be large in the fall and small or absent in the spring. Why? ————————————————

Male Reproductive System. A small, pale **testis** lies on the ventral side of each kidney (Figure 17.6A). Sperm pass from the testis into some of the kidney tubules and then are carried by the ureter to the cloaca and hence to the outside. Thus, the male ureters serve also as genital ducts. In leopard frogs, a small **vestigial oviduct** runs parallel to the ureter (this is absent in some species of *Rana*).

Female Reproductive System. **Ovaries** are attached by mesenteries to the dorsal wall of the coelom

(Figure 17.6B). In winter and early spring, the ovaries are distended with eggs. If the specimen was killed in summer or early fall, the ovaries will be small, pale, and fan-shaped. Convoluted **oviducts** widen anteriorly (dorsal to the lungs) into funnel-like **ostia** and posteriorly into **uteri,** which empty into the cloaca. Eggs are released from the ovary into the coelom, carried in coelomic fluid to the ostia, and then transported down the oviducts by ciliary action to the outside. At **amplexus** (the courtship embrace), the male clasps the female and fertilizes the eggs externally as they are laid in the water.

During breeding season, the thumbs of the male frogs of some species enlarge, presumably to assist in clasping the female during amplexus. Why would the male need to clasp the female so firmly? ——————— Who might be trying to displace the male? ———————
————————————————————————

Recently, research has found that a top-selling weed killer (atrazine) in the United States dramatically affects the sexual development of frogs in both the laboratory and the field. Male frogs exposed to the herbicide show indications of being hermaphrodites, developing both testes and ovaries. Does your frog show any of these signs?————————————————————————

Oral Report

Be able to demonstrate your dissection and state the functions of the structures you have studied.

Trace the route of food through the digestive tract, and tell what happens to it at each stage.

Trace the route of eggs and spermatozoa, and explain the anatomy and physiology of excretion and reproduction.

Understand the physiology of respiration, and tell how the mechanics of respiration in the frog differ from those in a human.

EXERCISE 17E
Circulatory System

Core Study

The shift from gill to lung breathing during the evolution of amphibians required important changes in circulation. With the elimination of gills, a major resistance to blood flow was removed. But two new problems arose. The first was to provide a blood circuit to the lungs. This was solved by converting the last pair of aortic arches (which in fishes carried blood through the last gill arch) into **pulmonary arteries** to serve the lungs. New **pulmonary veins** then developed to return oxygenated blood to the heart. The second problem was to separate this new **pulmonary circuit** from the rest of the body's circulation so that oxygenated blood from the lungs would be selectively sent to the body, and deoxygenated blood from the body would be sent to the lungs. This was achieved by partitioning the heart into a double pump, with each side serving each circuit. In this way, a **double circulation** comprising separate **pulmonary** and **systemic circuits** was formed (Figure 17.7). As you will see in your dissection of the frog, amphibians approached this necessary advance only partway: the atrium, which was single in fish, is completely divided into two atria, but the ventricle remains undivided. The task of complete separation had to await the evolution of the birds and mammals, which have completely divided hearts consisting of two atria and two ventricles.

To assist your study of the circulation, you will dissect a frog that has had its arterial system injected with a red or yellow latex and its venous system injected with blue latex. If only the arteries have been injected, the veins may be filled with dark, clotted blood.

☞ Dissect carefully and do not cut or injure the blood vessels. Using a probe, you may loosen the connective tissue that holds the vessels in place. Cut away a midsection of the pectoral girdle and pin back the arms so that the heart is fully exposed. Carefully remove the **pericardium,** the sac that contains the heart.

Identify the thick-walled conical **ventricle** (Figure 17.8); the thin-walled **left and right atria;** the **conus arteriosus,** arising from the ventricle and dividing to form the **truncus arteriosus** on each side; and on the dorsal side of the heart, the thin-walled **sinus venosus,** formed by the convergence of three large veins—two **precaval veins** and one **postcaval vein.**[3]

---

[3] "Caval" derives from the Latin *cavus,* meaning hollow, and refers to the sinus venosus into which both the precaval veins (also called the anterior vena cava) and the postcaval vein (= posterior vena cava) drain. The caval veins of tetrapod vertebrates replace the cardinal veins of fishes (see p. 211).

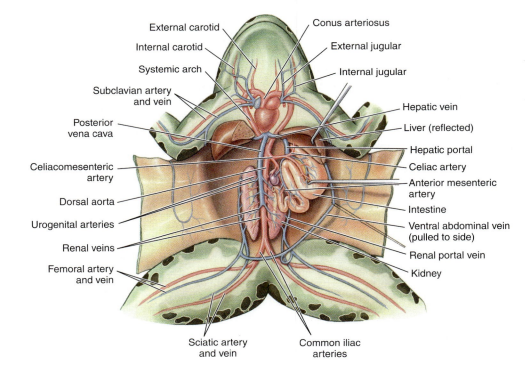

Figure 17.7
Circulatory system of the frog: arterial, red; venous, blue.

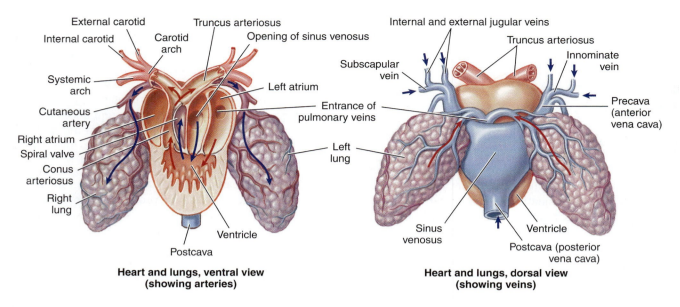

Figure 17.8
Structure of the frog heart. **Left,** Ventral view of frontal section. **Right,** Dorsal view. What structure collects deoxygenated blood returning to the heart and is formed by the convergence of three large veins? _____ What structure may help prevent mixing of oxygenated and deoxygenated blood as it leaves the heart (consult your textbook for help if necessary)? _____ What artery carries deoxygenated blood to the skin where the blood can be oxygenated? _____

Venous System

The **precava (anterior vena cava)** (Figure 17.8) is formed by the union of (1) the **external jugular** from the tongue and floor of the mouth, (2) the **innominate vein** (L. *in,* not, + *nomen,* named), made up of the **subscapular vein** from the shoulder and the **internal jugular vein** (L. *jugulum,* collarbone) from the brain, and (3) the **subclavian vein,** (L. *sub,* under, + *clavus,* key, meaning below the clavicle), which receives blood from the arm and dorsal body wall.

The **postcava (posterior vena cava)** extends from the sinus venosus through the liver to the region between the kidneys. It receives **hepatic veins** from the liver, **renal veins** from each kidney, and **ovarian** or **spermatic veins** from the gonads.

Pulmonary veins bring blood from the lungs to the left atrium. How does the oxygen content of the blood in these veins differ from that in any other vein? _____

Portal Systems

Ordinarily, veins carry blood directly from a capillary bed to the heart. This plan is interrupted in amphibians by capillary beds in two portal systems—the hepatic portal and the renal portal systems.

Hepatic Portal System. In the hepatic portal system, blood is carried to the capillaries of the liver by two veins: (1) The **ventral abdominal vein** in the ventral body wall collects from the pelvic veins, which are branches of the femoral veins; it empties into the liver. (2) The **hepatic**

portal vein receives the splenic, pancreatic, intestinal, and gastric veins. From the capillary bed in the liver, the blood is picked up by the **hepatic veins,** carried to the postcava, and so to the sinus venosus. The hepatic portal system, present in all vertebrates, is of great importance because it delivers nutrients absorbed from the gut directly to the liver. This guarantees that the liver will have first opportunity to store and process food materials before they are released into the general circulation. In this way, the blood leaving the liver remains relatively uniform regardless of any digestive activities underway.

Renal Portal System. Amphibians also inherited a renal portal system from their fish ancestors. Most of the blood returning from the hindlegs is interrupted in its journey toward the heart to be diverted into a network of capillaries in the kidneys. The **renal portal vein,** found along the margin of each kidney, is formed by the union of the **sciatic** and **femoral veins** (see Figure 17.7). Then, from the kidney, blood is collected by the **renal veins** and carried to the **postcava.** All vertebrates except the mammals have a renal portal system. You will see that both the hepatic and the renal portal systems fit the definition of a portal system as one that begins and ends in capillaries.

Arterial System

The **carotid, systemic,** and **pulmocutaneous arches** (known collectively as the **aortic arches**) arise from the **truncus arteriosus** (Figure 17.8).

Carotid Arch

The **common carotid artery** divides into (1) the **internal carotid,** which leads to the roof of the mouth, eye, brain, and spinal cord, and (2) the **external carotid (lingual)** to the floor of the mouth, tongue, and thyroid gland (see Figure 17.7).

Systemic Arch
Each systemic arch (one from each side) passes along the side of the esophagus to join middorsally to form the **dorsal aorta** (see Figure 17.7). Each gives off several arteries before joining; among these is the **subclavian artery** to the shoulder.

Find the **dorsal aorta** by lifting the kidneys. The dorsal aorta supplies all the body posterior of the head except the lungs and skin. Major branches of the dorsal aorta are:

1. **Celiacomesenteric artery,** which in turn gives rise to the **celiac** (Gr. *koilia,* belly) **artery** to the stomach, pancreas, and liver, and the **anterior mesenteric artery** to the spleen and intestines.

2. Six pairs of **urogenital arteries** to the kidneys, fat bodies, and gonads.

3. **Lumbar arteries** (not shown in Figure 17.7) to the muscles of the back.

4. **Common iliac** (L. *ilia,* flanks) **arteries,** formed by the division of the dorsal aorta. Each iliac gives off the **femoral artery** to the thigh, as well as branches to the urinary bladder, abdominal wall, and rectum. The iliac then continues into the hindleg as the **sciatic artery.**

Pulmocutaneous Arch
The third arch divides into the short **pulmonary artery** to the lungs and the longer **cutaneous artery** to the skin. Unlike the carotid and systemic arches, which carry oxygenated blood to the body, the pulmocutaneous arch carries *deoxygenated* blood to the lungs and skin, where the blood can be oxygenated.

Heart

👉 Make a frontal section of the heart (Figure 17.8), dividing it into dorsal and ventral valves.

Find the opening from the sinus venosus into the **right atrium** and the opening from the pulmonary veins into the **left atrium.** Why is the ventricle more muscular than the atria? _____

_____

The **conus arteriosus,** which receives blood from the **ventricle,** divides to form a left and right truncus arteriosus. Valves to prevent backflow of blood guard the entrances to the atria and the conus.

Even though the amphibian heart is three-chambered with two atria and one undivided ventricle, oxygenated and deoxygenated blood are effectively separated in the heart. Oxygenated blood from the lungs is sent preferentially to the body, whereas deoxygenated blood from the body is directed toward the pulmocutaneous arch. This partitioning is aided by a spiral fold inside the conus arteriosus and by pressure changes within the heart occurring with each heart contraction.

Written Report

✍ Fill in the lab report on frog circulation on pp. 236–237.

Oral Report

🗨 Demonstrate your dissection of the circulatory system to your instructor and be able to explain orally any phase of its anatomy or its functions. Be able to trace the flow of blood into the heart from body tissues, through the pulmonary circuit, and back to the body tissues.

Observing the Heartbeat

👉 Open a pithed frog as you did the preserved frog, but avoid cutting the abdominal vein. Cut carefully through the pectoral girdle, keeping the scissors well up to avoid injuring the heart. Pin back the forelimbs and keep the heart well moistened with frog Ringer solution.

Identify the ventricle, left and right atria, truncus arteriosus, and aortic arches. Watch the heartbeat and note the series of alternating contractions—first the two atria, then the ventricle, and finally the arterial trunk. Raise the ventricle carefully to view the contraction of the sinus venosus immediately before the contraction of the atria.

The frog is an **ectothermic** animal, meaning that its temperature is governed by the environmental temperature, rather than by internal means (endothermic). You can examine the effect of temperature changes on the heart rate by performing a simple experiment.

👉 Count and record the number of beats per minute at room temperature. Flood the abdominal cavity with ice-cold frog saline solution. Immediately count the beats again; then wait a little while until the maximum effect of the cold is achieved, and then count and record again.

Replace the cold saline solution with frog saline solution warmed to about 40°C. Count the beats immediately and record. Allow the warmth to take effect; then count and record again. Replace the warmed saline with saline at room temperature, and count the heart rate again. Record your results in the lab report on p. 237.

What effect does temperature change have on the heart rate? _____ What advantage would this reaction to cold have for the frog? _____ Why should the solution not be warmed to more than 40°C? _____ What is the effect of fever on a human's heart rate? _____

Name _____

Date _____

Section _____

Frog Circulation

1. Trace the shortest route a corpuscle could take on each of the following trips, underscoring each place where it would go through a **capillary bed.**

    a. Ventricle to lung and return _____

    _____

    _____

    b. Ventricle to brain and return _____

    _____

    _____

    _____

    c. Systemic arch to intestine to right arm _____

    _____

    _____

    _____

    d. Hindleg to left atrium by way of renal portal vein _____

    _____

    _____

    _____

2. What are the chief gains and losses that take place in the blood in these organs?

    a. Lung _____

    _____

    b. Kidney _____

    _____

    c. Intestinal wall _____

    _____

d. Liver _____

    _____

e. Muscles _____

    _____

3. Describe the shortest route from the **left atrium** to an **arm** and back to the **left atrium.**

    _____

    _____

    _____

    _____

    _____

4. Effect of temperature on heart rate.

    Rate of Contraction

    At room temperature   _____   /min (first count)

    Of cooled heart   _____   /min (first count)

    Of cooled heart   _____   /min (later count)

    Of warmed heart   _____   /min (first count)

    Of warmed heart   _____   /min (later count)

    At room temperature   _____   /min (later count)

The Nonavian Reptiles

EXERCISE 18
Painted Turtle

Core Study

Amphibians and nonavian reptiles are similar in that both are ectothermic tetrapods. However, they differ in several important characteristics. One of the most important differences is that nonavian reptiles have an amniotic egg, an egg that allows reproduction and development outside of water. Nonavian reptiles also have a tough, protective skin, better-developed lungs, a more efficient circulatory system, and a more complex nervous system.

Modern nonavian reptiles—lizards, snakes, crocodilians, and turtles—belong to the **diapsid** lineage, a group characterized by having a skull with two pairs of windowlike openings in the cheek (temporal) region. This is a derived condition that lightened the skull, furnished edges for jaw muscle attachment, and provided space that allowed jaw muscles to bulge when the jaw was closed. Interestingly, turtles lack these openings and are often considered living representatives of an early **anapsid** (no opening) group. Research published in the last 15 years using morphological and genetic evidence suggests that the skull openings were lost in early turtle evolution and that the current condition evolved secondarily from ancestors having temporal openings.

Turtles have changed little over the past 200 million years, and although they are highly specialized and much modified from the earliest known amniote fossils, they reveal several features that distinguish the reptilian lineages. A turtle's shell, the anatomical feature that makes it instantly recognizable, is undoubtedly one secret of its success, providing protection to an otherwise ungainly animal.

Painted Turtle

Phylum Chordata
    Subphylum Vertebrata
        Class Reptilia
            Order Testudines (Chelonia)
                Genus *Chrysemys*
                    Species *Chrysemys picta*

EXERCISE 18
Painted Turtle

Where Found

Painted turtles, *Chrysemys picta,* are familiar aquatic turtles in ponds, marshes, lake edges, and slow streams. They are widely distributed in the central and northern United States, as far south as Georgia and Louisiana, and north along the southern edge of the Canadian provinces. Painted turtles feed on aquatic vegetation, crayfish, snails, and insects. They are relatively common throughout most of their range; however, their numbers decrease whenever aquatic habitat (rivers, ponds, lakes) are modified or destroyed. Their numbers have declined in Canada to the point at which they are considered a species of concern.

External Structure

☞ If living turtles are available, study their locomotion and external features for adaptations that distinguish this group.

The anatomical form of the turtle is unusually broad and flattened compared to that of other reptiles, with the limbs extending laterally from the sides and bent downward at elbow and knee. As a result, the turtle's gait is slow and seemingly awkward, although not as inefficient as it might appear—it has, after all, served this successful group for some 200 million years while numerous, more agile reptiles have disappeared. The protective shell of the turtle, its exoskeleton, includes the dorsal **carapace** (Sp. *carapacho,* shell) and the ventral **plastron** (Fr., breastplate). Some turtles have hinged plastrons and can close their plastron tightly. Other species rely more on their snapping jaws for defense and have greatly reduced plastrons. The bones of the carapace are covered with horny **scutes** (homologous to the epidermal scales of other reptiles). Although the turtle's shell is confining, it is not responsible for the awkward gait since a sprawling posture with the body dragging on the ground was characteristic of

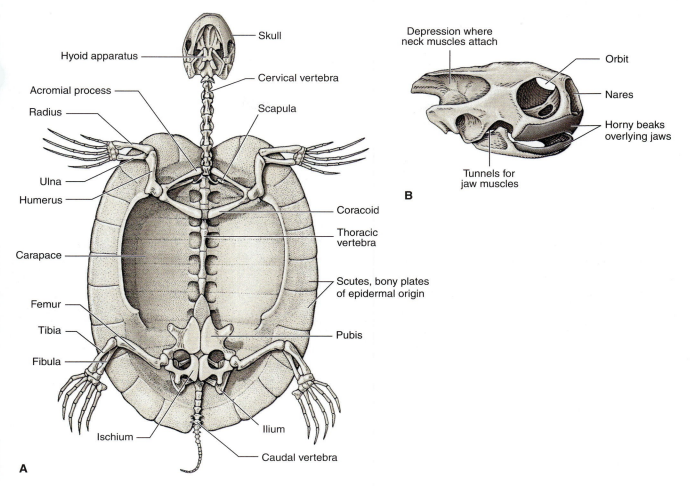

Figure 18.1
Skeleton of the painted turtle. **A,** Ventral view of entire skeleton. **B,** Lateral view of the skull.

primitive land-dwellers in general. However, the subsequent evolution of the shell did prevent any further refinement in locomotion.

The head is stoutly protected by bone. A peculiarity of turtles is their lack of teeth. Instead, the edges of the jaws are formed into sharp ridges covered with strong, horny beaks. Locate the **external nares** (nostrils). Note that they are set close together at the tip of the snout, enabling the turtle to breathe while remaining almost completely submerged. The eyes are well developed. (Turtles, like lizards, have an abundance of cones in the retina, providing them with a colorful view of the world.) The eyes have upper and lower eyelids and an additional third eyelid at the anterior corners, the transparent **nictitating** (L. *nictare,* to wink) **membrane,** that can be pulled across the eye to moisten and cleanse the cornea.

As in most turtles, the neck of the painted turtle is long and flexible, and it folds dorsoventrally to allow the head to be retracted within the protective confines of the shell. In a different suborder of turtles, distributed in the Southern Hemisphere, the neck folds sideways when the head is retracted.

Skeleton

☞ Examine the mounted skeleton of the painted turtle. Handle with care and use a probe or dissecting needle, not a pencil, for pointing.

Examine the carapace and plastron of the shell (Figure 18.1). The plastron, normally united to the carapace by bony bridges, will have been removed in the mounted skeleton. Note that the bones of the carapace are covered by horny scutes that do not coincide in number or position with the underlying bone. On the outside, there are five central scutes of epidermal origin, but on the inside there are eight central bony plates (of dermal origin) that are fused to the thoracic vertebra. Notice the growth lines in the scutes, formed as new keratin is laid down at the edges of the expanding scutes. However, these cannot be used to determine the turtle's age because new growth is influenced by many environmental factors.

As in other vertebrates, the **endoskeleton** comprises two divisions: the **axial skeleton** consisting of skull, vertebrae, and ribs, and the **appendicular skeleton** consisting of the pectoral and pelvic girdles and the limb bones.

The turtle skull is completely roofed over by the braincase (Figure 18.1B). Locate the openings for the eyes, nostrils, and ears. Note the wide excavations at the rear where the skull has been scalloped out to receive the neck muscles. Tunnels on either side at the rear of the upper jaw provide space for the powerful jaw muscles. These excavations for muscles in the anapsid skull of turtles are not true skull openings ("fenestrations") of the kind found in the diapsid skull of other reptiles. Beneath the lower jaw is the **hyoid** (Gr. *hyoeides,* Y-shaped) **apparatus,** a complex of bones, mostly fused, that support the tongue (Figure 18.1A). The hyoid apparatus is derived from remnants of the gill arches of fish ancestors.

The vertebral column consists of 8 **cervical vertebrae** that form the bony support of the turtle's flexible neck; 10 **thoracic vertebrae** fused to the carapace; 2 **sacral vertebrae** with expanded ends fused to the pelvic girdle (frogs and other amphibians never have more than one sacral vertebra); and 25 to 30 **caudal vertebrae.** Note the strong articulating processes (called zygopophyses) of the elongate cervical vertebrae. The articulations are arranged to allow the turtle to withdraw its neck into an S-shaped bend. The thoracic (trunk) vertebrae have elongate centra, and each gives rise to a rib on either side. Note that the central eight pairs of ribs extend laterally from the vertebrae like flying buttresses to fuse with, and lend strength to, the shell.

Turtles have stout pectoral and pelvic girdles to support the limbs and bear the animal's weight. Both girdles are much modified compared to those of other reptiles. The pectoral girdle lacks a sternum (because the plastron serves a sternum's function), and the strutlike bones of the pectoral girdle form an odd tripartite scaffold between the carapace and the plastron. An elongate **scapula** meets and articulates with the carapace dorsally. Fused to the scapula is the **acromial** (Gr. *akros,* summit, + *omos,* shoulder) **process** projecting anteriorly and at right angles to the scapula. Also lying at right angles to the two scapulas are the paddlelike **coracoid** (Gr. *korax,* crow, + *eidos,* form) **processes,** which are joined by cartilage at their medial margins. The pelvic girdle consists of the **ilia** (sing., **ilium**) attached to the ribs of two sacral vertebrae and, on either side, a broad **ischium** and **pubis.** The limbs of the turtle are stout but otherwise typical of reptiles and contain the same elements found in other vertebrate tetrapod pentadactyl limbs.

Internal Structure

Internal structure will be studied with preserved turtles.

Oral Cavity and Pharynx

☞ To open the mouth, first trim away some of the neck skin if necessary to expose the entire head. Cut through the temporal muscle on either side of the jaw, and then force a strong knife blade between the jaws. Force the jaws apart enough to

allow the angle of the jaws to be cut with heavy scissors or bone shears.

Open the mouth widely to expose the oral (buccal) cavity (Figure 18.2). The space enclosed by the jaws leads into the **pharynx,** a cavity extending from the posterior border of the tongue to the esophagus. Note the horny beaks covering the toothless jaws and the triangular **tongue** that is firmly attached throughout its length. The **glottis** is a narrow, slitlike opening posterior to the tongue that leads into the larynx. The external nares lead through the nasal passages to the **internal nares,** which open on the roof of the mouth. There is no secondary palate in turtles, but the roof of the mouth is vaulted and bears a pair of **palatal folds** that help direct the air toward the glottis. (The secondary palate, present in crocodilians and mammals, is a bony plate that completely separates the respiratory and food passages, thus providing a channel for the free passage of air from the nose to the pharynx; the secondary palate allows an animal to manipulate food in its mouth while breathing through its nose.) Locate the **eustachian (auditory) tubes** in the roof of the pharynx; these lead to the middle ear cavity on each side of the head. At the posterior margin of the pharynx is the **esophagus.**

Digestive System

☞ If not already done, cut through the bridges uniting the carapace and plastron with a bone saw or hacksaw. Strong bone shears are usually required to break through the anterior and posterior margins of these bony bridges to complete the separation. Remove the plastron by slicing carefully through

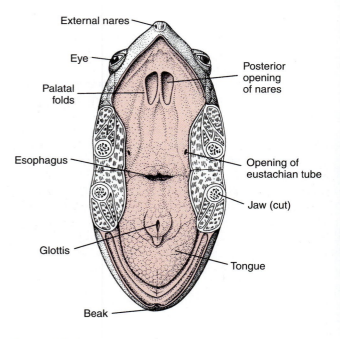

Figure 18.2
Oral cavity and pharynx of the painted turtle.

adhering tissue and muscle, holding the scalpel *close* to the undersurface of the plastron.

With the plastron removed, you will see much of the viscera covered by several muscles of the pectoral and pelvic girdles, with their origins on the plastron.

☞ Peel back and cut off the large, fan-shaped pectoralis major muscle near its insertion on the humerus. Cut through and remove the coracoid bone of the pectoral girdle and its muscles on both sides to expose the anterior viscera. Trim away the superficial muscles of the pelvic girdle.

The entire coelom is enclosed by a thin, tough, transparent membrane, the **parietal peritoneum.** Carefully trim away this membrane to reveal the **pleuroperitoneal cavity** containing most of the visceral organs. The peritoneum continues over the heart as the **pericardial sac,** which also should be trimmed away on the ventral surface to expose the heart.

☞ Break away the sides of the carapace with bone shears to expose more of the viscera. Wash out the pleuroperitoneal cavity with running water.

The brown **liver** extends across the body, passing dorsal to the heart (Figure 18.3). The left lobe covers the **stomach,** to which it is bound by a mesentery (gastrohepatic ligament). Use a probe or other blunt instrument to carefully free the stomach from the liver. The right lobe of the liver covers part of the **duodenum** (the initial segment of the small intestine) and part of the colon.

If the specimen is a female, the pleuroperitoneal cavity just posterior to the liver may be filled with yellow eggs of various sizes. The eggs are contained within the two **ovaries** and confined by a mesentery, the **mesovarium.**

Look beneath (dorsal to) the heart to find a tough membrane, the **transverse septum,** separating the pericardial cavity from the rest of the pleuroperitoneal cavity.

Now identify the components of the digestive system (Figure 18.3). Push the left lobe of the liver to

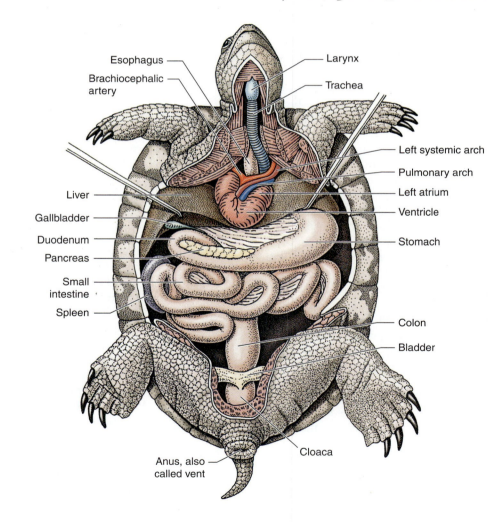

Figure 18.3
Internal anatomy of a painted turtle. In desert species, the bladder can be large and used as a water reservoir. Notice how tightly the organs sit in the body cavity. Why might turtles have physiological adaptations for dealing with extended periods of breath-holding? (Hint: how is the physiology impacted when its legs and head are withdrawn into the shell?)

the right and trace the ventral surface of the stomach forward to the **esophagus.** Following the gut posteriorly, find the duodenum; the **pancreas,** a slender, pale gland lying along the anterior border of the duodenum; and the **gallbladder,** a dark green sac embedded in the dorsal side of the right liver lobe. It may be necessary to break away the posterior margin of the right liver to expose the gallbladder. The gallbladder stores and concentrates bile from the liver; its green color is contributed by a variety of pigments present in the bile. The duodenum gives rise to the **small intestine** proper, which coils several times before emptying into the **colon** (large intestine). The colon terminates at the **cloaca,** which also receives ducts from the ovaries (oviduct) and kidneys (ureter, or metanephric duct).

Respiratory System

☞ With bone shears, cut through the acromial processes on both sides of the thoracic region and trim away the tissue in the neck region. Make a light midventral incision in the neck and retract tissue to expose the trachea.

Follow the **trachea** anteriorly, carefully scraping away muscle and connective tissue from the trachea and the hyoid bone beneath which the trachea disappears. Cut through the two horns of the hyoid bone on both sides, lateral to the trachea, and carefully free the hyoid from the underlying trachea and larynx (Figure 18.3).

The glottis, seen earlier, opens into the **larynx,** an enlarged vestibule that leads into the trachea. Note the complex of cartilages and associated muscles (the latter partly removed with the hyoid bone) that open and close the glottis. In many animals, the larynx contains vocal cords, but most turtles are voiceless, able to make little more than a hissing sound by rapidly expelling air from the lungs. If turtles cannot vocalize, how might they communicate with each other?

_____

_____

Follow the trachea posteriorly to the point where, with the esophagus, it drops middorsally between the large blood vessels emerging from the heart. At this point, the trachea divides into two **bronchi** (sing., **bronchus**). Trace the left bronchus to the lung. Reflect the left lobe of the liver and stomach to the right, and cut through the peritoneum to expose the **left lung.** The left bronchus will be seen lying between the left pulmonary artery and vein.

Slice open the lung with a lateral incision. The lung is basically saclike, although modestly subdivided into **alveolar pockets,** giving the inner wall a honeycomb appearance. The lung is mostly confined to a separate pleural cavity by an extension of the transverse septum, the **pleuropericardial membrane.** In mammals, these same membranes become incorporated into the muscular diaphragm. But turtles have no diaphragm and must employ other muscles to expand and compress the lungs. The volume of the body cavity is relatively fixed by the shell. What would happen to the internal volume if the turtle extended its head and legs? _____ _____ Would air come into the lungs or be pushed out? _____

Circulatory System. We will confine our study of the circulatory system to the heart and the major vessels entering and leaving it (Figure 18.3).

As noted earlier, the heart lies within a pericardial sac. Lift the ventricle to see the **sinus venosus,** which receives blood from the major **systemic veins,** the precaval and postcaval veins. Deoxygenated blood from the body drains from the sinus venosus into the **right atrium,** which empties into the right side of the ventricle. Oxygenated blood from the lungs enters the left atrium via **pulmonary veins** and is pumped from the left atrium into the left side of the heart.

The turtle ventricle is only partly divided by a septum (complete in the crocodilians), but very little admixture of oxygenated and deoxygenated blood occurs. From the ventricle, three large arterial trunks extend forward; these represent the subdivision of the conus arteriosus of fishes and amphibians.

1. The trunk farthest to the turtle's left is the **pulmonary arch,** which divides immediately into the right and left pulmonary arteries. The pulmonary arch takes deoxygenated blood from the right ventricular chamber and distributes it to the lungs.

2. The middle trunk is the **left systemic arch (left aorta),** which gives off branches to the stomach, pancreas, liver, and duodenum. It receives oxygenated blood from the left side of the ventricle.

3. The third trunk is the **right systemic arch (right aorta).** It is concealed from view by the **brachiocephalic** (Gr. _brachion,_ arm, + _kephale,_ head) **(innominate) artery,** a large branch that the right systemic arch gives off immediately after leaving the heart. The right systemic arch also receives oxygenated blood from the left ventricular chamber. Small **coronary arteries** spring from the base of the brachiocephalic artery and branch over the heart. The brachiocephalic divides at once into four arteries: **right** and **left subclavian arteries,** which supply the neck region, pectoral girdle, and shoulder muscles, and forelimbs; and **right** and **left carotid arteries,** which supply the head.

Although you will not trace the arteries farther, the left and right systemic arteries unite posteriorly in a V shape to form the **dorsal aorta.** This vessel continues posteriorly in the median dorsal line, giving off branches that supply most of the visceral organs and all of the muscles of the trunk and hindlimbs.

The Birds
Class Reptilia, Subclass Aves

EXERCISE 19

Pigeon

Core Study

Birds, described by the English zoologist Thomas Huxley as "glorified reptiles," do indeed bear the stamp of their reptilian heritage in many subtle ways. But birds have become so highly specialized into flying machines that the truth of Huxley's comment is not instantly evident when gazing casually at a bird.

The demands of flight, far more than anything else, have shaped the form and function of birds—and disguised their reptilian past. Virtually every adaptation found in flying birds focuses on two features: more power and less weight. Exploring these characteristics will be the central theme of this exercise.

Pigeon (Rock Dove)

Phylum Chordata
    Subphylum Vertebrata
        Class Reptilia
            Subclass Aves
                Order Columbiformes
                    Family Columbidae
                        Genus *Columba*
                            Species *Columba livia*

Where Found

The common pigeon (rock dove), so familiar to city dwellers, is of Old World origin but has been introduced throughout the world. It is not migratory, but it has excellent navigational proficiency. In fact, it is the homing pigeon that was used to carry messages in wars from the time of Caesar's conquest of Gaul through World War II. The pigeon is also one of the swiftest birds in flight. Many domesticated varieties have been developed over centuries of breeding; Charles Darwin was himself a pigeon fancier. The "city pigeon" is highly variable in coloration.

EXERCISE 19

Pigeon
  Pigeon (Rock Dove)

Feathers

☞ Examine a flight (contour) feather.

Identify the central **shaft** (also called the **rachis** [Gr. *rhachis,* spinel]), which is a continuation of the **quill** that is thrust into the feather follicle of the living bird. The shaft bears numerous **barbs,** which spread laterally to form the feather's expansive webbed surface, the **vane.**

Flex the feather in your hands, noting its resilience and toughness, despite its remarkably light weight. If you run your fingers down the vane toward the quill, you will separate some of the barbs that are normally linked together by tiny **barbules.** Note that considerable force is needed to separate the barbs. Because this happens in the course of the bird's daily activities, the bird spends time each day preening: zipping the barbs back together by drawing the feather through its bill. You can do the same with your fingers. Why would preening be so important for a bird? _____

☞ Examine a prepared slide of a contour feather, using a dissecting microscope or the low power of a compound microscope.

There are several types of feathers that may have been placed on display. **Contour feathers** (such as the one you just examined) give the bird its outward form; contour feathers used in flight are called **flight feathers. Down feathers** are soft tufts, lacking hooks, that are found on young birds or beneath the contour feathers of adult birds. Down feathers have excellent insulative value and function mainly to conserve heat. **Filoplume feathers** are hairlike feathers thought to play a sensory role. These feathers, consisting of a weak shaft with a tuft of short barbs at the tip, are the "hairs" visible on plucked fowl.

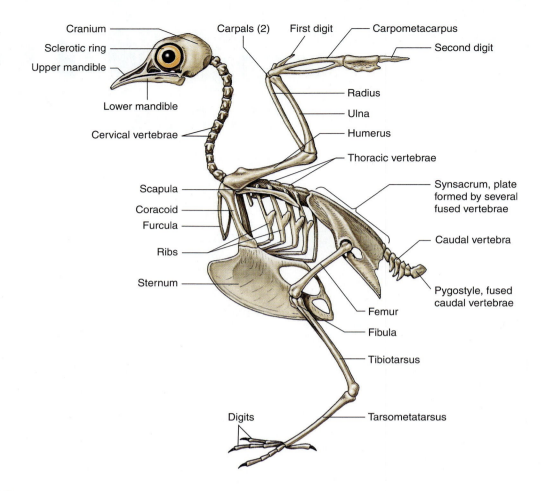

Figure 19.1

Skeleton of a pigeon. Why is the keel of the sternum so large? _____ What bone is formed by the fusion of 13 vertebrae? _____ What bone is sometimes called the "wishbone"? _____ How would you describe the fibula of the bird, as compared with that of other vertebrates (about the same, reduced, enlarged)? _____

Skeleton

☞ Study the mounted pigeon skeleton and unmounted bird bones on display.

The pigeon skeleton (Figure 19.1), like that of other birds, is a marvel of lightness combined with strength. It is even lighter than it looks because many of the limb and girdle bones are hollow. If a long bone such as the humerus has been broken or cut open for examination, note its tubular form and the internal struts that have developed where stresses must be borne. Many of the bird's bones are "pneumatized"—that is, penetrated by extensions of the air sac system—and thus they contain buoyant warm air rather than the bone marrow typical of mammalian bones. (Some bird bones do contain marrow, but much of a bird's red and white blood cell production occurs in the spleen and liver.)

A striking feature of the bird skeleton is its rigidity. Of the axial skeleton, only the neck remains flexible. The remaining vertebrae are fused together and with the pelvic girdle to form a stiff, boxlike framework to support the legs and provide rigidity for flight. The double-headed **ribs** are also mostly fused with the **thoracic vertebrae** and with the **sternum.** Unlike mammals, nearly all of which have 7 cervical vertebrae, the number of **cervical vertebrae** in birds varies from 8 to 24, depending on the species (long-necked birds have the most). Note the complex articulations of the cervical vertebrae of this most flexible part of the pigeon's body. How many cervical vertebrae does the pigeon have?

_____

The last rib-bearing thoracic vertebra is fused with five **lumbar,** two **sacral,** and five **caudal** vertebrae to form a thin, platelike structure, the **synsacrum.** The **ilium** of the pelvic girdle is also fused with the synsacrum. This very light, but stout, arrangement provides further rigidity to the body frame. Finally, note the short bony tail, consisting of five free **caudal vertebrae** that are followed by four caudal vertebrae fused into a **pygostyle,** which supports the tail feathers.

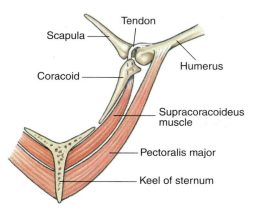

Figure 19.2
The major flight muscles of a bird are arranged to keep the center of gravity low in the body. Both the supracoracoideus and the pectoralis are anchored on the sternum keel. Contraction of the pectoralis muscle pulls the wing downward. As the pectoralis relaxes, the supracoracoideus muscle contracts and, acting as a pulley system, pulls the wing upward.

The pigeon skull is composed of individual bones that are completely united in the adult to form a single thin-walled, lightweight structure. Birds are descended from archosaurian reptiles, which belong to the diapsid lineage of amniotes; this lineage is characterized by skulls having two openings, or fenestrae, in the temporal region. Birds are so highly specialized, however, that it is difficult to see any trace of diapsid origin in their skulls. The large, bulging **cranium** encloses the brain, which is much larger, relative to body size, than the brain of a turtle because the complex movements of bird flight require far more central nervous coordination than does the plodding locomotion of a turtle. The pigeon's large eyes are housed in sockets and encircled in front by a protective ring of shingle-like bony plates, the **sclerotic ring.** The **beak** consists of a **lower mandible** hinged to the skull in a way that provides wide-gaping action. The **upper mandible** (also called the maxilla) is fused to the skull in pigeons, but some birds (parrots, for example) have kinetic skulls with movable bony elements that allow the upper mandible to tilt upward when the bird opens its mouth.

We will turn our attention now to the appendicular skeleton. Examine the pectoral girdle, composed of a tripod of paired bones: **scapula, coracoid,** and **furcula** ("wishbone"). The scapula is a thin, bladelike bone tied to the ribs by ligaments. The stout coracoid bone unites the scapula and the **sternum.** Describe the sternum _____ Why is it so large? _____
The muscles that depress the wing (pectoralis) and those that raise the wing (supracoracoideus) are both attached to the sternum. Where the scapula and coracoid unite, there is a hollow depression into which the ball of the chief bone of the wing, the **humerus,** fits. The supracoracoideus is attached by a tendon to the upper side of the humerus so that it pulls from below by an ingenious "rope-and-pulley" arrangement (Figure 19.2). In this way, muscle weight is kept below the center of gravity, providing greater flight stability.

The pelvic girdle, as we have seen, is a fused structure that is almost paper thin but is strengthened by bony ridges that can be seen by looking at the underside of the girdle. The **femur** (thighbone) is directed forward and is virtually buried in the flesh of the living bird. The **tibia** is the main bone of the shank ("drumstick"); the **fibula** is reduced to a thin splint. The ankle is greatly modified. Some of the pebblelike tarsal bones of the tetrapod limb are united with the tibia, and the others are fused with the metatarsals to form a single elongate **tarsometatarsus.** The pigeon, like most other birds, has four digits: three directed forward and the fourth directed backward.

The bones of the forelimbs are highly modified for flight. Note how the wing folds into a compact **Z** shape when the bird is at rest. Identify the **humerus,** and locate the expanded dorsal surface for the attachment of the pectoral muscles. The **radius** and **ulna** are longer than the humerus, and the ulna, the larger of the two, carries the secondary flight feathers. Most modified are the wrist and digits, which carry the primary flight feathers. Identify the two **carpals** (wrist bones) and the two elongate **carpometacarpals** (palm bones), so called because they are formed by the fusion of three carpal and three metacarpal bones. There are only three **digits** (fingers). The first digit, or "thumb," carries the feathers of the **alula.** The second finger is by far the largest; this, together with the palm bones, carries the primary flight feathers. The third digit, like the first, is reduced to a small bone (not shown in Figure 19.1) that carries a single outermost flight feather.

The Mammals
Fetal Pig

Fetal Pig

Class Mammalia
   Subclass Theria
      Infraclass Eutheria
         Order Artiodactyla
            Genus *Sus*
               Species *Sus domesticus*

Mammalia are those animals whose young are nourished by milk from the breasts of the mother. Mammals have a muscular diaphragm, a structure found in no other class, and a four-chambered heart. Most of them are covered with hair. Their nervous system is especially well developed. Their eggs develop in a uterus, with placental attachment for nourishment (except in the monotremes, which lay eggs, and in the marsupials, which have only a weakly and briefly developed placental attachment).

The order Artiodactyla includes the even-toed, hoofed mammals, such as deer, sheep, cattle, and camels. These usually have two toes, but some, such as hippopotamuses and pigs, have four toes.

Fetal pigs are an especially desirable laboratory example of a mammal. They are easy to obtain, relatively inexpensive, and easily stored in individual plastic bags. Fetal pigs are obtained from the uteri of sows slaughtered for market. Because they are unborn, their bones are still largely cartilaginous, which makes the specimens pliable and easy to handle. They have an umbilical cord by which they were attached to the placenta in the uterus.

The embryo depends on maternal blood to bring it nutrients and oxygen and to carry off the waste products of metabolism because its own organs cannot serve these functions until birth. This exchange of materials between fetal blood and maternal blood takes place within the placenta of the mother's uterus. The difference between the fetus and the adult is largely physiological, but there are also a few morphological differences, especially in the circulatory system, which you will observe in your specimen.

The period of gestation in a pig is 16 to 17 weeks, compared with 20 days in a rat, 8 weeks in a cat, 9 months in a human being, 11 months in a horse, and 22 months in an elephant. Pig litters average 7 to 12 piglets, but may have as many as 18. The pigs are about 30 cm long at birth

---

Fetal Pig
External Structure

EXERCISE 20A
Skeleton
  Axial Skeleton
  Appendicular Skeleton

EXERCISE 20B
Muscular System
  Organization of Skeletal Muscles
  Dissection of Fetal Pig Muscles

EXERCISE 20C
Digestive System
  Head and Throat
  Abdominal Cavity
  Thoracic Cavity and Neck Region
  Digestive Tract

EXERCISE 20D
Urogenital System
  Urinary System
  Male Reproductive System
  Female Reproductive System

EXERCISE 20E
Circulatory System
  Heart
  General Plan of Circulation

---

and weigh from 1 to 1.5 kg (about 2 to 3 pounds). The age of a fetus may be estimated from the length of its body:

At 3 weeks, the fetus is about 1.3 cm long.
At 7 weeks, the fetus is about 3.8 cm long.
At 14 weeks, the fetus is about 23 cm long.
At full term, the fetus is about 30 cm long.

External Structure

☞ Before proceeding with the regular exercises, look at the external structure of your pig.

On the head, locate the **mouth** with its fleshy lips and the **nostrils** at the tip of the snout. The snout has a tough rim for rooting and bears **vibrissae,** stiff sensory hairs (whiskers). Each eye has two lids and a small

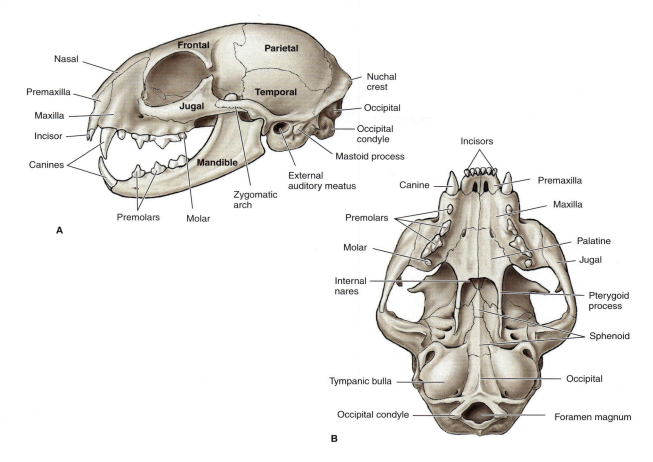

Figure 20.1

A, Skull of a cat, lateral view. **B,** Skull of a cat, ventral view.

membrane in the medial corner, which represents the nictitating membrane. The fleshy pinnae, or ear flaps, contain the external auditory opening.

On the trunk, locate the **thorax,** supported by ribs, sternum, and shoulder girdle with forelimbs attached; the **abdomen,** supported by a vertebral column and muscular walls; the **sacral** region, comprising the pelvic girdle with hindlimbs attached; the **umbilical cord;** five to eight pairs of **mammae,** or nipples, on the abdomen; and the **anus** at the base of the tail.

What is the sex of your specimen? _____ In the **male,** the **urogenital opening** is just posterior to the umbilical cord; the **scrotal sacs** form two swellings at the posterior end of the body (the **penis** can sometimes be felt under the skin as a long thin cord passing from the urogenital opening back between the hindlegs). In the **female,** the urogenital opening is just ventral to the anus and has a fleshy tubercle projecting from it.

Examine the cut end of the umbilical cord. Note the ends of four tubes in the cord. These represent an umbilical vein, two umbilical arteries, and an allantoic duct, all of which during fetal life are concerned with transporting food, oxygen, and waste products to or from the placenta of the mother's uterus.

Notice that the entire body of the fetal pig is covered with a thin cuticle called the **periderm.**

EXERCISE 20A
Skeleton
Core Study

The skeleton of an adult pig, although different in size and proportion, is nonetheless quite similar to the skeletons of other mammals, such as a dog, a cat, or a human. The bones are homologous, and the origins and insertions of muscles are usually comparable.

Because of its immature condition, the skeleton of a fetal pig is unsuitable for classroom study. However, if you are planning to dissect the muscles of a fetal pig, it is essential to be familiar with the bones. Skeletons of a cat or dog can be used quite satisfactorily for this purpose. Figures 20.1 and 20.2 depicting a cat skeleton will help you identify the various bones.

☞ As you study the mounted skeleton of the dog or cat, compare the parts with those of the fetal pig skeleton (Figure 20.3). Then try to locate these parts and

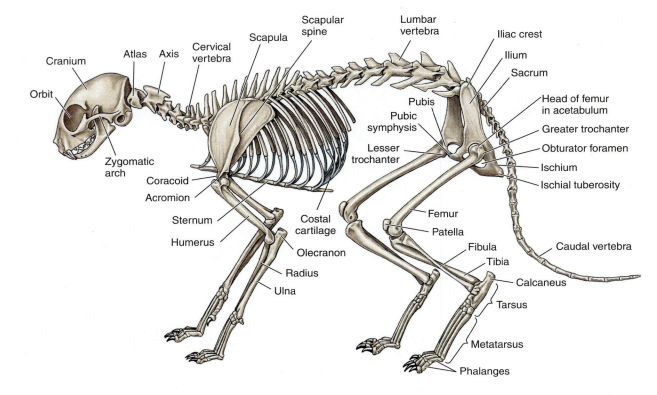

Figure 20.2

Skeleton of a cat.

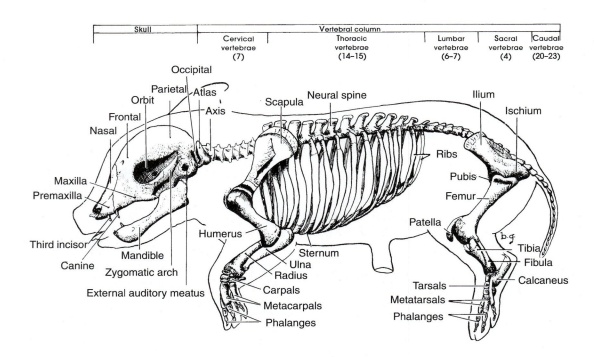

Figure 20.3

Skeleton of a fetal pig.

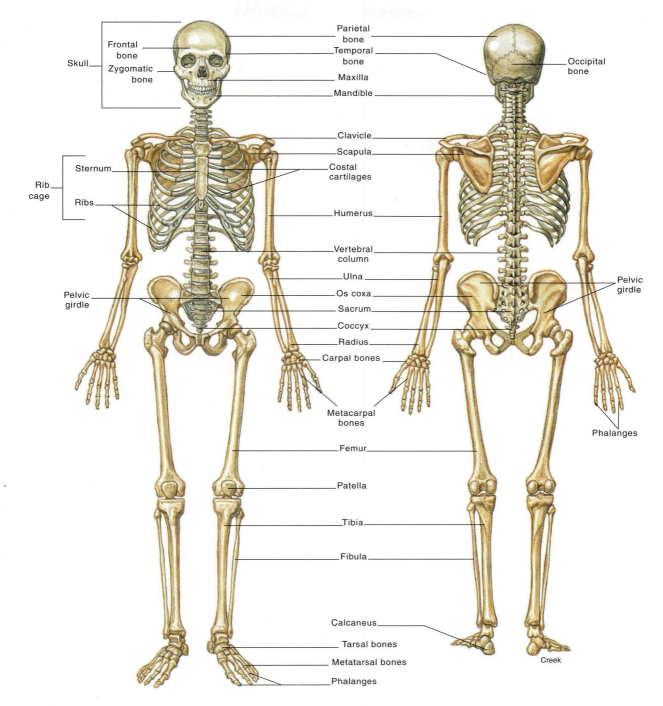

Figure 20.4

Human skeleton.

visualize their relationships within the flesh of the preserved pig. As you work, notice the similarities with the human skeleton (Figure 20.4).

As in a frog, the pig skeleton can be divided into the **axial skeleton** (skull, vertebral column, ribs, and sternum) and the **appendicular skeleton** (pectoral and pelvic girdles and their appendages).

Axial Skeleton

Skull

The skull can be divided into a **facial region,** containing the bones of the eyes, nose, and jaws, and a **cranial region,** which houses the brain and ears. The smooth, rounded occipital condyles (Gr. *kondylos,* knuckle) of the skull articulate with the ring-shaped first cervical vertebra (called the **atlas,** for Atlas of Greek mythology who was condemned to hold the heavens on his shoulders for all eternity). The foramen magnum ("great opening") is the opening at the posterior end of the braincase where the spinal cord emerges. Many of the important bones of the skull can be identified with the help of Figure 20.1.

Vertebral Column

Note the five types of vertebrae: **cervical,** in the neck; **thoracic,** bearing the ribs; **lumbar,** without ribs but with large transverse processes; **sacral,** fused together to form a point of attachment for the pelvic girdle; and **caudal,** the tail vertebrae. In humans, three to five vestigial caudal vertebrae are fused to form the coccyx of the tail bone. All of the vertebrae are built on the same general plan but with recognizable individual differences (Figures 20.2 and 20.3).

Ribs

Observe the structure of a rib and its articulation with a vertebra. Each rib articulates with both the body of the vertebra and a transverse process. The **shaft** of the rib ends in a **costal** (L. *costa,* rib) **cartilage,** which attaches to the sternum or to another costal cartilage and so indirectly to the sternum. The cat has one pair of free, or floating, ribs. The pig has 14 or 15 pairs of ribs, of which 7 pairs attach directly to the sternum and 7 or 8 attach indirectly. How many does the human have? _____

Sternum

The sternum is composed of a number of ossified segments, the first of which is called the **manubrium,** and the last of which is called the **xiphisternum.** Those in between make up the **body** of the sternum.

Appendicular Skeleton
Pectoral Girdle and Its Appendages

The pectoral girdle comprises a pair of triangular **scapulae,** each with a lateral **spine** and a **glenoid fossa** at the ventral point for attachment with the head of the humerus. The **forelimb** (Figures 20.2 to 20.4) includes (1) the **humerus;** (2) the two forearm bones—a shorter, more medial **radius** and a longer, more lateral **ulna,** with an **olecranon** (ol-ek′re-non) **process** at the proximal, or elbow, end; (3) the **carpus,** consisting of two rows of small bones; (4) the **metacarpals** (five in the cat and human and four in the pig); and (5) the **digits,** or toes, made up of **phalanges.**

Pelvic Girdle and Its Appendages

The pelvic girdle in adult mammals consists of a pair of **innominate** (L., without name) **bones,** each formed by the fusion of the **ilium** (L., flank), the **ischium** (Gr. *ischion,* hip), and the **pubis** (L., mature). A lateral cavity, the **acetabulum** (L., vinegar cup), accepts the head of the femur. The pubic bones and the ischial bones on opposite sides unite at **symphyses,** and the ilia articulate with the sacrum so that the innominates and the sacrum together form a complete ring, or **pelvic canal.** Each pelvic appendage includes (1) the **femur,** or thigh bone; (2) the larger **tibia** and more slender **fibula** of the shank; (3) the ankle, or **tarsus,** comprising seven bones, and the fibular tarsal bone (**calcaneus;** cal-ka′nee-us; L. *calx,* heel), forming the projecting heel bone; (4) the **metatarsals** (five in the cat, of which the first is very small, and four in the pig); and (5) the **digits,** or toes (four in both the cat and the pig), composed of **phalanges.**

Structure of a Long Bone

Longitudinal and transverse sections through a long bone (Figure 20.5) show that it consists of a shell of **compact bone** surrounding a type of **spongy bone (cancellous bone),** the spaces of which are filled with **marrow.** The shaft (also called the **diaphysis;** di-af′-uh-sis; Gr. *dia,* through, + *phyein,* to grow) is usually hollowed to form a **marrow cavity.** In the young animal, there is only red marrow, a blood-forming substance, but this is gradually replaced in the adult with yellow marrow, which is much like adipose tissue. The ends of the bone, or **epiphyses** (ee-pif′uh-sees; Gr. *epi,* upon, + *phyein,* to grow), usually bear a layer of **articular cartilage.** The rest of the bone is covered with a membrane, the **periosteum** (pear-ee-os′te-um; Gr. *peri,* around, + *ostrakon,* shell). Arteries, veins, nerves, and lymphatics pass through the compact bone to supply the marrow and cells responsible for bone formation and maintenance.

Further Study
Growth of a Bone

The skeleton of the embryo consists of cartilage and fibrous tissue, in which the bones develop by a process of ossification. Bones that develop in fibrous tissue—namely, some of the bones of the cranium and face—are called membranous bones. Most of the bones of the body develop from cartilage and are designated as endochondral ("within cartilage") bones.

A typical long bone usually has three primary centers of ossification—one for the diaphysis, or shaft, and one for each epiphysis, or extremity (Figure 20.5). As long as this cartilage persists and grows, new bone may form, and the length may increase.

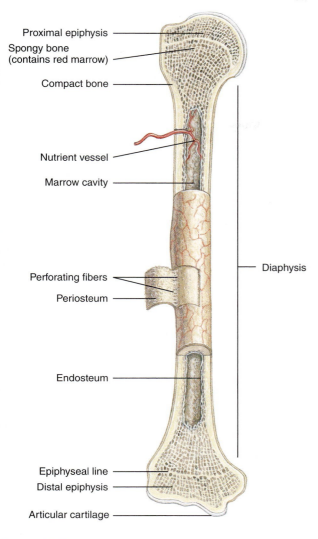

Proximal epiphysis

Spongy bone
(contains red marrow)

Compact bone

Nutrient vessel

Marrow cavity

Diaphysis

Perforating fibers

Periosteum

Endosteum

Epiphyseal line

Distal epiphysis

Articular cartilage

Figure 20.5
Diagram of longitudinal section of a long bone.

Articulations

An articulation, or joint, is the union of two or more bones or cartilages by another tissue, usually fibrous tissue or cartilage or a combination of the two.

Three types of joints are recognized:

A **synarthrosis** (sin-ar-thro′sis; Gr. *syn,* with, + *orthron,* joint) is an immovable joint. Interlocking margins of the bones are united by fibrous tissue. Example: sutures of the skull.

A **diarthrosis** (Gr. *dis,* twice, + *arthron,* joint) is a movable joint. The ends of articulating bones are covered with cartilage and enclosed in a joint capsule of fibrous tissue. The capsule contains a joint cavity lined with a vascular synovial membrane that secretes a lubricating fluid. Example: most of the joints—knee, elbow, and others.

An **amphiarthrosis** (am-fee-ar-thro′sis; Gr. *amphi,* both, + *arthron,* joint) is a slightly movable joint. The bones are joined by a flattened disc of fibrocartilage and bound together by ligaments. These tough bands or sheets are composed mostly of white fibrous tissue and are pliable but not elastic (except for the nuchal ligament at the back of the neck). Examples: the pubic symphysis and the joints between the vertebrae.

EXERCISE 20B
Muscular System

Because the muscles of a fetal pig are softer and the separations of the muscles less evident than in a lean cat or frog, the pig has been less frequently used for muscle dissection in beginning classes. However, if full-term pigs (30 cm or more in length) are used, and careful attention is given to the dissection, even beginning students can demonstrate a great many of the muscles along with their origins and insertions. Uninjected pigs are quite satisfactory for this work; in fact, in some ways they are easier to work with than injected pigs.

The human being has approximately 700 identified and named skeletal muscles. A pig probably has fewer (since it lacks our five-fingered manual dexterity), but still possesses hundreds of muscles. Here only about 40 of the largest and most conspicuous will be studied. Even so, many introductory courses lack sufficient laboratory time in which to dissect these 40, so your instructor may be selective, or some of the dissection may be done on an extra-credit basis.

Organization of Skeletal Muscles

Skeletal muscle is under voluntary control—that is, it is innervated by motor fibers. Skeletal muscle can contract effectively to about 30% of its resting length. However, the force that a muscle develops is not the same throughout its shortening length. If a muscle is stretched out fully, little force can be developed. Similarly, when a muscle is fully contracted, its force again diminishes. You know from your own experience that it is difficult to lift a heavy object with your forearm fully extended, but it becomes easier as your arm approaches a 90° angle. Again, when the muscle is maximally shortened, the force you can exert declines.

Functional Bundles

Muscle fibers are bound together by a fibrous connective tissue called **fascia** (fa′shē-uh; pl., **fasciae,** fa′shē-ē; L., bundle) into bundles called **fasciculi** (fa-sick′yu-li; L., *fasciculus,* small bundle). The fasciculi are in turn organized in various ways into an entire muscle. The most common type of muscle is the **parallel muscle,** in which the fasciculi are arranged side by side parallel to the long axis of the muscle, but other muscles are arranged in sheets with broad attachments; still others are arranged in a circle, such as those that form sphincters around orifices.

How Muscles Are Connected

Skeletal muscles are connected to cartilage, bone, ligaments, or skin either directly by their investing fascia or indirectly by means of tendons or aponeuroses, never by the actual muscle fibers themselves. If the muscle fibers come very close to the bone, we say the muscle has a "fleshy attachment." A **tendon** is a narrow band of tough, fibrous connective tissue, and an **aponeurosis** (Gr. *apo,* from, + *neuron,* sinew) is a broad, thin sheet of tough connective tissue that connects the muscle to its place of attachment.

Origin, Insertion, and Action. A muscle begins at its **origin,** which in general is the stationary end. It ends at its **insertion,** which is the end that moves. The movement produced by the muscle is its **action.** For example, the biceps brachii of the human arm originates on the scapula and inserts on the radius of the forearm. Its action is to flex the forearm. Sometimes, these rules for origin and insertion do not easily apply. If, for example, a muscle extends between a broad aponeurosis and a narrow tendon, the aponeurosis is considered the origin and the tendon the insertion. Some muscles have several tendons at one end and only one at the other end; in this case, the muscle has multiple origins but only one insertion. The multiple origins in such instances are called **heads.** Even these rules will not apply to every muscle, and in the end it is better to know what a muscle does (its action) than to try to know every origin and insertion.

Naming the Actions

Anatomists have long used special terms to describe dynamic motion; such terms prevent having to use complicated notations such as "bends the foreleg toward the body" or "raises the arm toward the shoulder." **Flexion** (flek′shun) moves a distal part of a limb toward the next proximal part—for example, bending the fingers or the elbow. Flexion may also refer to bending the head or trunk toward the ventral surface. **Extension** is the opposite movement; it *increases* the angle between articulating elements. Straightening the arm or the fingers is an extension movement. Muscles causing these movements are called **flexors** and **extensors.**

Adduction (L. *ad,* to, + *ducere,* to lead) refers to a movement of the distal end of a bone—for example, the humerus or the femur—that brings it closer to the ventral median line of the body. In **abduction** (L. *ab,* from), such a bone is moved farther away from the ventral median line, as in raising the arm out to one side. Muscles that cause these movements are **adductors** and **abductors,** respectively. (It will be helpful if you can remember the meanings of the prefixes *ad* [meaning *to*] and *ab* [meaning *from*]).

Protractors in a tetrapod (an animal with four legs) may be defined as muscles that move the distal ends of bones—for example, the femur or humerus—forward longitudinally; **retractors** move them backward. A **depressor** may cause a part, such as the mandible or an eyelid, to be lowered; a **levator** would raise such a part. **Rotation** is the turning of a part—for example, the rotation of the radius on the ulna or the first vertebra (atlas) on the second (axis). Human anatomists use several other muscle action terms as well (for example, inversion, eversion, pronation, and supination), but learning the actions defined here will suffice for our study of the fetal pig.

Naming the Muscles

The names of muscles are descriptive and may be derived from one or more of the following characteristics: (1) **position** (the brachialis is an arm muscle, the pectoralis is a chest muscle, and the cutaneous lies just under the skin); (2) **action** (adductor longus, depressor rostri); (3) **shape** (deltoid); (4) **direction** (transversus abdominis); (5) **number of divisions** (the biceps is two-headed, the digastric is two-bellied); and (6) **attachments** (the sternomastoid is attached to the sternum and to the mastoid process of the skull). Many names combine two or more of these descriptive elements. For example, the extensor carpi obliquus is an extensor attached to the carpus, which it extends, and its fibers run in an oblique direction.

Dissection of Fetal Pig Muscles

Before beginning to dissect the muscles, be sure you are familiar with the external structure of the animal as given in the introduction to this chapter (see pp. 248–249).

It is absolutely essential to study bones and muscles together as functional units. The shape, attachments, and actions of the muscles have meaning only in connection with the bones they cover, hold in place, and move. Refer often to the skeleton as you dissect the muscles. Feel for and identify the bones underlying the muscles you are dissecting.

If your pig has been injected, there will be an incision in the neck through which one of the jugular veins was injected. *Use the side opposite this incision to dissect the lateral muscles.* In the neck region, as you dissect the musculature, refer also to the exercise on the salivary glands (see p. 262), and identify the glands and blood vessels mentioned there and shown in Figure 20.11 (see p. 262).

The animal must first be skinned. Make a longitudinal middorsal incision through the skin from head to tail (making sure the cut is through the skin only). Make a midventral incision from chin to groin and circular incisions through the skin around the neck, the chest, and the groin. Then, starting with a corner of skin in the dorsal neck region, lift the skin with forceps and use a probe or the *blunt edge* of the scalpel to push back the muscle underneath. Continue removing pieces of skin until you have uncovered the entire body and legs (unless the instructor asks you to do only one side or only the forequarter).

The outer layer of muscle is a thin, superficial layer of **cutaneous muscle,** many fibers of which are attached directly to the skin. *Try not to remove this muscle layer with the skin.*

After the pig has been skinned, there is usually a great deal of fat and connective tissue still covering the muscles. Much of this can be removed by gentle rubbing with a paper towel or by careful scraping with a scalpel.

Identify the outer layer of cutaneous muscle. In the neck and face region, this cutaneous layer is called the **platysma** (pla-tiz'ma; Gr., flat piece); in the trunk region, it is called the **cutaneous maximus.** A pig uses these muscles in twitching the skin to shake off insects, dirt, or other irritants. This thin layer of muscle is *not*

shown in the illustrations. After identifying this layer, *remove it carefully* to identify the superficial muscles underneath. Remember that these muscles are very thin; remove them carefully so as not to destroy the other muscles you are to identify.

To locate the borders of muscles, scrape off the overlying connective tissue and fascia and look for the direction of the muscle fibers. A muscle edge may be seen where the fibers change direction. Try to slip the flat handle of the scalpel between the layers of muscle at this point. *Do not cut the muscles or tear them with a dissecting needle.* Try to loosen each muscle and find out where it is attached but *do not cut* a muscle unless instructed to do so. When you are told to cut a muscle to locate deeper muscles, cut through the belly of the muscle but leave the ends attached for identification.

Figure 20.6 shows the more superficial muscles after removal of the cutaneous and platysma layer.

If, after skinning the animal, you find the muscles are still too soft to separate, exposure to air will help harden them in a few hours. Dipping the pig in preservative or sponging a little preservative over its surface and keeping it overnight in a plastic bag should make it easier to handle.

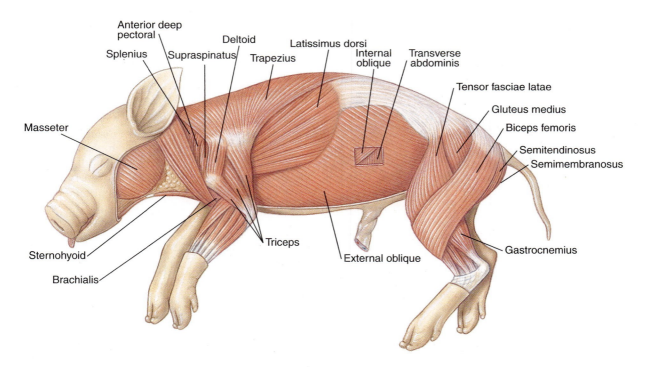

Figure 20.6
Superficial muscles of the fetal pig, lateral view. What muscle is most responsible for chewing? _____ Clench your teeth and feel your own muscle that closes your jaw. What muscle elevates the shoulder? _____ What anterior superficial thigh muscle flexes the hip joint and extends the knee? _____

The following discussion loosely divides the muscles into two groups—muscles of the forequarter and muscles of the hindquarter. The first group contains some of the muscles of the face and neck as well as those of the shoulder, chest, and forelimb. The second group includes muscles of the back, abdomen, and hindlimb. This is not an exhaustive list, but it will include the chief superficial muscles and many of the muscles of the second layer.

Muscles of the Forequarter

Muscles of the Face, Neck, Chest, and Shoulder (Table 20.1). Beginning with the throat muscles, locate the most ventral pair, the **sternohyoids,** marking the ventral midline of the neck and covering the larynx (Figure 20.7). These muscles retract and depress the hyoid and the base of the tongue, as in swallowing. Immediately internal to the sternohyoid on each side is the long **sternothyroid,** which retracts the larynx. Note that each has a lateral and a medial branch; separate these muscles to locate the small, dark, compact **thyroid gland** lying on the ventral side of the trachea.

The **digastric** muscle is the major depressor of the mandible—that is, it acts to open the jaw. It originates by a strong tendon from the base of the skull. Stretching ventrally between the mandibles as a thin transverse sheet is the **mylohyoid,** which compresses the floor of the mouth and assists in swallowing. The large muscle of the cheek is the **masseter.** This muscle and another (the temporal, not shown in Figure 20.8) are the major muscles that elevate the jaw and close the mouth.

TABLE 20.1			
Muscles of the Forequarter			
Muscle	**Origin**	**Insertion**	**Action**
Muscles of the Face, Neck, Chest, and Shoulder			
Sternohyoid	Anterior end of the sternum	Hyoid bone	Retracts and depresses hyoid and base of tongue as in swallowing
Sternothyroid	Sternum	Larynx	Retracts larynx
Digastric	By a tendon from the mastoid process of the skull	Medial surface of mandible	Depresses mandible
Mylohyoid	Medial surface of the mandibles	Hyoid bone	Raises the floor of mouth and hyoid bone
Masseter	Zygomatic arch	Lateral surface of mandible	Elevates jaw and closes mouth
Sternocephalic	Anterior end of sternum	Mastoid process of skull	Turns head; the two muscles together depress head
Brachiocephalic	Two origins: nuchal crest and mastoid process of skull	Proximal end of humerus and fascia of shoulder	Singly, inclines head; when head is fixed, draws limb forward; together, extend head
Superficial pectoral	Sternum	By a broad aponeurosis on medial surface of humerus	Adducts humerus
Posterior deep pectoral	Posterior half of sternum and cartilages of fourth to ninth ribs	Proximal end of humerus	Retracts and adducts forelimb
Anterior deep pectoral	Anterior part of sternum	Scapular fascia and aponeurosis that covers dorsal end of supraspinatus	Adducts and retracts limb
Trapezius	Nuchal crest of skull and neural spines of first 10 thoracic vertebrae	Spine of scapula	Elevates shoulder
Latissimus dorsi	Some of the thoracic and lumbar vertebrae and the four ribs preceding the last rib	Medial surface of humerus	Draws humerus upward and backward and flexes shoulder
Deltoid	Scapular aponeurosis	By an aponeurosis on proximal end of humerus	Flexes shoulder and abducts arm

TABLE 20.1 (*continued*)

Muscles of the Forequarter

Muscle	Origin	Insertion	Action
Muscles of the Face, Neck, Chest, and Shoulder			
Rhomboideus	Second cervical to ninth or tenth thoracic vertebrae	Medial surface of dorsal border of scapula	Draws scapula mediodorsally or rotates it
Rhomboideus capitis	Occipital bone	Dorsal border of scapula	Draws scapula forward and rotates shoulder
Splenius	First four or five thoracic neural spines	Occipital and temporal bones and first few cervical vertebrae	Singly, inclines head and neck to one side; together, elevate head and neck
Ventral serratus	Cervical part on transverse processes of last four or five cervical vertebrae; thoracic part on lateral surfaces of the last eight or nine ribs	Medial surface of scapula	Singly, cervical part draws shoulder forward and thoracic part backward; together, shift weight to the limb of contracting side; both sides together form an elastic support that suspends trunk between scapulas; raise thorax
Supraspinatus	Anterior and dorsal portion of scapula and scapular spine	Proximal end of humerus	Extends humerus
Infraspinatus	Lateral surface and spine of scapula	Lateral surface of proximal end of humerus	Abducts and rotates forelimb
Muscles of the Foreleg			
Triceps brachii	Long head: posterior border of scapula Lateral head: lateral side of proximal end of humerus Medial head: medial surface of proximal end of humerus, covering insertion of teres major	All three heads insert on the medial and lateral surfaces of olecranon process of ulna	Extend forearm
Brachialis	Proximal third of humerus, ventral to lateral head of triceps	Medial surface of distal end of radius and ulna	Flexes elbow
Biceps brachii	Ventral surface of scapula near glenoid fossa	Proximal ends of radius and ulna	Flexes elbow

Locate the **sternocephalic,** a flat muscle band that passes diagonally across the throat posterior to the submaxillary gland and beneath the parotid gland; it turns the head. When both sternocephalics contract together, the head is depressed. Posterior to the sternocephalic is the **brachiocephalic,** a large, band-shaped muscle that originates on the skull and inserts on the shoulder. This muscle raises the head, or if the head is fixed in position by other muscles, it draws the forelimb forward. It originates on two different processes (nuchal crest and mastoid process) of the skull.

Now examine the muscles that position the shoulder girdle and move the forelimb (Figure 20.7). The **superficial pectoral** (equivalent to the pectoralis major of humans) adducts the humerus. The **posterior deep pectoral** and the **anterior deep pectoral** both retract and adduct the forelimb. The most superficial muscle of the back (after removal of the cutaneous maximus) is the thin, triangular **trapezius** (Figure 20.8). This muscle elevates the shoulder and moves the scapula. The **latissimus dorsi** (see Figure 20.6) is a large, broad muscle that fans out across the back; it flexes the shoulder and is a major retractor of the humerus.

Now carefully cut through the trapezius at its insertion on the spine of the scapula and separate it from the

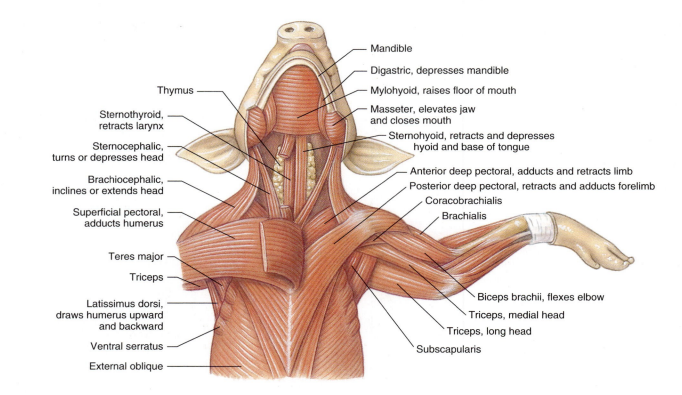

Figure 20.7
Muscles of the ventral thoracic region of a fetal pig. The pectoral muscle has been removed from the pig's left side to reveal the underlying musculature.

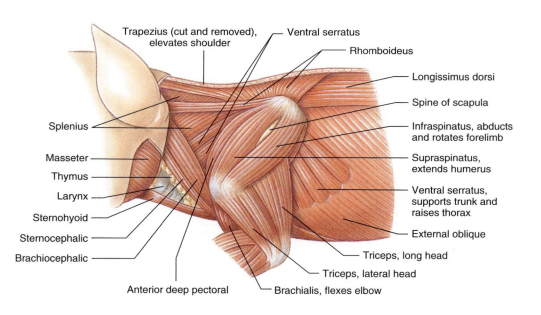

Figure 20.8
Muscles of the neck and shoulder of a fetal pig, lateral view. The trapezius muscle has been removed to reveal underlying musculature.

underlying muscles. A triangular muscle just beneath the trapezius is the **rhomboideus.** Anterior to this is the straplike **rhomboideus capitis.** Both of these muscles act upon the scapula to rotate it or draw it forward or dor-

sally. Beneath the rhomboideus capitis, locate the triangular **splenius,** which helps elevate and turn the head.

Cut the latissimus dorsi at its origin along the spine and remove most of the muscle, leaving only a centimeter or two at its insertion on the humerus. Locate the **ventral serratus** (Figure 20.8), an extensive, fan-shaped chest muscle that originates on the cervical vertebrae and several ribs and inserts on the scapula beneath the insertion of the rhomboideus muscles. It acts to shift the scapula forward and backward and serves as a muscular support to sling the weight of the trunk.

Arising near the insertion of the trapezius on the scapula is the **deltoid.** It inserts on the humerus and acts to protract (move forward) the upper foreleg. Carefully trim away the deltoid muscle at both the origin and the insertion. Beneath it lies the **infraspinatus,** which abducts and rotates the forelimb. Anterior to this is the **supraspinatus,** a fleshy muscle on the anterior surface of the scapula; it acts to extend the humerus.

Muscles of the Foreleg (Table 20.1). The largest muscle of the foreleg is the **triceps brachii,** an extensor of the forearm (Figures 20.7 and 20.8). It arises from three heads: a triangular-shaped long head at the posterior border of the scapula, a lateral head from the lateral surface of the humerus, and a medial head from the medial surface of the humerus. Two smaller muscles lying on the anterior and ventral surfaces of the humerus are the **brachialis** and the **biceps brachii;** both act upon the elbow to flex the forearm.

Muscles of the Hindquarter

Muscles of the Abdomen, Back, and Hip (Table 20.2). Three thin sheets of muscle lie in the lateral abdominal wall. The most superficial is the **external oblique** lying immediately beneath the cutaneous layer of

TABLE 20.2			
Muscles of the Hindquarter			
Muscle	**Origin**	**Insertion**	**Action**
Muscles of the Abdomen, Back, and Hip			
External oblique	Lateral surface of last 9 or 10 ribs and lumbodorsal fascia	Linea alba, ilium, and femoral fascia	Compress abdomen, arch back; singly, it flexes trunk laterally
Internal oblique	Similar to external oblique	Similar to external oblique	Similar to external oblique
Transverse abdominal	Similar to external oblique	Similar to external oblique	Similar to external oblique
Rectus abdominis	Pubic symphysis	Sternum	Constricts abdomen
Longissimus dorsi	Sacrum, ilium, and neural processes of lumbar and thoracic vertebrae	Transverse processes of most vertebrae and lateral surfaces of the ribs except the first	Singly, flexes spine laterally; together, extend back and neck; rib attachments may aid in expiration
Tensor fasciae latae	Crest of ilium	Fascia over knee, patella, and crest of tibia	Flexes hip joint and extends knee joint
Biceps femoris	Lateral part of ischium and sacrum	By a wide aponeurosis to patella and fascia of thigh and leg	Abducts and extends limb; may also flex knee joint
Gluteus medius	Fascia of longissimus dorsi, ilium, and sacroiliac and sacrosciatic ligaments	Proximal end of femur	Abducts thigh
Muscles of the Hindleg			
Quadriceps femoris, a large muscle group consisting of			
1. Rectus femoris	Ilium	Patella and its ligament	Extends shank
2. Vastus lateralis	Proximal end of femur	Patella and its ligament	Extends shank
3. Vastus medialis	Proximal end of femur	Patella and its ligament	Extends shank
4. Vastus intermedialis	Proximal end of femur	Patella and its ligament	Extends shank
Gracilis	Pubic symphysis and ventral surface of pubis	Patellar ligament and proximal end of tibia	Adducts hindlimb

continued

TABLE 20.2

Muscles of the Hindquarter—*continued*

Muscle	Origin	Insertion	Action
Muscles of the Hindleg—*continued*			
Sartorius	Iliac fascia and tendon of psoas minor (external iliac vessels lie between the two heads)	Patellar ligament and proximal end of tibia	Adducts hindlimb and flexes hip joint
Semimembranosus	Ischium	Distal end of femur and proximal end of tibia, both on medial side	Extends hip joint and adducts hindlimb
Semitendinosus	First and second caudal vertebrae and ischium	Proximal end of tibia and calcaneus	Extends hip and tarsal joint and flexes knee joint
Adductor	Ventral surface of pubis and ischium and tendon of origin of gracilis	Proximal end of femur	Adducts hindlimb and extends and rotates femur inward
Pectineus	Anterior border of pubis	Medial side of shaft of the femur	Adducts hindlimb and and flexes hip
Iliacus	Ventral surface of ilium and wing of sacrum	Proximal end of femur together with psoas major	Flexes hip and rotates thigh outward
Psoas major	Ventral sides of transverse processes of lumbar vertebrae and last two ribs	With the iliacus on proximal end of femur	Flexes hip and rotates thigh outward

muscle (see Figure 20.6). Beneath this is the **internal oblique,** which may be revealed by cutting a window high up on the external oblique. Then, by separating fibers of the internal oblique, you will see the **transverse abdominal,** the deepest layer and the thinnest of the three muscles (and the most difficult to see). All three muscles insert on the **linea alba,** a tendinous band that extends from the pubis to the sternum (Figure 20.9). Together, these muscles support the abdominal wall and compress the viscera during expiration and defecation. Beneath the external oblique and extending between the pelvic girdle and ribs on each side of the midventral line is a longitudinal band of muscle, the **rectus abdominis** (Figure 20.9). It also supports and constricts the abdomen.

Dorsal and lateral to the vertebral column, locate the **longissimus dorsi,** a very long muscle extending from the sacrum to the neck (see Figure 20.8). Acting together, these muscles extend the back and neck; acting singly, each flexes the spine laterally.

The most anterior superficial thigh muscle is the **tensor fasciae latae** (see Figures 20.6 and 20.10); it acts to flex the hip joint and extend the knee joint. Posteriorly, the most superficial thigh muscle is the **biceps femoris** (see Figures 20.6 and 20.10). Its action is complex, acting across both the hip and knee joints to retract the thigh and flex the shank.

Between the tensor fasciae latae and the biceps femoris, and partially covered by them, is the **gluteus medius** (see Figures 20.6 and 20.10). It acts to abduct the thigh.

Muscles of the Hindleg (Table 20.2). The **quadriceps femoris** is a large muscle group covering the anterior and lateral sides of the femur. It comprises four muscles: **rectus femoris,** a thick muscle on the anterior side of the femur; **vastus lateralis** (Figure 20.10), lateral to the rectus femoris and partly covering it; **vastus medialis** (see Figure 20.9), on the medial surface of the rectus femoris; and **vastus intermedialis,** a deep muscle lying beneath the rectus femoris. All four of these muscles converge on the patella (kneecap) and then continue as the patellar ligament to insert on the tibia. These are the extensors of the shank.

The posteromedial half of the thigh is covered with a thin, wide muscle, the **gracilis** (see Figure 20.9). It adducts the thigh and flexes the shank. Just anterior to the gracilis is the **sartorius,** a thin band of muscle that covers the femoral blood vessels; it is delicate and easily destroyed if not identified. Cut through the gracilis and sartorius to reveal the large **semimembranosus** muscle in the medial portion of the thigh (Figure 20.9). It extends the hip joint and adducts the

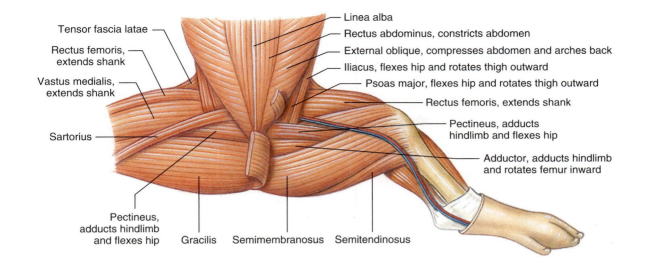

Figure 20.9
Muscles of the hindlimb of a fetal pig, ventral view. The gracilis and sartorius have been cut and removed from the pig's left leg.

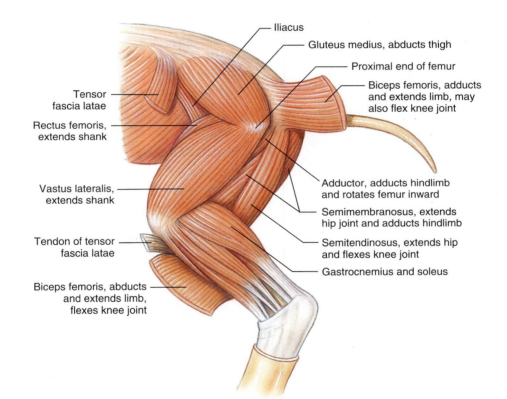

Figure 20.10
Muscles of the hindlimb of a fetal pig, lateral view.

hindlimb. Just posterior to the semimembranosus is the thick, band-shaped **semitendinosus;** it acts mainly to extend the hip. The semimembranosus and semitendinosus, together with the biceps femoris, are the hamstring muscles of humans.

The **adductor,** lying anterior to the semimembranosus and covered by the gracilis, is a triangular-shaped

muscle that, as its name suggests, adducts the femur— that is, draws it toward the midline.

Also on the medial side of the thigh are three smaller muscles: the triangular-shaped **pectineus,** an adductor of the thigh; the **iliacus,** which flexes the hip and rotates the thigh outward; and the **psoas major,** which acts the same as the iliacus.

Some of the shank muscles are shown in Figure 20.10 but are not described in this exercise.

Identification

Be able to identify and explain the origin, insertion, and action of as many of the foregoing muscles as your instructor has assigned.

EXERCISE 20C
Digestive System
Core Study

Head and Throat

Salivary Glands

Three pairs of salivary glands produce a continual background level of fluid secretions containing lysozymes and immunoglobulins that flush the teeth and mouth cavity and help keep bacterial growth under control. During meals, much larger quantities of saliva are produced containing lubricating glycoproteins called **mucins** and a large amount of **salivary amylase** (α-amylase) that begins the breakdown of complex carbohydrates such as starch.

☞ On the right side of the face, neck, and chin, carefully remove the skin if you have not already done so. A muscle layer will tend to adhere, but push this layer back into place gently so as not to destroy the glands beneath. Now carefully remove the thin muscles back of the angle of the jaw and beneath the ear to uncover the **parotid gland.** Do not destroy any large blood vessels.

The triangular parotid (pa-rot'id; Gr. *para,* beside, + *ous,* ear) gland is broad, thin, and rather diffused, extending from almost the midline of the throat to the base of the ear (Figure 20.11). Do not confuse the salivary glands, which are choppy and lobed in appearance, with the lymph nodes, which are smoother and shiny. The **parotid duct** comes from the deep surface of the gland and follows the ventral border of the masseter (cheek) muscle along the external maxillary vein to the corner of the mouth (Figure 20.11).

The **submaxillary (mandibular) gland** lies under the parotid gland and just posterior to the angle of the jaw. It is darker, compact, and oval. Its duct comes from the anterior surface of the gland and passes anteriorly, medial to the mandible, and through the sublingual gland to empty into the floor of the mouth. This duct is very difficult to trace.

☞ To find the **sublingual glands,** remove the mylohyoid muscle and the slender pair of geniohyoid muscles immediately beneath it.

On each side of the head, a whitish, elongated sublingual gland is located between the digastric muscle, which lies inside the mandible, and the genioglossus, which is one of the muscles at the base of the tongue. A sublingual artery and vein will be seen along the ventral side of each gland. The sublingual glands empty by way of several short ducts to the floor of the mouth.

Mouth Cavity and Pharynx

☞ Using scissors, cut through the angle of the mouth on both sides. Cut posteriorly, pulling open the mouth as you proceed. Follow the angle of the tongue and do not cut into the roof

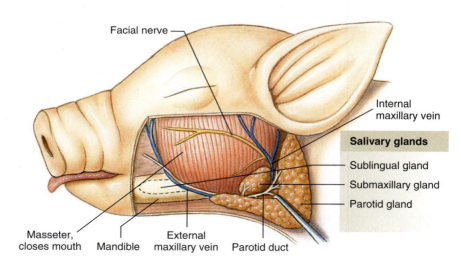

Figure 20.11
Dissection of the head and neck of a fetal pig to show some superficial veins, nerves, and salivary glands.

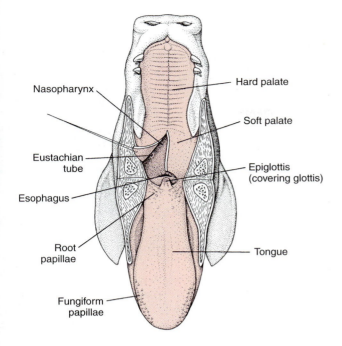

Nasopharynx

Hard palate

Soft palate

Eustachian tube

Epiglottis (covering glottis)

Esophagus

Root papillae

Tongue

Fungiform papillae

Figure 20.12
Oral cavity of the fetal pig.

of the mouth. Continue the cuts to the esophagus to fully expose the oral cavity and pharynx (Figure 20.12).

Teeth may not be erupted yet, although the canines and third pair of incisors may be seen in older fetuses. (The third incisors and the canines are the first to erupt; the second incisors are the last.) The young pig will have three incisors, one canine, and four premolars on each side of each jaw.

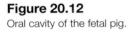 Remove the flesh along the right jaws and carefully cut away enough of the jawbone to expose the buds of the embryonic teeth.

The mouth cavity is roofed by a narrow, bony, **hard palate,** sheathed ventrally with mucous membranes that are ridged into transverse folds. Extending posteriorly from the hard palate is the **soft palate** composed of thick membrane. The hard and soft palates of mammals completely separate the oral cavity from the air passages above, an innovation that allows a mammal to chew a mouthful of food at leisure while breathing freely through its nose. Among other vertebrates, only crocodilians have a hard palate; the soft palate is unique to mammals.

Open the mouth wide, drawing down the tongue, to locate at the posterior end of the soft palate the opening into the **nasopharynx** (fair′inks; Gr. *pharyngx,* gullet), the space above the soft palate. It connects with the nasal passages from the nostrils.

To expose the nasopharynx, make a midline incision of the soft palate. Locate the small openings on either side of the roof of the nasopharynx.

From these openings, the **eustachian tubes** (named after B. Eustachio, an Italian physician) lead to the middle ear.

Posterior to the nasopharynx is the **laryngeal pharynx,** which connects the oral cavity with the **esophagus.** Both nasal and the laryngeal pharynx are derived from the pharynx of ancestral chordates, which evolved as a filter-feeding apparatus. Pharyngeal (gill) pouches develop in this region in all vertebrate embryos. In fishes, these pouches break through to develop into gill chambers, but in tetrapods they become transformed into other structures: middle ear cavity and glandular tissue (thyroid, parathyroid, and thymus).

The **larynx** lies in the floor of the laryngeal pharynx. Locate the flaplike **epiglottis,** which folds up over the **glottis** (the open end of the larynx) to close it when food is being swallowed. Note that in the mouth the air passages are *dorsal* to the food passage. In the throat, however, the air is carried through the larynx and trachea, which are *ventral* to the food passage (esophagus). These passageways cross in the pharyngeal cavity (Figure 20.13). When the animal is respiring, the epiglottis fits up against the opening into the nasopharynx, allowing air into the larynx but preventing saliva or food from entering the mouth. During swallowing, the larynx is pushed forward, causing the epiglottis to fold over the glottis, thus opening the food passage while closing off the air passage.

Continue your dissection of the neck region by making a midventral incision down the neck.

Parting the skin and clearing away some tissue around the larynx will expose the **thymus,** a large, soft, irregular mass of glandular tissue lying lateral to the sternohyoid muscles (Figure 20.14). It is an extensive gland in the fetus and young animal, but after puberty it decreases in size although continuing to function throughout life. The thymus extends caudally under the sternum with its posterior portion overlying the heart. The thymus is part of the body's lymphatic system and is filled with lymphocytes of all sizes, especially T cells, which are important in immunological responses.

As you clear tissue from the larynx, find the **trachea** (windpipe) extending caudally from it. The trachea is stiffened by a series of C-shaped cartilage rings, which are incomplete dorsally where the trachea lies against the esophagus.

Tracing the trachea posteriorly, you will see the **thyroid gland,** a small, dark red, oval gland lying on the trachea beneath the sternothyroid muscles (Figure 20.14). The thyroid is an endocrine gland that produces thyroxin and triiodothyronine, two hormones that promote growth and development and regulate the metabolic rate.

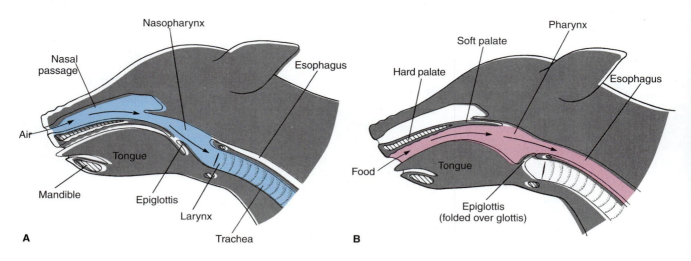

Figure 20.13
Relationship of the respiratory passage to the mouth and esophagus in breathing and swallowing. **A,** During breathing, the glottis is open to receive air from the nostrils and is protected from food and saliva by the epiglottis. **B,** For swallowing, the larynx is pushed anteriorly, causing the epiglottis to fold over the glottis, thus closing the air passage to the lungs. Feel your Adam's apple (larynx) as it moves up when you swallow.

Abdominal Cavity

Directions for Dissection

To proceed further, it is necessary to expose the organs of the abdominal cavity. Place the pig ventral side up in the dissecting pan. Tie a cord or rubber band around one forelimb, loop the cord under the pan, and fasten it to the other forelimb. Do the same to the hindlegs.

☞ With a scalpel, make a midventral incision through the skin and muscles but not into the body cavity, continuing the incision already made in the neck posteriorly to within 1 cm of the umbilical cord (incision 1, Figure 20.15). Cut around each side of the cord (2). If your specimen is female, continue on down the midline from the cord to the anal region (3). If it is male, make two incisions, one on each side of the midline, to avoid cutting the penis, which lies underneath (3a). Now, in either sex, deepen the incisions you have made in the abdominal region through the muscle layer to reach the body cavity, taking care not to injure the underlying organs. With scissors, make two lateral cuts on each side, one just anterior to the hindlegs (4) and the other posterior to the ribs (5); then turn back the flaps of the body wall (6). Flush out the abdominal cavity with running water.

All visceral organs are invested in mesentery and held in place with connective tissue. Loosen this tissue carefully to separate organs, tubes, and vessels, being careful not to cut or tear them. *Do not remove any organs unless you are specifically directed to do so.* Be careful in all your preliminary dissection not to destroy blood vessels or nerves; keep them intact for later dissection of the circulatory and nervous systems.

It is important to remember that instructions referring to the "right side" mean the animal's right side, which will be on your left as the animal lies ventral side up in the dissection pan.

Notice that the umbilical cord is attached anteriorly by a tube, the **umbilical vein.**

☞ Tie a string around the umbilical vein in two places and sever the vein between the two strings.

The strings will identify this vein later. Lay the umbilical cord between the hindlegs, and identify the following parts of the abdominal cavity.

The **body wall** consists of several layers: (1) tough external **skin,** (2) two layers of **oblique muscle** and an inner layer of **transverse muscle** (try to separate the layers and determine the direction of the fibers), and (3) an inner lining of thin, transparent **peritoneum.**

The **diaphragm** is a muscular, dome-shaped partition separating the peritoneal cavity (abdominal cavity) from the thoracic cavity, which together constitute the coelom. *Do not remove the diaphragm.*

The peritoneum is the smooth, shiny membrane that lines the abdominal cavity and supports and covers the organs within it. That which lines the body walls is called the **parietal peritoneum.** It is reflected off the dorsal region of the body wall in a double layer to form the **mesenteries,** which suspend the internal organs, and then continues on around the organs as a cover, where it is called the **visceral peritoneum.**

The **liver** is a large, reddish gland with four main lobes lying just posterior to the diaphragm. The greenish,

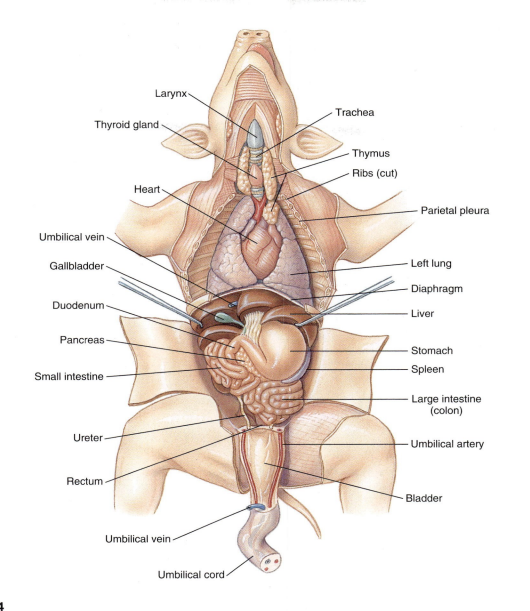

Larynx

Thyroid gland

Trachea

Thymus

Ribs (cut)

Heart

Parietal pleura

Umbilical vein

Gallbladder

Left lung

Diaphragm

Duodenum

Liver

Pancreas

Stomach

Spleen

Small intestine

Large intestine (colon)

Ureter

Umbilical artery

Rectum

Bladder

Umbilical vein

Umbilical cord

Figure 20.14

Internal anatomy of the fetal pig, ventral view. The first part of the small intestine (just past the stomach) is called the _____ . The liver has how many lobes? _____ What long, reddish organ serves as an important lymphatic organ and also has an important immunological function? _____

saclike **gallbladder** may be seen under one of the central lobes (see Figure 20.14).

The **stomach** is nearly covered by the left lobe of the liver. The **small intestine** is loosely coiled and held by mesenteries. Note the blood vessels in the mesentery that supports the digestive tract. The **large intestine** is compactly coiled on the left side posterior to the stomach.

The **spleen** is a long, reddish organ attached by a mesentery to the greater curvature of the stomach. The spleen contains one of the largest concentrations of lymphatic tissue in the body. It functions to phagocytize spent blood components and salvage the iron from

hemoglobin for reuse. It is also immunologically important in initiating immune responses by B cells and T cells.

The **umbilical arteries** are two large arteries extending from the dorsal wall of the coelom to and through the umbilical cord (see Figure 20.14).

The **allantoic bladder,** the fetal urinary bladder, is a large sac lying between the umbilical arteries. It connects with the allantoic duct in the umbilical cord.

The **kidneys** are two large, bean-shaped organs attached to the dorsal wall dorsal to the intestines. They are *outside* the peritoneal cavity in the **cisterna magna** and are separated from the other abdominal organs by the peritoneum.

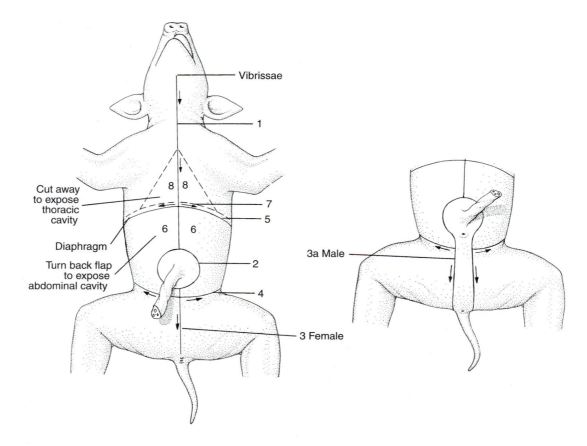

Figure 20.15

Cutting diagram. The numbers indicate the order in which each incision is to be made. *1* to *6* expose the abdominal cavity; *7* and *8* expose the thoracic cavity.

Digestive Tract

The digestive system consists of the alimentary canal, extending from mouth to anus, and glands (such as salivary glands, liver, and pancreas) that assist in its function of converting food into a form that can be assimilated for growth and energy requirements. You have already studied the anterior portions of the alimentary canal: the mouth cavity and the salivary glands. We will now consider the digestive tract proper, beginning with the esophagus.

The **esophagus** is a soft, muscular tube that leads from the pharynx to the stomach. Locate it in the neck region posterior to the larynx, where it is attached to the dorsal side of the trachea by connective tissue. Find the esophagus in the thoracic cavity posterior to the lungs, and in the abdominal cavity find where it emerges through the diaphragm at the cardiac end of the stomach. The muscles at the anterior end of the tube are striated (voluntary), gradually changing to smooth muscle. How does this affect swallowing?

The **stomach** (see Figure 20.14) is a large, muscular organ that breaks up food and thoroughly mixes it with gastric juice. Identify its **cardiac end** near the

heart, its **pyloric end** that joins the duodenum, its **greater curvature** where the spleen is attached, and its **lesser curvature.** The **fundus** (L., bottom) is the anterior blind pouch. The contents of the fetal digestive tract, made green by pigments in the bile salts, are called **meconium** and contain epithelium sloughed from the mucosa lining, sebaceous secretions, and amniotic fluid swallowed by the fetus. Open the stomach longitudinally, rinse it out, and find (1) the **rugae,** or folds, in its walls; (2) the opening from the esophagus; and (3) the **pyloric** (Gr. *pylōros,* gatekeeper) **valve,** which regulates the passage of food into the duodenum.

The **small intestine** includes the **duodenum** (doo-uh-dē′num; L., "twelve-each," referring to its length in humans, which equals about 12 finger-widths), or first portion, which lies next to the pancreas and receives the common bile duct and pancreatic duct, and the **jejunum** and **ileum,** indistinguishable in the fetal pig, which make up the remainder of the small intestine.

Remove a piece of the intestine, open it, and examine it *under water* with a hand lens or dissecting microscope. Observe the minute, fingerlike **villi** (sing., **villus;** L., shaggy hair), which greatly increase the absorptive surface of the intestine.

Most digestion and absorption take place in the small intestine.

The **large intestine** includes the long, tightly coiled **colon** and the straight, posterior **rectum,** which extends through the pelvic girdle to the **anus.** Its primary function is to absorb water and minerals from the liquified chyme that enters it.

Find the **cecum,** which is a blind pouch of the colon at its junction with the ileum. In humans and the anthropoid apes, the cecum has a narrow diverticulum (a tube, blind at its distal end) called the **vermiform** (L., worm-shaped) **appendix.** Open the cecum (on its convex side opposite the ileum), and note how the entrance of the ileum forms a ring-shaped **ileocecal valve.** The posterior end of the rectum will be exposed in a later dissection.

Digestive Glands

The **liver,** a large, brownish gland posterior to the diaphragm, has four main lobes: the left and right lateral lobes and the left and right central lobes. One of the many important functions of the liver is to produce **bile,** a fluid containing bile salts, which are steroid derivatives responsible for the emulsification of fats. Bile is stored and concentrated in the **gallbladder,** a small, greenish, oval sac embedded in the dorsal surface of the right central lobe of the liver. The liver is connected to the upper border of the stomach by a tough, transparent membrane, the **gastrohepatic ligament,** in which are embedded blood vessels (in the left side) and ducts (in the right side). Carefully loosen the gallbladder and note its tiny **cystic duct.** This unites with **hepatic ducts** from the liver to form the **common bile duct,** which carries bile to the duodenum. Probe the gastrohepatic ligament and adjoining liver tissue carefully to find these ducts. Do not injure the blood vessels lying beside them.

The **pancreas** is a mass of soft glandular tissue in the mesentery between the duodenum and the end of the stomach. Push the small intestine, except the duodenum, to the left to explore the gland. Its pancreatic juice is carried by a **pancreatic duct** to the pyloric end of the duodenum. The pancreas is a double gland having both endocrine and exocrine portions. Its endocrine portion produces two hormones, insulin and glucagon, that are of great importance in carbohydrate and fat metabolism. The exocrine portion secretes the pancreatic juice—a mixture of water, electrolytes, and enzymes. The enzymes include carbohydrases, which digest sugars and starches; lipases, which split lipids; and proteases, which break down proteins. The pancreas is the only source of lipases in the digestive system.

Other digestive juices are secreted by the **mucosa** lining the stomach and the small intestine.

Histological Study of the Intestine

Examine a cross section of human or other mammalian small intestine, and compare it with the histological structure of amphibian intestine. Examine slides of a mammalian liver and a pancreas.

Further Study

Thoracic Cavity and Neck Region

☞ With scissors, begin just anterior to the diaphragm and cut along the midventral line through the sternum, to a point midway between the forelegs. Keep the lower blade of the scissors up to keep from injuring the heart underneath. Now make a lateral cut on each side just anterior to the diaphragm (see incision 7, Figure 20.15). This exposes the thoracic cavity but leaves the diaphragm in place.

The **mediastinal septum,** which separates the right and left lung cavities, is a thin, transparent tissue attached to the sternal region of the thoracic wall (Figure 20.16).

☞ Carefully separate the mediastinal septum from the body wall. Now lift up one side of the thoracic wall and look for the small **internal thoracic artery** and **vein** (also called sternal or mammary) embedded in the musculature of the body wall. Carefully separate these vessels on each side and lay them down over the heart and lungs for future use. Now you may cut away some of the

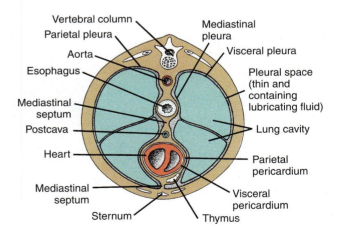

Figure 20.16
Diagrammatic transverse section through the thorax in the region of the ventricles to show the relationships of the pleural and pericardial membranes, the mediastinum, and the lung cavities.

ventral thoracic wall (see incision 8, Figure 20.15) to allow a better view of the thoracic cavity containing the left and right **lungs** and the **heart.**

The **pleura** (Gr., side) is the name given to the peritoneum that lines each half of the thoracic cavity and covers the lungs (Figure 20.16). The peritoneum lining the thoracic cavity is the **parietal** (L. *paries,* wall) **pleura;** the part applied to the lungs is the **visceral** (L., bowels) **pleura.** The small space between is the **pleural cavity,** which contains lubricating **pleural fluid** to prevent friction. The portions of the parietal pleurae on the medial side next to the heart are called the **mediastinal pleurae.** The **mediastinum** is the region between the mediastinal pleurae. It contains the pericardium and heart and the roots of the big arteries and veins, as well as the trachea, esophagus and part of the thymus.

The double-walled **pericardium** enclosing the heart is made up of an outer **parietal pericardium** and an inner **visceral pericardium** applied to the heart, with pericardial fluid in the space between.

Identification

Be able to locate and give the functions of the parts of the digestive system.

EXERCISE 20D
Urogenital System

Urinary System

☞ Read the directions carefully and dissect cautiously. Do not tear or remove any organs, blood vessels, or ducts. Instead, separate them carefully from the surrounding tissues. You will dissect the urogenital system of only one sex, and then exchange your dissected specimen with another student who has dissected the opposite sex. Therefore, prepare your specimen with the care you would give to a demonstration dissection.

The urinary system consists of a pair of kidneys, a pair of ureters, a urinary bladder, and a urethra (shared with the reproductive system in the male).

The fetal **urinary bladder** is the **allantoic** (Gr. *allas,* sausage) **bladder,** a long sac located between the umbilical arteries (Figure 20.17). It narrows ventrally to form the **allantoic duct,** which continues through the umbilical cord and is the fetal excretory canal. The bladder narrows dorsally to empty into the **urethra,** the adult excretory canal. The urethra will be dissected later. After birth, the allantoic end of the bladder closes to form the urinary bladder.

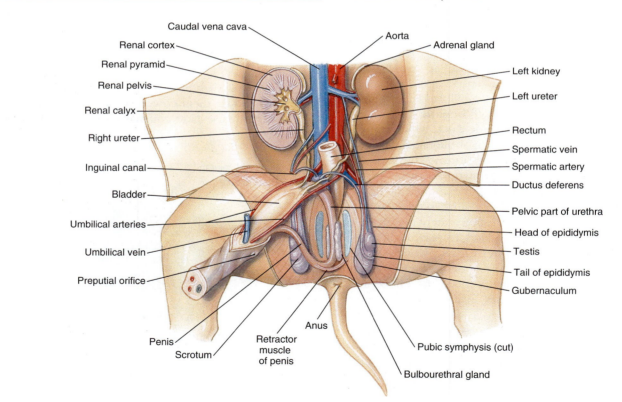

Figure 20.17
Male urogenital system of the fetal pig. The right kidney has been longitudinally sectioned to show its internal structure.

The **kidneys** are dark and bean-shaped. They lie outside the peritoneum on the lumbar region of the dorsal body wall. Uncover the right kidney carefully. A depression on the median side of each kidney is called the **hilus.** Through it pass the renal blood vessels and the **ureter,** or excretory duct. Follow the left ureter posteriorly to its entrance into the bladder. Be careful of small ducts and vessels that cross the ureter. (Note the small **adrenal gland,** an endocrine gland lying close to the medial side of the anterior end of the kidney and embedded in fat and peritoneum.)

☞ Slit open the right kidney longitudinally, cutting in from the outer border. Remove the ventral half of the kidney and lay it in a dish of water.

Study the kidney section with a hand lens or dissecting scope. Identify the **cortex** (L., bark), or outer layer, containing the microscopic renal corpuscles; the **medulla** (L., marrow, pith), or deeper layer, containing the radially arranged blood vessels and collecting tubules; and the **renal pyramids,** which contain groups of collecting tubules coming together to empty through **papillae** into the **pelvis.** The pelvis is a thin-walled chamber that connects with the ureter. Divisions of the pelvis into which the papillae empty are referred to as **calyces** (ka'luh-sez; sing, **calyx**) (Figure 20.17).

☞ For a description of the anatomy and physiology of excretion, read your textbook. Examine the demonstration specimens of sheep kidneys on display. You will be expected to understand the structure and function of the kidney and its functional unit, the nephron.

Male Reproductive System

The location of the testes depends on the fetal pig's stage of development. Each testis originates in the abdominal cavity near the kidney. During the development of the fetus, a prolongation of the peritoneum, the **processus vaginalis,** grows down into each half of an external pouch, the **scrotum.** Later, the testis "descends" into the scrotum through the **inguinal canal,** where it lies within the sac, or sheath, formed by the processus vaginalis. The inguinal canal is the tubular passage connecting the abdominal cavity with the scrotal sac (Figure 20.17). The descent of both testes into the scrotum is usually completed shortly before birth. Part of the cutaneous scrotum can be seen ventral to the anus.

Lay the umbilical cord and allantoic bladder between the legs, and locate the **urethra** (from Gr. *ouron,* urine) at the dorsal end of the bladder. The urethra bends dorsally and posteriorly to disappear into the pelvic region. Find the sperm ducts (pl., **vasa deferentia;** sing., **vas deferens**), two white tubes that

emerge from the openings of the **inguinal** (L., groin) **canals,** cross over the umbilical arteries and ureters, and come together medially to enter the urethra. Also emerging from each inguinal ring are the spermatic artery, vein, and nerve. Together with the vas deferens, these make up the **spermatic cord,** which leads to the testis.

Now lay the bladder up over the abdominal cavity and locate the thin, hard, cordlike **penis** under the strip of skin left posterior to the urogenital opening. The penis lies in a sheath in the ventral abdominal wall, ending at the **urogenital opening.**

☞ Carefully separate the penis from surrounding tissue, and then cut away the skin and muscle that covered and surrounded it. Complete the removal of any skin remaining on the inside of the thigh and rump, and carefully separate away the underlying fascia. The thin-walled scrotal sac extends posteriorly across the ventral surface of the high muscles toward the cutaneous scrotum. Free the left scrotal sac from the surrounding tissues.

Pass a probe through the inguinal canal into the processus vaginalis, which houses the **testis.** The testis is a small, hard, oval body containing hundreds of microscopic **seminiferous tubules** in which the sperm develop.

☞ Cut open the left scrotal sac to expose the testis.

The seminiferous tubules of the testis unite into a much-coiled **epididymis** (Gr. *epi,* upon, + *didymos,* testicle). The epididymis begins as a whitish lobe on the anterior surface of the testis and passes posteriorly around one side of the testis to its caudal end, where it unites with the **vas deferens** (sperm duct). The vas deferens (L. *vas,* vessel, + *deferre,* to carry off) passes cranially through the inguinal canal, loops over the ureter, and enters the urethra, as already seen. A fibrous cord attaches the testis and the epididymis to the posterior end of the processus vaginalis. It is called the **gubernaculum** (L., rudder). A narrow band of muscle (the cremaster) runs along the lateral and posterior part of the processus vaginalis parallel with the vas deferens.

☞ Separate the tissues on each side of the penis in the pelvic region; then, being careful not to injure the penis or cut too deeply, use a scalpel to cut through the cartilage of the pelvic girdle.
    Spread the legs apart to expose the **urethra** and its connection with the penis.

The urethral canal extends throughout the length of the penis and serves as a common duct for both sperm and urine.

Now, beginning at its juncture with the bladder, follow the urethra posteriorly and locate these **male glands:**

Seminal vesicles. The seminal vesicles are a pair of small glands on the dorsal side of the urethra. These glands contribute a secretion to the semen that is rich in fructose, a six-carbon sugar that stimulates previously inactive but mature spermatozoa to become highly motile. In humans, the secretion of the seminal vesicles makes up more than 60% of the volume of the semen.

Prostate gland. The prostate gland is poorly developed in the fetus; it lies between and often partly covered by the seminal vesicles, but may be difficult to find. The alkaline secretions of the prostate assist in neutralizing acids normally present in the urethra, as well as in the vagina of the female. In humans, the prostate secretion is known to contain a compound that may help prevent urinary tract infections in males.

Bulbourethral glands (Cowper glands). The bulbourethral glands are a pair of narrow glands about 1 cm long on each side of the urethra near its junction with the penis (Figure 20.17). These glands add a thick, sticky alkaline mucus to the semen that has lubricating properties.

Identification

Be able to follow the paths of urine and sperm to the outside. Compare the male urogenital system of the pig with those of a frog and a human.

Female Reproductive System

The **ovaries** are small, pale organs lying just posterior to the kidneys (Figure 20.18). Each is suspended by a mesentery, the **mesovarium** (Gr. *mesos,* middle, + L. **ovarium,** ovary), which can be seen extending between the kidney and the ovary. The **uterus** (L., womb) is Y-shaped. The **horns of the uterus** (the arms of the Y) extend from the ovaries to unite medially at the **body of the uterus,** which leads to the **vagina** (L., sheath). It is in the horns of the uterus, not the body of the uterus, that the fetal pigs develop. (Most mammals have a similar Y-shaped uterus, called **bicornuate** ["double-horned"], but in the higher primates, including humans, the two uterine horns are completely fused, a condition called **simplex.**) Each uterine horn is suspended by a mesentery, the **broad ligament.** Follow the uterine

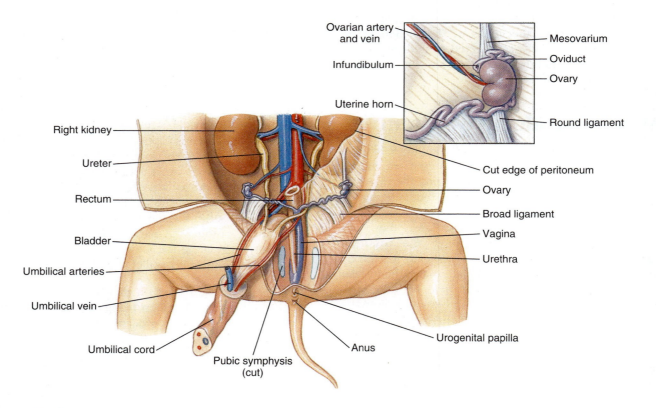

Figure 20.18
Female urogenital system of the fetal pig. What structures serve as excretory ducts for the kidneys? _____ What structure drains the bladder? _____ Embryonic pigs develop in what paired structures? _____

horn anterior to the ovary where it gives rise to a highly convoluted **oviduct** (also called **fallopian tube** after G. Fallopio, Italian anatomist). The oviduct coils around the ovary and terminates at a wide, ciliated funnel, the **infundibulum** (L., funnel). In adult pigs, eggs released from the ovary at ovulation are swept into the opening **(ostium)** of the infundibulum by ciliary currents.

To expose the rest of the reproductive system, the pelvis must be cut open.

☞ Lay the allantoic bladder anteriorly over the abdominal viscera. Cut through the muscle medially between the legs and through the pelvic girdle. Be careful not to cut through the urethra or vagina. Spread the legs apart and separate the urethra from the surrounding tissue. (Note where the ureters join the dorsal end of the bladder.)

Lay the bladder and urethra to one side. Follow the body of the uterus posteriorly to a slight constriction called the **cervix.** From here, the tube widens and is called the **vagina.** The vagina and urethra soon join to form a **urogenital sinus,** which is a short common passageway for the two systems.

The **vulva** is the external opening of the urogenital sinus, ventral to the anus. The ventral side of the vulva extends out to form a pointed **genital papilla.** A small, rounded **clitoris** may be seen extending from the ventral floor of the urogenital sinus, but is not always evident.

In adults, at copulation, the male penis places the spermatozoa, contained in seminal fluid, into the vagina. The sperm must pass through the uterus to the oviduct to fertilize the egg. After fertilization, the zygote passes down to a horn of the uterus to develop. Note how the horns are adapted for carrying a litter. How does this compare with the human uterus? _____ How is the developing fetus nourished? _____ How are waste products from the fetus disposed of? _____

The placenta of the pig is known as a **chorioallantoic,** or **diffuse, placenta** (Figure 20.19). In the uterus of the dog and some other carnivores, the placenta is called a zonary placenta because the chorionic villi are located in a girdlelike zone, or band, around the middle of each pup rather than over the whole surface of the chorion as in the pig. The human placenta is disc-shaped.

Identification ◀

Be able to trace the path of the unfertilized egg from the ovary to the uterus. Compare it with the path in a frog.

Be able to trace the path of urine in the female. Are any parts shared by both the urinary and reproductive systems?

_____

If you have not studied the male system, trade your specimen for a male, and make a thorough study of the male system.

Oral Report

Be able to identify and state the functions of the reproductive organs of both the male and female pig.

Written Report

On separate paper, write a comparison of the mammalian and amphibian reproductive systems as illustrated by your study of the pig and frog. How is each adapted to its own type of reproduction?

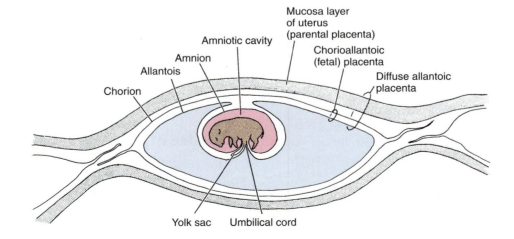

Figure 20.19
Diagram of a pig fetus in utero, showing the relationship of fetal and parental membranes in a diffuse placenta.

EXERCISE 20E
Circulatory System

Core Study

The circulatory system of the pig is quite similar to that of other mammals, including humans. In contrast to the single-circuit system of the fish, with its two-chambered heart, and the incomplete double circuit of the three-chambered amphibian heart, the mammal has an effective four-chambered heart, which allows the blood two complete circuits: a **systemic circuit** through the body, followed by a **pulmonary circuit** to the lungs for oxygenation. The right side of the heart receives the oxygen-poor blood returning from the body tissues and pumps it to the lungs; the left side receives the oxygen-rich blood returning from the lungs and pumps it to the body tissues.

The study of fetal pig circulation has the added advantage of illustrating not only typical mammalian circulation but also typical **fetal circulation.** The changes in circulation necessary for the transition from a non-breathing, noneating fetus to an independent individual with a fully independent circulatory system are both crucially important and elegantly simple.

Uninjected vessels will contain only dried blood (or they may be empty) and therefore will be fragile, flattened, and either brown or colorless. If injected, the arteries will have been filled with latex or a starchy injection mass through one of the arteries in the cut umbilical cord and will be firm and pink or yellow. The veins will have been injected through an external jugular vein in the neck and will be blue. Sometimes, however, the injection medium does not fully penetrate the vasculature. The lymphatic system will not be studied.

You should uncover and separate the vessels with a blunt probe and trace them as far into the body as possible, but be careful not to break or remove them. Nerves often follow an artery and vein and will appear as tough, shiny, white cords. Do not remove them. You may in fact find it efficient to identify the major nerves at this time as you find them. As you identify a vessel, separate it and carefully remove investing muscle and connective tissue, taking care not to break the vessel or destroy other vessels that you have not yet identified. Do not remove any body organs unless specifically directed to do so.

Because it is often difficult to trace the arterial system without damaging the venous system, which lies above it, you will study the veins first. However, since corresponding arteries and veins usually lie side by side, it is often convenient to study both systems at the same time.

Keep in mind that there is considerable variation among individual pigs in the points of vessel bifurcation, especially in the venous system of the neck and shoulder region. *The venous arrangement in your pig almost certainly will not look exactly like the manual illustrations*. Make notes or sketches of any variations that you find in your specimen. With careful dissection, both veins and arteries can be left intact.

Heart

☞ Note carefully the shape and slope of the diaphragm and how it forms the posterior boundary of the thoracic cavity. Then cut the diaphragm away from the body wall to make entrance into the thoracic cavity more convenient.

Open the pericardial sac and examine the heart. It has two small, thin-walled atria and two larger muscular-walled ventricles.

Right Atrium (Anterior and Ventral)

Lift the heart (Figure 20.20) and see the precaval and postcaval veins that empty into the right atrium. The **postcava** (posterior vena cava) from the abdominal region emerges through the diaphragm; the **precava** (anterior vena cava) comes through the space between the first ribs (Figure 20.20).

Left Atrium (Anterior and Dorsal)

The left atrium receives the **pulmonary veins** on the dorsal side. Note the conspicuous earlike right and left auricles (L. *auricula,* ear) lying on each side of the heart. The term "auricle" is often incorrectly used as a synonym for "atrium"; however, the term should be reserved for the earlike flap that protrudes from each atrium.

Right Ventricle

The right ventricle is large and thick-walled. The **pulmonary artery** leaves the right ventricle and passes over the anterior end of the heart to the left, where it divides back of the heart to go to the lungs.

Left Ventricle

The left ventricle is larger and covers the apex of the heart (posterior). It gives off the large **aorta,** which rises anteriorly just behind (dorsal to) the pulmonary artery. The **coronary sulcus** is the groove on the surface of the heart between the right and left ventricles. It contains the **coronary artery** and **vein,** which supply the tissues of the heart itself.

Structure of the Heart

External View

☞ Use fresh or preserved pig or sheep hearts and compare with the heart of the fetal pig.

Locate the right and left **atria** and the right and left **ventricles.** Find the **coronary sulcus** and the **coronary artery** and **vein,** which supply the muscles of the heart.

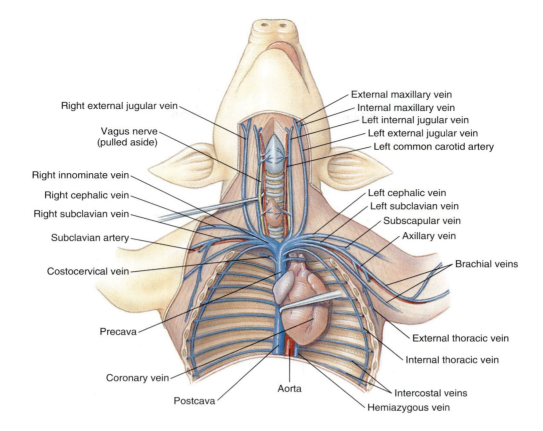

Figure 20.20
Veins of the head, shoulders, and forelimbs of the fetal pig. The internal thoracic veins, shown laterally on the opened thoracic cavity, actually lie ventral to the heart.

Now identify on the ventral side the large, thick-walled **pulmonary trunk** leaving the right ventricle and the large **aorta** leaving the left ventricle. On the dorsal side, find the large, thin-walled **precava** and **postcava** entering the right atrium and the **pulmonary veins** entering the left atrium.

Frontal Section

☞ Now make a frontal section, dividing the heart into dorsal and ventral halves. Start at the apex and direct the cut between the origins of the pulmonary and aortic trunks. Leave the two halves of the heart attached at the top. Wash out the heart cavities (or, if injected, carefully pull out the latex filling).

The cavities of the heart are lined with a shiny membrane, the **endocardium.** Identify the chambers of the heart as well as the valves at the entrance to each chamber that prevent backflow of blood.

Right Atrium. The right atrium is thin-walled, with openings from the precava, postcava, and hemiazygous veins. Find the entrance of these veins. Now, in the dorsal half of the fetal pig heart, probe for an opening between the two atria. This is the **foramen ovale,** one of the fetal shortcuts.

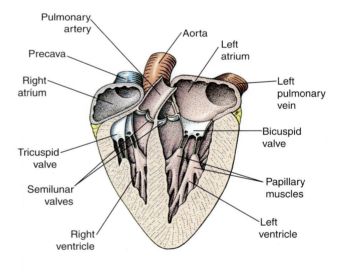

Figure 20.21
Diagrammatic frontal section of a sheep heart.

Right Ventricle. The right ventricle is thick-walled. An atrioventricular valve, called the **tricuspid valve,** prevents backflow of blood to the atrium (Figure 20.21). The valve is composed of three **cusps,** or flaps of tissue, that extend from the floor of the atrium and are connected by fibrous cords, the **chordae tendineae** (ten-din'ee-ee), to **papillary muscles** projecting from the walls of the ventricle.

In the ventral half of the heart, the opening into the pulmonary artery is guarded by **semilunar valves.**

☞ Cut the pulmonary artery close to the heart and remove the latex. Slit the vessel for a short distance into the ventricle and look into it to see the three cusps, or pockets of tissue. Determine how they would allow passage into the vessel but prevent return of blood into the ventricle.

Left Atrium. The left atrium is thin-walled. Find the entrance of the pulmonary veins that return oxygen-rich blood to the heart.

Left Ventricle. The left ventricle is the most muscular chamber of the heart, for it must send the freshly oxygenated blood at high pressure to all the tissues of the body. An atrioventricular valve, the **bicuspid valve** (also called the mitral valve), guards the entrance from the left atrium. Find the cusps of this valve. Push the valve open and closed to see how it works.

Find the three-cusped **semilunar valve,** which prevents backflow from the aorta (Figure 20.21).

Notice the difference in the thickness of the walls of arteries and veins. Of what are these walls composed? _____ Would you find the same kind of muscle in the heart as in the vessels? _____ What is the pacemaker of the heart? _____ Consult your text.

General Plan of Circulation

In mammalian circulation (after birth), the systemic and pulmonary systems are separate. Blood from all parts of the body except the lungs returns by way of the large **precava** and **postcava** to the **right atrium.** From there it goes to the **right ventricle** to be pumped out through the **pulmonary arteries** to the lungs to be oxygenated. The blood returns through the **pulmonary veins** to the **left atrium** and then to the **left ventricle,** which sends it through the **aorta** to branches that carry the blood finally to the capillaries of all parts of the body. The venous system returns it again to the right side of the heart to begin another circuit. The mammal has a **hepatic portal system** but no renal portal system as in the amphibian.

Before birth, when the lungs are not yet functioning, the fetus depends on the placenta for nutrients and oxygen, which are brought to it through the **umbilical vein.** Therefore, the general plan of circulation is modified in the fetus so that most of the pulmonary circulation is short-circuited directly into the systemic bloodstream to be carried to the placenta instead of the lungs. Only enough blood goes to the lungs to nourish the lung tissue until birth. These modifications will be mentioned as they arise in later descriptions.

Veins of the Head, Shoulders, and Forelimbs

☞ Carefully remove the remainder of the sternum and the first rib without damaging the veins beneath.

You will find considerable variation in the veins of individual specimens.

The precaval division of the venous circuit carries blood from the head, neck, thorax, and forelimbs to the heart.

Begin by locating the **internal jugular vein** found lying along the trachea and adjacent to the common carotid artery and the vagus nerve (Figure 20.20). It drains the brain, larynx, and thyroid.

Trace the internal jugular vein posteriorly to its confluence with the **external jugular vein** and **subclavian vein;** these join to form the **innominate vein.** Follow the two innominate veins (right and left) through the opening at the level of the first rib into the chest cavity. Here they join to form the **precaval,** which enters the right atrium. The precava also receives at this level a pair of **internal thoracic veins** lying on either side of the sternum. This is the vein that, along with the accompanying artery, you detached from the ventral muscle wall of the chest cavity (see Exercise 20C). You will also find its continuation in the abdominal muscle wall.

Another vessel entering the precava dorsolaterally close to the heart is the **costocervical trunk.** It receives several veins from the neck and dorsal thorax.

Lift the heart and left lung and push them to the right. The unpaired **hemiazygous (azygous) vein** lies in the chest cavity along the left side of the aorta and receives blood from the left and right intercostal veins. Follow the hemiazygous anteriorly to the point at which it crosses the aorta and goes under the heart. Lift the heart up to see the vein cross the pulmonary veins and enter the right atrium along with the postcava.

Return now to the neck region and locate the **external jugular vein.** This vein was probably used to inject the venous system and will be tied and severed on one side of the neck. The external jugular is formed by the union of the **internal maxillary** (dorsal) and the **external maxillary** (ventral) near the angle of the jaw. These vessels drain superficial parts of the head.

Joining the base of the external jugular (or sometimes the subclavian) is the **cephalic vein,** a large superficial vein from the arm and shoulder. It lies just beneath the skin.

The **subclavian vein** extends from the shoulder and forelimb. It is a deeper vein that follows the subclavian artery. (There may be from one to three subclavian veins. If more than one, considerable anastomosing [networklike branching and rejoining of vessels] may exist among them.) The subclavian, as it continues into the arm, is called the **axillary** (L. *axilla,* armpit) in the armpit and the **brachial** (Gr. *brachion,* upper arm) in the upper arm. Large nerves of the brachial nerve plexus overlie these veins.

Arteries of the Head, Shoulders, and Forelimbs

Arteries carry blood away from the heart, and with the exception of the pulmonary arteries, all branch from the main artery, the aorta.

The **aorta** begins at the left ventricle (ascending aorta), curves dorsally (aortic arch) behind the left lung, and extends posteriorly along the middorsal line (descending, or dorsal, aorta). The first two branches are the **brachiocephalic trunk** and, just to the left of it, the **left subclavian artery** (Figure 20.22).

Brachiocephalic Trunk. The brachiocephalic is a large single artery that extends anteriorly a short distance and branches into the carotid trunk and the right subclavian artery.

The **carotid trunk** may extend anteriorly for as much as a centimeter, or it may divide at once into the **left** and **right common carotid arteries,** which form a Y over the trachea and extend up each side of it. Each common carotid artery follows an internal jugular vein and the vagus nerve toward the head, giving off branches to the esophagus, thyroid, and larynx. Near the head, each common carotid gives off an **internal carotid,** a deep artery passing dorsally to the skull and brain, and an **external carotid,** which is a continuation of the common carotid that supplies the tongue and face.

The **right subclavian artery** arises from the brachiocephalic trunk and continues into the right arm as the **axillary** in the armpit, the **brachial** in the upper arm, and the **radial** and **ulnar** in the lower arm. The **subscapular** extends to the deep muscles of the shoulder and branches off the axillary part of the artery.

Left Subclavian Artery. The left subclavian arises directly from the aorta. Otherwise, it is similar to the right subclavian artery.

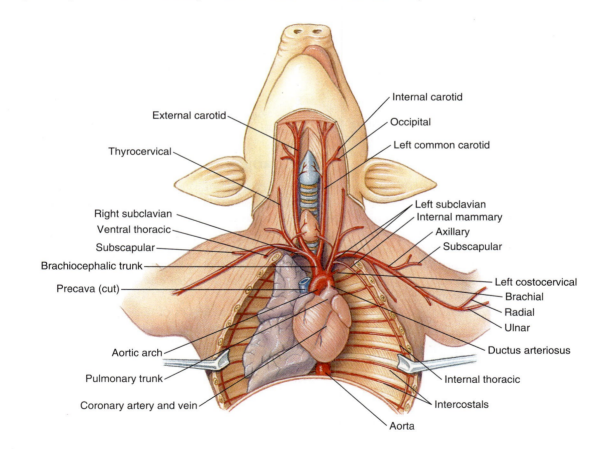

Figure 20.22

Arteries of the head, shoulders, and forelimbs of the fetal pig. The internal thoracic arteries, shown laterally, actually lie ventral to the heart. What pair of arteries supply blood to the tongue and face? _____ What artery supplies blood to the thyroid and parotid glands? _____ Near the head, the common carotid artery splits into what two arteries? _____

Each of the subclavian arteries gives off several branches, which include the **internal thoracic artery** and the **thyrocervical (internal cervical) artery.** The internal thoracic artery (also called the mammary or sternal artery and, in humans, an artery used in coronary bypass operations) is the artery you detached earlier. It supplies the ventral muscular wall of the thorax and abdomen.

The thyrocervical artery arises from the subclavian at the same level as the internal thoracic. It supplies the thyroid and parotid glands and some of the pectoral muscles.

Pulmonary Circulation

Pulmonary Trunk. The pulmonary trunk arises from the right ventricle, ventral to the aorta. Follow it as it branches into **right** and **left pulmonary arteries,** which carry oxygen-poor blood to the lungs. Scrape away some lung tissue to find branches of these vessels.

Pulmonary Veins. Pulmonary veins empty oxygen-rich blood into the left atrium.

☞ Probe gently under the left atrium to expose the veins.

You should be able to find the large trunk entering the heart just under the hemiazygous vein and a pair of vessels servicing each lobe of the lungs.

Fetal Shortcuts

Foramen Ovale. The foramen ovale is a fetal opening in the wall between the right and left atria. Part of the blood from the right atrium can pass through the foramen ovale directly to the left atrium, where it can go to the left ventricle, to the aorta, and back into the systemic circulation without going to the lungs at all (Figure 20.23). The remainder of the blood enters the pulmonary trunk.

Ductus Arteriosus. The ductus arteriosus is the short connection between the pulmonary trunk and the aorta. Trace this connection, which begins where the smaller pulmonary arteries branch off toward the lungs. Part of the blood from the right ventricle goes to the lungs, and part is shunted through the ductus arteriosus to the aorta (Figure 20.23).

Ductus Venosus. A third fetal shortcut is the ductus venosus, which connects the umbilical vein with the postcava, passing through a channel in the liver tissue. During fetal life, the anterior part of the postcava carries a mixture of oxygen-poor blood from body tissues and oxygen-rich (and nutrients) from the placenta by way of the umbilical vein (the vein tied and cut earlier). After birth, both the umbilical vein and the ductus venosus degenerate.

☞ You will be able to see the arrangement of the vessels in the liver tissue by using a probe to

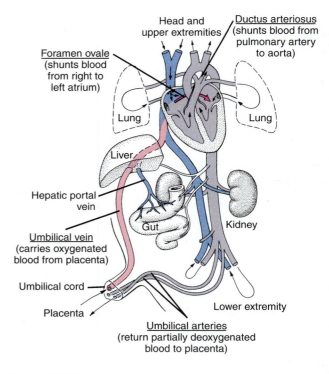

Figure 20.23
Scheme of the fetal circulation.

scrape or "comb out" the liver tissue to separate it from the vessels. Wash out the loose tissue.

Some of the vessels you see will probably be hepatic ducts.

Changes in Circulation at Birth

Two crucially important events happen at the moment of birth: (1) the placental bloodstream upon which the fetus has depended is abruptly cut off, and (2) the pulmonary circulation immediately assumes the task of oxygenating the blood. One of the most extraordinary aspects of mammalian development is the perfect preparedness of the circulatory architecture for this event.

When the newborn piglet (or human) takes its first breath, the vascular resistance through the lungs is suddenly lowered as the lungs expand. The ductus arteriosus functionally constricts almost immediately, and the lungs receive full blood circulation. Blood now returns from the pulmonary veins to the left atrium, raising the pressure in this chamber enough to close the flaplike valve over the foramen ovale. With blood no longer passing between the right and left atrium, the foramen ovale gradually grows permanently closed, leaving a scar called the **fossa ovalis.** The ductus arteriosus also closes permanently, becoming a ligament between the pulmonary artery and the aortic arch.

With severing of the umbilical cord, flow through the umbilical vein and arteries ceases immediately, and these vessels are eventually reduced to fibrous cords.

Postcaval Venous Circulation

The **postcava** carries blood to the heart from the hindlimbs and trunk. Follow it from the right atrium through the intermediate lobe of the lung and through the diaphragm to the liver (Figure 20.24). Push the intestines to the left side, clean away most of the dorsal portion of the liver, and note where the postcava emerges and continues posteriorly. Clean away the peritoneum that covers the postcava, and find the following veins that enter it on each side.

One or two **renal veins** extend from each kidney. Do both the right and left veins enter at the same level? The narrow, band-shaped adrenal gland (suprarenal gland) lying against the anteromedial border of each kidney is drained by a small **adrenal vein** that may enter the renal vein or may empty directly into the postcava. It is accompanied by the adrenal artery. Just posterior to the adrenal vein and at about the same level as the renal veins, find the **parietal vein** extending to the body wall, accompanied by the parietal artery. The **genital vein** and **artery** serve the testes and ovaries. In the male, the vein is called

the **spermatic vein** and is incorporated together with the spermatic artery into the spermatic cord to the testis. In the female, it is called the **ovarian vein;** it lies suspended in the mesovarium along with the ovarian artery.

Follow the postcava posteriorly to its bifurcation into the **common iliac veins.** These in turn divide into the **internal** and **external iliac veins.** The internal iliac vein (hypogastric) is the medial branch extending dorsally and posteriorly into the deep tissue and draining blood from the rectum, bladder, and gluteal muscles. The external iliac vein is the lateral branch and the largest vein entering the common iliac. It drains the foot and leg and is formed by the union of the large **femoral vein,** which extends ventrally toward the knee, and the **deep femoral vein,** which extends dorsally and posteriorly into the deep muscles of the thigh. It also receives a lateral branch, the **circumflex iliac vein,** which drains the muscles of the abdomen and the upper thigh.

Hepatic Portal System

The hepatic portal system is a series of veins that drain the digestive system and spleen. The blood is collected in the **hepatic portal vein,** which carries it to the liver capillaries. As the blood laden with nutrients from the

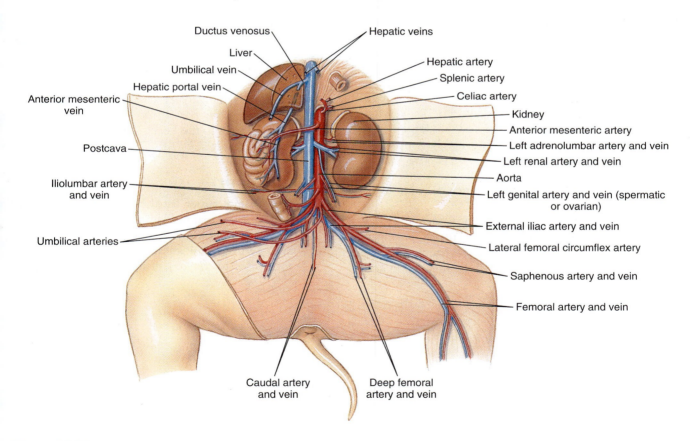

Figure 20.24
Circulation in the trunk and hindlegs of the fetal pig.

intestine passes through the liver, the liver may store excess sugars in the form of glycogen, or it may give up sugar to the blood if sugar is needed. Some of the amino acids formed by protein digestion may also be modified here. Blood from the liver capillaries is collected by the **hepatic veins,** short veins that empty into the postcava.

The hepatic portal system in your specimen may not be injected. Unless these vessels are injected or are filled with dry blood, they may be difficult to locate. Try to examine an injected specimen.

Lift up the liver, stomach, and duodenum and draw the intestine posteriorly to expose the pancreas. Loosen the pancreatic tissue from the hepatic portal vein that runs through it. Now lay the duodenum over to your right and see how the vein enters the lobes of the liver near the common bile duct (in the gastrohepatic ligament). Lay the small intestine over to your left and fan out the mesenteries of the small intestine to see how the veins from the intestines are collected into the **anterior mesenteric vein** (Figure 20.24). This joins with the small **posterior mesenteric vein** from the large intestine and with veins from the stomach and spleen to form the **hepatic portal vein.**

Arteries of the Trunk and Hindlegs

Descending Aorta. The descending aorta follows the vertebral column posteriorly, first lying dorsal and then ventral to the postcava (Figure 20.24). After passing through the diaphragm, the aorta gives rise to a single large artery, the **celiac artery.** To find it, clip away the diaphragm and remove the tissue around the aorta at the anterior end of the abdominal cavity. The celiac artery divides almost immediately into arteries serving the stomach, spleen, pancreas, liver, and duodenum.

Just posterior to the celiac, find another unpaired artery, the **anterior mesenteric artery.** This vessel must be dissected from the surrounding tissue. It sends branches to the pancreas, small intestine, and large intestine.

Next locate the paired **parietal arteries, renal arteries,** and **spermatic** or **ovarian arteries.** Posterior to these is the single **posterior mesenteric artery,** which divides and sends branches to the colon and rectum. It will probably have been broken off during earlier dissection of the abdominal cavity.

At its posterior end, the aorta now divides into two large lateral **external iliac arteries** to the legs, two medial **internal iliac arteries** to the sacral region, and a small continuation of the aorta into the tail, the **caudal artery.** Trace one of the external iliac arteries. It first gives off a lateral **circumflex artery** that supplies some of the pelvic muscles, and then penetrates the peritoneal body wall to enter the leg as the **femoral artery.** Note that nearly all of the caudal arterial circulation is accompanied by corresponding venous circulation, already described. Follow the femoral artery to the point where, on its median side, it gives off a **deep femoral artery.** This artery serves the deep muscles of the thigh.

Now locate the origin of the **internal iliac arteries.** Each internal iliac gives off at once a large **umbilical artery** and then, as a smaller vessel, continues dorsally and posteriorly into the sacral region beside the internal iliac vein, giving off branches to the bladder, rectum, and gluteal muscles.

In the fetus, the umbilical arteries are major vessels that return blood from the fetus to the placenta by way of the umbilical cord. After birth, they become smaller and serve the urinary bladder.

Identification

Be familiar with the direction of blood flow throughout the body. Be able to describe how fetal circulation differs from postnatal circulation. Be able to trace blood from any part of the heart to any part of the body and back to the heart.

Written Report

On separate paper, list the two chief differences between the mammalian and amphibian circulatory systems as illustrated by your study of the pig and the frog. How are these differences adaptive to an all-terrestrial or half-aquatic life?

APPENDIX A

Instructor's Resources for Implementing Exercises

Exercise 1: Ecological Relationships of Animals

Exercise 1: A Study of Population Growth, with Application of the Scientific Method

Materials

For each group of four students:
  5 half-pint, wide-mouth mason jars
  5 screw-on rings
  5 pieces of aluminum window screen, cut to fit snugly into the rings
  5 pieces of paper towel, same size as the screen
Resource, consisting of 5 pounds stoneground whole-wheat flour mixed with 1/4 pound powdered brewer's yeast (both available at health food stores); this will supply enough resource for 30 groups of 4 students each
Inert filler; vermiculite works well (available at garden stores); sawdust seems to inhibit growth; vermiculite should be screened with window screen or finer mesh to remove large particles
250 adult beetles; these can be purchased from scientific supply houses or secured from a culture kept in the laboratory; fill several quart jars 3/4 full with flour/yeast mixture (95:5), add some beetles, cover with a screen and paper towel top; as the medium is used up, add more; every year or so, discard half the medium (and included beetles) and replace with fresh medium

Notes

1. **This is a project exercise requiring most of a term to complete.** Because it is an excellent demonstration of the scientific method, which is usually introduced at the beginning of a zoology or biology course, we suggest setting aside part of the first laboratory period to get it underway. Only about an hour is required for each student group to separate the beetles into jars, but students will need more reading time to understand the background of the exercise and to become properly acquainted with the methodology. Consequently, it is best to require the students to read the exercise before coming to the laboratory. A promise of a quick quiz at the beginning of the laboratory period concerning the scientific method and the procedures is usually sufficient to ensure compliance.

2. **This lab is a good place to discuss basic population biology** and/or show a film on population biology.

3. **One way to help separate flour from beetles** is to fashion a lid by putting three pieces of screen (and no paper towel) in one screw-on ring, screw it onto the jar, and then gently shake out the flour. The screens must be closely opposed to keep beetles from slipping through. The adults, pupae, and larger larvae will be caught on the three layers of screen and remain in the jar (a few may escape through the screening, so students should watch for this during the separation). At the end of the experiment, you may want to combine and save all beetles and flour to add to your culture for next term.

4. **The report.** This project is a good subject for a scientific paper of approximately three to five pages in length. Assign it to be written using the standard format for scientific articles: introduction, materials and methods, results, discussion, and literature cited. You may want to provide a handout that fully describes your expectations. Encourage students to look at their data in original ways. Also encourage them to explore the literature. You may wish to place the article by Peters and Barbosa (1977) on reserve at the library.

5. **Tips.** There are several ways to simplify this laboratory. One is to count only adults. Another is to use only three jars per group, either containing 3, 10, and 50 g of resource with filler, or the same without filler.

Exercise 2: Introduction to Animal Classification

Exercise 2A: Use of a Taxonomic Key for Animal Identification

Materials

Animal specimens representing several phyla and classes, suitable reference texts or field guides

All references to papers and publications mentioned in Appendix A are available on the website that accompanies *Animal Diversity,* sixth edition. Visit www.mhhe.com/hickmanad6e for more information.

Notes

1. **Cladogram exercise.** For the exercise on phylogeny reconstruction and how to make a cladogram, no additional materials are required. The instructor may choose to require students to submit a brief written report showing the character matrix and resulting cladogram. Students sometimes struggle with the concept that the interpretation of a branching pattern is critical to interpretation of a tree. The idea that branches can rotate within a tree without changing the tree's meaning is difficult. However, constructing simple model trees using ball and socket toys, sticks and Styrofoam balls, or similar items is often extremely helpful to students. Manipulating a physical model of a cladogram reaches students with a wide range of learning styles.

2. **Selection of specimens.** For Exercise 2A, select specimens that represent a wide range of animal taxa. Although students will recognize by common name many of the specimens easily available to you, they usually will not be able to classify them correctly without the aid of the key. Specimens that will challenge students include: hydroids (slide mounted), sea pansy or sea whip, ctenophore, freshwater turbellarians (slide), tapeworm, liver fluke, *Ascaris,* spiny-headed worm (acanthocephalan), brachiopod, sea cucumber, chiton, marine segmented worm such as *Arenicola* or *Aphrodite,* chaetognath, horseshoe crab, spider, tunicate, and amphioxus. Less challenging, perhaps, but still providing experience in the use of a taxonomic key are a variety of vertebrates and insects, as well as a sponge, octopus, bivalve, snail, sea star or brittle star, earthworm, and marine crab or freshwater crayfish. All of these are available as preserved material from biological supply houses.

3. **The report.** The written report requires students to verify the identification of each insect order by consulting a reference, such as an insect field guide or other reference. This exercise acquaints students with the reference, introduces them to the variety of forms within a taxon, and emphasizes the importance of confirming a tentative identification.

Exercise 3: The Microscope

Exercise 3A: Compound Light Microscope

Materials

Compound microscopes
Microscope lamps
Blank slides and coverslips
Prepared slides
    Typewritten letters (e, a, h, or k)
    Colored threads
Salt or sugar crystals
Distilled water
Materials suitable for wet mounts
Pipette
Gum arabic solution
Pond water

Notes

Koehler illumination. Students find it easier to learn to focus properly than to adjust the illumination properly. To avoid problems later, we spend a few minutes talking the students through the procedure. Correct adjustment of the illumination system is commonly referred to as Koehler (or Köhler) illumination. Here in more detail are the steps involved:

a. Focus on a specimen (preferably a stained section) with low power.

b. Move the condenser with its focus knob up toward the top of its range. Close down the iris diaphragm; then move the condenser up and down until a sharp image of the diaphragm is seen. Center the circle of light using the condenser centering screws.

c. Once the condenser is centered, open the iris diaphragm until the dark edges are just outside the field of view; that is, the entire field of view should be filled with light.

d. Remove the ocular (or one of the oculars, if using a binocular microscope) and look into the tube. Adjust the iris diaphragm so that the circle of light covers 2/3 to 3/4 of the illuminated area on the objective lens.

e. Replace the eyepiece. With a bit of practice, students learn how to use the iris diaphragm effectively without having to remove the ocular each time they switch objectives. Students should be advised *not* to change the condenser position to get more or less contrast. Lowering the condenser tends to reduce resolution. Instead, they should be encouraged to use the diaphragm to readjust illumination.

Exercise 4: Protozoan Groups

Exercise 4A: Phylum Amoebozoa— *Amoeba* and Others

Materials

Amoeba culture
Other sarcodine cultures as available
Slides, ringed or plain
Coverslips
Petrolatum
Thread
10% nigrosine (for demonstration #3)
Radiolarian shells, preserved or on permanent slides
Prepared slides
    Amoeba
    Entamoeba
    Other sarcodines as available
Microscopes

Notes

1. **Finding amebas in the culture.** Amebas will be found on the bottom of the shipping container or petri dish into which they are poured for the laboratory. Advise students not to agitate the container by squirting the unused contents of a pipette back into the culture. It helps to pour off most of the culture water before the lab, leaving only a few millimeters in the container.

2. **Subculturing amebas.** Amebas are rather easily subcultured. Several methods are described in Needham (1959), all using some variation of the classic hay or rice infusion. To see what the Society of Protozoologists is recommending for culture establishment, see pp. 2–3 in Lee et al. (1985).

3. **Nigrosin vital stain,** 10%, is a good relief stain against a background of which ameboid movement and contractile vacuole activity show up well.

4. **Demonstration of osmoregulation.** How would the osmoregulatory requirements of marine protozoans compare with those of freshwater forms? This might be demonstrated by placing a drop of rich *Amoeba* culture in each of four deep-well slides. Fill three of the wells with culture water that has been made up to 2%, 4%, and 6% saline (by adding 2, 4, or 6 g of salt/100 ml of

water); fill the fourth well with plain culture water. After a few minutes, place a drop from the bottom of each well on a slide and examine. Time the rate of discharge of contractile vacuoles in specimens from each osmotic concentration. If there is a difference, can you explain it?

Exercise 4B: Phyla Euglenozoa and Plantae—*Euglena, Trypanosoma,* and *Volvox*

Materials

Euglena culture
Volvox culture
Stained slides of *Euglena, Volvox,* and *Trypanosoma*
Other flagellate cultures or slides as available
Methylcellulose, polyvinyl alcohol, Protoslo, or Detain
Methyl violet or Noland's stain
Microscopes
Slides and coverslips

Notes

1. ***Euglena gracilis*** and ***Euglena viridis,*** the two most commonly supplied species of genus *Euglena* differ in several ways. *E. gracilis,* which is 35 to 65 μm in length, has 6 to 12 shield-shaped chloroplasts, each bearing a prominent central pyrenoid region (containing enzymes that catalyze the conversion of photosynthetically produced glucose into starch and paramylon). The pyrenoid of each chloroplast is covered with a watch-glass-shaped complex of starch grains and paramylon bodies (paramylon, also spelled paramylum, is an unbranched nutrient polysaccharide related to starch, but the glucose molecules are β-linked rather than α-linked as they are in starch). In the slightly larger *E. viridis* (40 to 80 μm in length), the chloroplasts form ribbons that radiate from a single, large, central pyrenoid body. *E. gracilis* is a smoother and more rapid swimmer, with pronounced euglenoid movement when swimming stops; the swimming and euglenoid movements of *E. viridis* are more jerky than those of *E. gracilis.*

2. **How to concentrate *Euglena*.** *Euglena* may be concentrated in a culture by shining a bright light through a narrow slit in a piece of cardboard placed against the shaded culture container. After a few minutes, the positively phototropic *Euglena* will be concentrated in a narrow band at the container's edge, where they can be removed easily with a pipette. (Do not use *too* bright a light; *Euglena* become negatively phototactic in intense light.)

3. ***Volvox globator,*** the species usually offered by supply houses, is found in nitrogen-rich freshwater habitats. Most colonies range from 350 to 500 μm in diameter and contain from 5000 to 15,000 cells. This species is monoecious, so both microgametes and macrogametes are formed by the same colony, as described in the manual. *V. aureus,* sometimes supplied commercially for laboratory use, has both dioecious and monoecious varieties. The somewhat smaller *V. tertius,* a dioecious species used in experimental work, is not ordinarily supplied commercially for use in teaching laboratories.

    Remind students to look for *Volvox* colonies in the bottom of the container—they are easily seen—rather than aimlessly pipetting from the culture container.

All references to papers and publications mentioned in Appendix A are available on the website that accompanies *Animal Diversity,* sixth edition. Visit www.mhhe.com/hickmanad6e for more information.

4. **Avoiding cross-contamination of cultures.** If several different protozoan species are used in the same laboratory over a period of several days, be careful to use separate pipettes for each to avoid cross-contamination of cultures. *Euglena* especially will quickly build up populations in other cultures if accidentally introduced.

Exercise 4C: Phylum Apicomplexa—*Plasmodium* and *Gregarina*

Materials

Mealworms (*Tenebrio* larvae)
    or
Cockroaches (*Periplaneta, Blatta*)
Watch glasses
Invertebrate Ringer solution, or 0.65% saline solution
Stained slides of gregarines
Stained human blood smears containing *Plasmodium* stages
Slides and coverslips
Teasing needles
Pipettes
Microscopes

Notes

1. **Preparing insect saline.** Insect physiological saline especially balanced for cockroaches is prepared by dissolving 10.9 g NaCl, 1.6 g KCl, 0.8 g CaCl$_2$, and 0.17 g MgCl$_2$ in 1 liter of distilled water.

2. **Prepared slides of *Plasmodium* and gregarines.** Prepared slides of *Plasmodium* are also available from biological supply houses. You may give students a blood smear containing various red blood cell stages (trophozoites and merozoites) for study; other stages, such as sporozoites, gametocytes, and a cross section of a mosquito gut showing oocysts, can be placed on display with accompanying explanatory legends. Mounted slides of male and female *Anopheles* mosquitos are also available as a demonstration.

    Figure 4.12 in the manual illustrates *Plasmodium falciparum,* considered the most virulent of the *Plasmodium* species in humans. The ring stage is smaller than in either *P. malariae* or *P. vivax,* and the crescent shape of the gametocytes is distinctive for *P. falciparum.* If species other than *P. falciparum* or *P. vivax* are displayed, you may wish to provide color plates of each of the red cell stages (found in most parasitology texts, such as Roberts and Janovy, 2005, *Foundations of Parasitology*) so that the students can see the species differences. Excellent color plates of these and other *Plasmodium* species are also found in Coatney et al. (1971).

    If living material for the study of gregarines is not available, prepared slides may be obtained from some biological supply houses (see Appendix B for suppliers of prepared microslides).

Exercise 4D: Phylum Ciliophora—*Paramecium* and Other Ciliates

Materials

Stained slides of paramecia—normal, undergoing binary fission, and in conjugation
Paramecia cultures
Vorticella, Stentor, Spirostomum, or other ciliate cultures
Protoslo or 10% methylcellulose
Congo red–yeast mixture or Congo red–milk mixture

Acidified methyl green
0.25% NaCl solution
Dilute picric acid
Weak acetic acid
Salt crystals
Cotton
Toothpicks or pins
Slides and coverslips
Microscopes

Notes

1. **Preparing Congo red–stained yeast.** Boil some dry yeast in a small amount of water in a test tube; then cool. Add a 1% solution of Congo red and allow it to diffuse into the yeast for a few minutes. Then decant most of the dye solution to concentrate the yeast cells.

2. **Preparing acidified methyl green.** Combine 1 g methyl green, 100 ml distilled water, and 1 ml glacial acetic acid.

3. **Slowing paramecia for study.** In our experience, students have better luck using cotton fibers than methylcellulose (Protoslo) to slow movement. Some of the paramecia will become trapped beneath or between the fibers, where they can be observed. This is especially important for the study of contractile vacuoles and feeding. However, some instructors may elect to have the students make their initial observations of swimming behavior using Protoslo.

4. **Watching feeding in paramecia.** When offered yeast, paramecia will eat rapidly and soon become sated. Encourage students to begin their observations as soon as possible (preferably within a few seconds) after adding Congo red–stained yeast in order to watch actual feeding.

   Congo red turns blue in an acid environment. Food vacuoles become acidic by fusion with acidosomes as digestion progresses, but seeing this happen by watching for a color change (which is subtle at best) may require more time and patience than most students are willing to invest.

   Be certain to emphasize to students the importance of using a *very small* amount of Congo red–stained yeast when examining feeding in *Paramecium*.

Experimenting in Zoology: Effect of Temperature on the Locomotor Activity of *Stentor*

Materials

Cultures of *Stentor* (each student pair will require 3 or 4 *Stentor*)
Shallow depression slides and coverslips
Finely divided graph paper (millimeter grid) and cellophane tape
Culture dish (4 1/2 inch or 8 inch) or plastic dish for water bath
Vaseline® and Q-tips®
Pasteur pipette (glass)
Supply of crushed ice for each student pair
Thermometer
Optional: hand tally counter for each student pair

Notes

1. **Preparations.** This experiment is not difficult to conduct and can be completed in about 1 1/2 hours if students are given depression slides with the *Stentor* already selected. Allow another hour if students are to do all preparations (prepare slides, transfer *Stentor,* and set up viewing assembly). Results are predictable and usually turn out with gratifying uniformity

among different student pairs in a class. Refrigerator dishes also serve nicely for the water bath, as do standard 4 1/2 inch culture dishes. Each student pair will require a dish of crushed ice, a beaker of warm water, and a thermometer.

   Tape a 20 mm by 20 mm grid (cut from graph paper) under the depression slide, using clear packaging tape that is waterproof. Trim excess tape from the bottom of the slide with a razor blade.

   Be certain that the depression slides are perfectly clean before the *Stentor* are added.

2. **Selecting the *Stentor*.** Each student pair needs about 3 *Stentor* so that 1 can be chosen that is actively swimming. Sometimes a *Stentor* becomes trapped between the coverslip and the edge of the depression slide concavity; others are just not inclined to swim. The *Stentor* must be removed from the culture and transferred to the depression slide individually with a glass Pasteur pipette (the 9 inch disposable variety are a little easier to use than the 5 3/4 inch), but with a little practice, this is accomplished quickly. If time is short, this step should be completed in advance by the instructor or laboratory assistant; however, we find that students do it themselves nearly as well with little practice. *Stentor* are rather easily seen with the naked eye (over a white background) and can be drawn up individually with a Pasteur pipette and bulb. The vaseline must be added carefully in an unbroken circle just outside the depression, and the coverslip must be pressed firmly down over the slide to form a perfect seal.

3. **Potential problems.** If too much fluid is placed in the depression, the *Stentor* may be washed out when the coverslip is added. If too little water is used, a bubble will be left in the field of view. Sometimes the *Stentor* get trapped in the vaseline. At the highest temperatures, the *Stentor* often slow down or stop altogether; this happens when the temperature approaches the incipient lethal temperature for the organism. Students will often note that the *Stentor* change shape at the higher temperature, becoming elongate (trumpet-shaped); they may continue to swim or they may attach themselves to the glass surface. The thermal history of the culture will determine the maximum temperature that can be used—usually either 25° or 30°C.

4. **Swimming rate** can be counted easily without hand tally counters, but if you have them, students enjoy using them, perhaps because it provides an aura of precision to the procedure.

5. **Temperature equilibrium.** Students should be advised to stabilize the water temperature for 2 to 3 minutes at each measurement point. However, the *Stentor* come into temperature equilibrium quickly, probably in less than a minute at a new temperature.

6. **Converting proportional swimming speeds to true swimming speeds.** It is a simple matter to convert all the measured rates (lines crossed per minute) to true swimming speeds by setting up proportionalities between the two—for example:

$$\frac{\text{TSS at } 30°}{\text{LCM at } 30°} = \frac{\text{TSS at } 20°}{\text{LCM at } 20°}$$

and

$$\text{TSS at } 20° = \frac{\text{TSS at } 30° \times \text{LCM at } 20°}{\text{LCM at } 30°}$$

where TSS = true swimming speed in mm/sec and LCM = lines crossed per minute. If true swimming speed is measured at the lowest temperature at the start (where the slow movement makes accurate measurements easier), students can convert all the lines-crossed-per-minute values to true swimming speeds as they progress through the experiment.

Exercise 5: The Sponges

Exercise 5: Class Calcarea—*Sycon*

Materials

Preserved or living *Sycon* (= *Scypha, Grantia*) Examples of asconoid and leuconoid sponges
Prepared slides
  Sycon, transverse sections
  Leucosolenia, transverse sections
Spicule stew
Single-edged razor blade
Chlorine bleach (sodium hypochlorite)
Microslides
Coverslips
Watch glasses
Hand lenses or dissecting microscopes
Compound microscopes

Notes

Preparation of spicule samples. Calcareous and siliceous sponges are identified principally from their spicules. To prepare a spicule sample, place a small piece (2- to 3-mm-square block) of the sponge in a tube and add about 2 ml of Clorox (sodium hypochlorite solution). Allow an hour or so for the organic matter to dissolve and the spicules to settle. Pipette off the Clorox, add water to wash, allow the spicules to settle, and then remove the water with a Pasteur pipette. Repeat the washing with water, transfer the spicules by pipette to a microscope slide, and add a drop of water and a coverslip. Examine at about 100× or more.

Permanent spicule mounts are easily prepared by following the water washes described with one or two washings with 95% alcohol. After the final wash, transfer the spicules with a pipette to a microscope slide and allow the alcohol to evaporate. Apply a drop of mounting medium, add a coverslip, and label.

Exercise 6: The Radiate Animals

Exercise 6A: Class Hydrozoa—*Hydra, Obelia,* and *Gonionemus*

Materials

Living material
  Hydras
  Artemia larvae, *Daphnia,* or enchytreid worms
  Marine hydroids, if available

Preserved material
  Obelia
  Gonionemus
Prepared slides
  Stained hydras
  Budding hydras
  Male and female hydras
  Cross sections of hydras
  Obelia colonies
  Obelia medusae
Clean microscope slides
Coarse thread
Watch glasses and depression slides
10^{-4} reduced glutathione
Bouin's fluid
Compound microscopes
Dissecting microscopes or hand lenses
Coverslips

Notes

1. **Green vs. brown hydras for class use.** The green hydras, *Chlorohydra viridissima,* appear green because their cells are packed with the symbiotic green algae of genus *Chlorella.* A single hydra may harbor 150,000 algal cells contained within vacuoles of the hydra's cells. The algae provide the hydra with photosynthetic products, such as maltose, which the hydra rapidly converts to glucose for its own metabolic needs. The algae also supply oxygen, a by-product of their photosynthesis. In return, the hydra provides the algae with amino acids and nucleotides, which the algae use for protein and nucleic acid synthesis. Green and brown hydras share the same habitat and normally multiply at similar rates. If food is scarce, however, green hydras have the advantage.

   Green hydras make an interesting demonstration for the class. Brown hydras, however, are larger and more suitable for class study of movement and feeding behavior.

2. **Living *Obelia* colonies** on seaweed fronds may be purchased from suppliers of marine material (e.g., Marine Biological Laboratory, Woods Hole, Mass.; see Appendix B for address).

   In our experience, the polyps of living *Obelia* are rapidly cropped by several predators on the seaweed, especially small nudibranchs. If not examined within 2 or 3 days after collection, little may be left but the stem. Numerous other marine forms are present on the colonies (polychaetes, microcrustaceans, ectoprocts, etc.), which will be of great interest to students. Often the *Obelia* colonies are covered with diatoms.

   Other colonial hydroids (e.g., *Eudendrium, Pennaria, Tubularia, Bougainvilla*) are available as living material from marine suppliers. Of these, we have found *Tubularia* especially suitable for classroom work. *Tubularia crocea,* known as the pink-hearted hydroid, is common on wharves, pilings, and bridges on both the Atlantic and Pacific coasts. This species is mentioned briefly in a delightful book by Richard Headstrom (1984), along with comments on many other tubularian species. **T. crocea** is described in detail by T. H. Waterman (1950).

3. **Raising brine shrimp.** Brine shrimp *(Artemia)* are excellent for demonstrating feeding behavior in hydras. Brine shrimp eggs can be obtained from any biological supply company and hatched in a day or two. They are widely used as food for aquarium animals. Use a shallow, rectangular pan or tray of glass, plastic, or enamel (not metal). Cut a divider of glass, Plexiglas, or wood that will extend across the width of the pan, fitting snugly against the sides of the pan but lacking 1/2 to 1 inch of reaching the bottom. Fill the pan three-quarters full of salt water (natural or artificial seawater, or simply a tablespoon of sodium chloride dissolved in a quart of tap water). The divider should extend above the surface of the water.

All references to papers and publications mentioned in Appendix A are available on the website that accompanies *Animal Diversity,* sixth edition. Visit www.mhhe.com/hickmanad6e for more information.

Add about 1/4 to 1/2 teaspoon of dry brine shrimp eggs to the water in one side of the pan; they will float on top of the water and be confined to that side of the pan. Place an opaque cover over the pan except for 1 inch or so at the end farthest from the eggs. The hatched nauplii will be attracted to the light, swim under the divider, and cluster at the open end, where they may be easily siphoned or drawn off into a brine shrimp net or finger bowl. At 23.8° to 26°C (75° to 80°F), they will hatch in 24 to 48 hours. Two or three trays started daily on a rotating basis will keep a constant supply available for daily feeding.

If larvae are to be fed to freshwater forms, they should be rinsed well first.

If larvae are to be held alive for a period, they should be removed to another container of salt water where they are not crowded and are fed small amounts of yeast suspension or green one-celled algae daily. The water should be changed often to prevent fouling. Although aeration is not essential, it is helpful in keeping successful cultures.

Additional information on rearing *Artemia* is available in the classic compendium edited by J. G. Needham (1937).

Exercise 6B: Class Scyphozoa— *Aurelia,* a "True" Jellyfish

Materials

Preserved *Aurelia*
Living jellyfish, if available
Hand lenses or dissecting microscopes
Finger bowls
Ladle

Notes

1. **Obtaining living scyphozoans.** These are most readily obtained from suppliers of marine materials during spring and summer months; few are available during fall and winter, when most zoology courses are offered.

   The **scyphistoma** of *Aurelia,* however, can be obtained from the Marine Biological Laboratory (see Appendix B for address) October through April. These animals are relatively hardy and can be maintained in cooled seawater aquaria and even in aerated culture dishes. They are fed ground shrimp or particles of meat. The Gulf Specimen Company offers *Chrysaora* (stinging nettle jellyfish), and other species may be available upon inquiry. Additional references are provided in Brown, J. G. (ed.), 1937, *Culture Methods for Invertebrate Animals,* Comstock Publishing Company.

2. **The "upside-down jellyfish."** *Cassiopeia* (order Rhizostomeae) is common in the shallow waters around Florida. They usually can be obtained from Florida marine supply houses and can be kept for several days or weeks in a marine aquarium. This interesting jellyfish has eight thick, gelatinous oral lobes that are fused in such a manner as to obliterate the central mouth and form numerous canals with small openings in the oral lobes. The oral lobes also bear many small tentacles with nematocysts. There are 16 rhopalia, but no tentacles around the scalloped margin.

   Although *Cassiopeia* can swim by rhythmic pulsations of the bell, it spends much of its time lying oral side up on the bottom of lagoons and tidal pools, anchored by the suckerlike action of its aboral surface (hence the name "upside-down jellyfish"). Here, as it pulsates, it draws water over its oral lobes, bringing in food and oxygen. Small organisms are paralyzed by nematocysts, entangled by mucus, and swept into the canals by flagellar action.

Cassiopeia can be cultured in marine tanks of artificial seawater. The scyphistomae reproduce readily by two asexual methods—budding off young medusae or budding off small planuloid larvae that detach, swim about, and finally settle down and develop into scyphistomae.

Exercise 6C: Class Anthozoa— *Metridium* and *Astrangia*

Materials

Metridium, preserved
Living sea anemones and corals, if available
Finger bowls and/or dissecting pans
Dried corals
Astrangia, preserved

Notes

Sea anemone behavior. We find that 4 or 5 sea anemones will serve a class of 20 students. Allow 45 minutes for the behavioral study. At least a day before the laboratory, we place the sea anemones in clear refrigerator boxes if the anemones are large, or in finger bowls if they are small. These are kept submerged in the marine aquarium until the laboratory period, when they are lifted out and gently placed on the lab benches at the beginning of the period. Students are warned not to touch the anemones until they begin the behavioral study, since once an anemone strongly contracts from clumsy handling, it may not relax for a long time.

Touching an anemone with saliva on a coverslip may cause it to contract, so that should be done only near the end of the exercise.

Nematocysts discharged on a coverslip can be observed more easily if a drop of methylene blue or acid fuchsin is added.

The demonstration with acontia and nematocyst discharge is dramatic; this result alone is worth doing the exercise. The demonstration may be done with a single anemone if living material is in short supply. Prodding the anemone strongly to stimulate acontia discharge will not injure the animal, but it will remain strongly contracted for some time thereafter. The acontia are equipped with cilia, which cause the acontia to move slowly in a snakelike fashion on the slide.

Nematocysts of *Metridium senile* acontia are grainlike capsules, measuring 6×10^{-2} mm in length and about 5×10^{-3} mm in width. They are visible at low power but must be viewed at high power to reveal the discharged threads.

Experimenting in Zoology: Predator Functional Response: Feeding Rate in *Hydra*

Materials

Brine shrimp eggs and shrimp hatchery	55 mm disposable Petri dishes
	Disposable pipettes
Deep depression slides	Dissecting microscope (low-power)
Pond water	Clock or stopwatch
Brown hydra (enough for about 10 per student) that have not been fed for at least 24 hours	

Note

1. The most convenient prey to use for this experiment is *Artemia*, the brine shrimp. Brown hydra readily feed on the first stage nauplii of this arthropod but may have trouble handling them as they increase in size during later development. Therefore, brine shrimp should be hatched no earlier than 1–2 days before they are needed. Functional response rate will be recorded as the number of prey eaten during a 10-minute period as prey density increases.

Exercise 7: The Flatworms

Exercise 7A: Class Turbellaria— Planarians

Materials

Live planarians
Prepared slides
    Planaria, stained whole mounts
    Bdelloura, stained whole mounts
    Planaria, cross sections
Powdered carmine or talc
Black paper
    Snap-cap vials
    Fine artist's brushes
    Small mirrors
Small flashlights (pocket-size) or narrow-beam electric lamps
Modeling clay
Petri dishes
Spring, pond, or well water, or dechlorinated tap water (not distilled water or demineralized water)
Raw liver (or cut-up mealworms)
Ice cubes
Watch glasses or depression slides
Dissecting microscopes or hand lenses

Notes

1. **Sources of planaria for class study.** Biological supply houses such as Carolina, Ward's, and Connecticut Valley offer "brown planaria" (*Dugesia trigrina*), probably the best species for routine class study as well as the regeneration experiment. Most supply houses additionally offer "black planaria" (usually *Phagocata gracilis*) and semitransparent "white planaria" (usually *Procotyla fluviatilis*), which have highly visible digestive tracts.

    Planarians may be kept for some days in the plastic shipping containers *if* the water is changed daily and the inside surfaces of the containers are wiped clean of accumulated slime.

2. **Collecting and keeping planarians.** Planarians also may be collected on pieces of raw beef tied to stones or plants near the water's edge and can be kept in the laboratory if fed once a week on small amounts of fresh beef liver of hard-boiled egg yolk, pieces of earthworm, crushed snails, etc. Remove any uneaten food after an hour or so. The water should be changed frequently to prevent pollution.

3. **Reactions to stimuli.** When observing response to food by smearing a bit of liver on a coverslip inverted over a depression slide, caution students that the amount of liver must be

very small. If more than just a spot of meat is placed on the coverslip, the water will quickly become clouded and permeated with liver particles; with the odor of meat everywhere, the planarians then seem unable to locate the meat.

Experimenting in Zoology: Planaria Regeneration Experiment

This may be done as an extra-credit exercise. We recommend using a tissue-covered ice cube rather than other methods for quieting and extending planarians; cuts can be made with precision under the dissecting microscope, and the animals are not harmed by chilling; 30-ml snap-cap plastic vials make excellent containers and are reusable. Punch a single small hole in the lid to allow air exchange. Alternatively, use the screw-cap specimen jars in which protozoans and other invertebrates are shipped from biological supply houses. Partial longitudinal cuts will almost always tend to grow back together and must be recut two or even three times during the first 24 hours for success (using a tissue-covered ice cube). Even so, some will rejoin even after repeated resectioning.

Keep planarians in separate containers after cutting to prevent whole specimens from cannibalizing cut specimens.

Brown planarians (*Dugesia tigrina*) are best for the regeneration exercise. Fortunately, since students are often forgetful, planarians will live for weeks without much attention.

There are excellent discussions of planaria regeneration in Pearse et al. (1987, pp. 214–221) and in Buchsbaum et al. (1987, pp. 171–179). This material should be made available to students to assist them in their written reports.

Exercise 7B: Class Trematoda— Digenetic Flukes

Materials

For study of *Clonorchis:*
    Stained slides
        Clonorchis
        Miracidia
        Cercariae
Compound microscopes (dissecting microscopes useful for viewing entire whole mount)
For study of *Schistosoma:*
    Prepared slides
        Schistosoma mansoni in copula
        Schistosoma eggs
    Compound microscopes
For observations of living flukes:
    Pithed leopard frogs *(Rana pipiens)*
    Live snails (*Physa, Certhidea,* others)
    Dissecting tools
    Normal saline solution (0.65%)
    Syracuse watch glasses
    Microscope slides
    Compound and dissecting microscopes

Notes

Prepared microslides. See Appendix B for a list of suppliers. Most suppliers offer whole mounts of *Clonorchis* and *Schistosoma mansoni.* Miracidiae, sporocysts (in section of snail tissue), and cercariae are also available and make useful demonstrations to supplement the schistosome exercise.

All references to papers and publications mentioned in Appendix A are available on the website that accompanies *Animal Diversity,* sixth edition. Visit www.mhhe.com/hickmanad6e for more information.

Exercise 7C: Class Cestoda— Tapeworms

Materials

Preserved tapeworms
Prepared slides of tapeworms showing scolex and immature, mature, and gravid proglottids
Microscope

Notes

1. **Choice of tapeworm.** This exercise has been written for use with either *Taenia pisiformis* or *Dipylidium caninum,* depending on the instructor's preference. If new slides are being purchased (see Appendix B for a list of microslide suppliers), it is prudent to check slide quality carefully before accepting an order; many commercially available tapeworm slides are of poor quality, suffering from a variety of problems such as overstaining, understaining, or selection of mature segments that do not clearly reveal reproductive structures.

2. **Measly meat demonstration.** In earlier editions of this manual, we suggested studying a piece of "measly" meat infected with tapeworm cysts or bladder worms. However, we have been unable to locate commercial suppliers of measly meat. Nevertheless, measly meat makes a dramatic demonstration, and it may be worth the effort to try to obtain a sample of formalin-preserved infected meat through a veterinarian.

Exercise 8: Nematodes and Four Small Protostome Phyla

Exercise 8A: Phylum Nematoda— *Ascaris* and Others

Materials

Preserved *Ascaris* specimens
Prepared slides
    Stained *Ascaris* cross sections
    Trichinella cysts in pork muscle
    Trichinella, male and female
    Necator americanus
    Enterobius vermicularis
    Turbatrix
Living *Turbatrix*
Soil samples, if desired
Slides and coverslips
Razor blades
Dissecting pans and pins
Microscopes

Notes

1. **A caution.** The eggs of female *Ascaris* may remain viable for many weeks in formalin-preserved females. Although the chances of infection are remote, students should be cautioned to wash their hands after dissecting the worms.

2. **How to quiet *Turbatrix*.** We find that *Turbatrix* can be quieted for study most effectively by *gently* warming the slide over a lamp. Others prefer to add a drop of 1 N HCl to the culture. With this alternative, the worms will remain active for several minutes before becoming quiescent. An advantage to

this alternative is that students sometimes see the birth of living juvenile worms as the mother slowly succumbs to the acid.

    Turbatrix may be weakly stained for study with Nile blue sulfate in 70% alcohol.

    Some simple experiments with *Turbatrix* are suggested by Goldstein and Metzner (1971, pp. 161–165).

Exercise 8B: A Brief Look at Some Other Protostomes

Materials

Living materials
    Philodina, or mixed rotifers
    Chaetonotus, or mixed gastrotrichs
    Gordius, or other nematomorphs, if available
Preserved or plastic-mounted material
    "Horsehair worms" (such as *Gordius*)
    Spiny-headed worms (such as *Macracanthorhynchus*)
Prepared slides of any of the previously listed types
Slides, depression slides, and coverslips
Microscopes

Exercise 9: The Molluscs

Exercise 9A: Class Bivalvia (= Pelecypoda)—Freshwater Clam

Materials

Living bivalves in an aquarium
Living or preserved clams for dissection
Clean, empty bivalve shells
Dissecting pans
Pasteur pipettes
Glass rod
Carmine suspension

Notes

Living clams are recommended for this exercise. They are odorless, the tissues retain their natural color, and students can observe ciliary action and the beating heart. It is best for the instructor to open the clams; with a bit of practice, this can be done rapidly and with minimal damage to internal structures. Heating the clams in warm (not hot!) water causes them to relax and facilitates insertion of the knife between the shells. Use a stout, strong-bladed knife that can be forced with safety between the valves, while holding the clam against a firm surface. Scalpels should never be used to open clams.

    After opening a living clam, it should be covered with pond water or dechlorinated tap water.

Exercise 10: The Annelids

Exercise 10B: Class Oligochaeta— Earthworms

Materials

Living earthworms
7% ethanol for anesthetizing living worms for dissection
    (immerse worms in anesthetic 30 minutes before class use)

Stained cross section of earthworms
Paper towel
Glass plates
Dissecting microscopes (or hand lenses)

Notes

1. **Earthworm behavior.** Worms respond better to a strong light (negative phototaxis) if they are kept in subdued light first.

   If a worm is not responding as expected to moisture and light, try covering it with a second paper towel also moistened in the center, so that the worm is gently "sandwiched" between the two towels. It will better sense the difference between moist and dry areas of the towels.

2. **Using living earthworms.** This exercise applies to living earthworms but may be easily adapted to preserved worms. However, because living worms are actually cheaper than preserved worms, it makes little sense to use the preserved. Living worms offer several advantages: organs have their natural color and are more easily identified; the circulation is prominent and easily traced; and the gut, especially the gizzard, reveals its natural peristaltic movements.

   Take care that worms are fully anesthetized before students begin dissections. Ethanol *cannot* be added to the saline once the worms are opened because it kills them.

   For those who may object to using living animals for dissection, but who want students to see the natural internal pigments of the living animal, living earthworms may be killed by immersion for 30 minutes in 10% to 15% alcohol just before the laboratory.

3. **Saline for dissection.** Living earthworms should be covered with near-isotonic saline after they are opened. Frog saline (0.6% NaCl, or 6 g NaCl per liter of solution) almost perfectly matches the osmolarity of earthworm body fluid. Several liters will be needed for a laboratory. We find it convenient to place the saline in a 5-gallon jug that has a spigot at its base. Elevate the jug and place a pan beneath the spigot to catch spilled saline.

   We use dissecting microscopes for this laboratory. The dissecting pan with the pinned-out and saline-covered earthworm is supported on plywood "bridges" that span the microscope stage.

Exercise 10C: Class Hirudinea—Leeches

Materials

Living leeches
Preserved *Hirudo*

Notes

1. **How to see the nephridiopores.** Gently squeezing an alcohol-narcotized (7% ethanol) specimen will cause a little fluid to be exuded from the nephridial bladders.

2. **Medicinal leeches.** Medicinal leeches are available from Carolina Biological, from Ward's, and from Leeches U.S.A. (see Appendix B). These are large specimens, easily maintained in pond or dechlorinated water in large stacking dishes; the dishes must be kept covered because these leeches show a strong propensity to wander. Students find them fascinating.

However, *use caution* in allowing these leeches to fasten themselves for any length of time to students' hands or arms. If hungry (and all that we have received are), they will waste little time in slicing into the skin with their triradiate, sawlike, chitinous jaws to initiate blood flow. The salivary glands of medicinal leeches contain a complex mixture of anticoagulant (hirudin, an antithrombokinase), anesthetic (leeching is nearly painless), vasodilators that cause tissues surrounding the wound to swell, and other compounds that liquefy coagulated blood (hence the effectiveness of medicinal leeches in plastic surgery for reestablishing circulation in replanted appendages or skin-flaps that are failing due to blood engorgement).

We have had students willingly "leech" themselves in our laboratory to the utter captivation of all. But this should be approached with caution for two reasons: (1) the Y-shaped wound will continue to bleed for several hours, and (2) there is a marginal danger of infection from the bacterium *Aeromonas hydrophilia,* which is present in the saliva of leeches (Whitlock et al., 1983). Wound seepage can be stopped easily with a compression bandage, which must be kept in position for at least 24 hours. Although the chances of infection appear to be remote (see, for example, Rao et al., 1985), the possibility of wound infection will doubtless dissuade most instructors in our increasingly litigious American society from subjecting students to "leeching."

Exercise 11: The Chelicerate Arthropods

Exercise 11: The Chelicerate Arthropods—Horseshoe Crab and Garden Spider

Materials

Living materials, if available
   Limulus, in marine aquarium
   Spiders, in terrarium
Preserved materials
   Limulus
   Golden garden spiders
Dissecting microscopes or hand lenses

Notes

1. **Time required for exercise.** Study of the external anatomy of both the horseshoe crab and the garden spider requires little more than 1 hour. We combine this exercise with Exercise 12 for a 4-hour laboratory period.

2. **Large golden garden spiders,** *Aurelia aurantia,* are available from Connecticut Valley Biological. This is the best source we have found. Those offered as *Argiope* by Carolina are a different species, much less suitable for this exercise.

3. **Suggestions for spider dissection.** Some instructors prefer to have students keep the spider covered with water during dissection, but the spiders may be examined in air without damage if glycerol is added to the alcohol preservative (about 5 cc per 100 ml of 70% ethanol) to reduce drying. Deep Petri dishes (e.g., 100 × 20 mm) containing a thin layer (about 0.5 cm thick) of dissecting pan wax make excellent containers. The spider can be secured with insect micropins. This exercise is much more successful if the students have good dissecting microscopes (rather than hand lenses) and good illumination from a bright focusing spotlight.

All references to papers and publications mentioned in Appendix A are available on the website that accompanies *Animal Diversity,* sixth edition. Visit www.mhhe.com/hickmanad6e for more information.

Another excellent way to hold spiders in any position for study is to place them in a Syracuse dish containing some washed sand. Cover with alcohol. The spider is held in the desired position by pushing it gently into the sand.

Exercise 12: The Crustacean Arthropods

Exercise 12: Subphylum Crustacea—Crayfish, Lobsters, and Other Crustaceans

Materials

Living materials
  Crayfish or lobsters
  Cultures of developing *Artemia* in different developmental stages
      (directions in "Notes")
  Other crustaceans, such as *Eubranchipus, Daphnia, Cyclops,*
      ostracods, barnacles, and crabs, as available
Preserved materials
  Crayfish or lobsters (may be injected)
  Barnacles, crabs, and the like, as available
Bowls or small aquaria
Dissecting pans
Slides and coverslips
Compound microscopes

Notes

1. **Time required for exercise.** Allow 3 hours to complete the entire exercise, with additional time as needed for individual oral demonstrations. For shorter laboratory periods, we suggest abridging the study of crayfish appendages.

2. **Mounting appendages.** We use 4 × 10 inch pieces cut from white Bristol board. Students glue the appendages to the board in the same order they appear in Figure 12.3 using Elmer's glue.

3. **Latex-injected crayfish.** Usually the heart is destroyed in such preparations, and perfusion of the arteries with latex is frequently poor. Uninjected specimens should be used to demonstrate the anatomy of the undamaged crayfish heart.

4. **Raising brine shrimp:** See notes for Exercise 6A.

Experimenting in Zoology: The Phototactic Behavior of *Daphnia*

Materials

Daphnia
Stopwatch
Open-ended glass column (2 to 3 cm diameter, 25 to 30 cm long)
Ring stand and two clamps
Dissecting scope illuminator or other small light
Pipette
Wax pencil

Notes

1. **Sources of *Daphnia*.** Living *Daphnia* may be obtained from any biological supply house (see Appendix B for listings). They also can be collected by using a plankton net in shallow ephemeral pools that do not contain fish.

2. **Light source.** Dissecting scope illuminators work well because they can be mounted on the ring stand and have a narrow beam. If the illuminator has a switch for variable beam strength, make sure that all students are using the same intensity and that they do not vary it from trial to trial. Other lamps can be used as long as they have a narrow beam that can be directed through the glass toward the rubber stopper.

3. **Drawing conclusions about the data.** After the class data have been compiled, a chi-square test can be used on the control and lighted experiments to see if the *Daphnia* are demonstrating phototaxis.

Exercise 13: The Arthropods: Myriapods and Insects

Exercise 13A: Myriapods—Centipedes and Millipedes

Materials

Living materials, if available
  Centipedes
  Millipedes
Preserved materials
  Centipedes, such as *Scolopendra* or *Lithobius*
  Millipedes, such as *Spirobolus* or *Julus*
Small terrariums or jars for living materials
Dissecting pans and instruments
Dissecting microscopes

Notes

Time required for this exercise. An examination of centipede external anatomy requires 25 to 30 minutes. Allow another 15 minutes if examining both a centipede and a millipede. This exercise can be combined with either Exercise 13B or 13C for a 2-hour laboratory period. For a 3- or 4-hour period, combine this exercise with both Exercises 13B and 13C.

Exercise 13B: Insects—Grasshopper and Honeybee

Materials

Living grasshoppers, if available
Preserved materials
  Grasshoppers, such as *Romalea*
  Apis, workers
  Apis, queen, drones, larvae, pupae (for demonstration only)
Dissecting pans or dishes (see note 4 for Exercise 13C)
Dissecting microscopes

Notes

1. **Demonstration mounts.** Demonstration mounts of insect parts, such as legs, wings, antennae, and mouthparts, can be prepared very simply. Clean a microslide with acetone, spray on a very thin coat of clear lacquer or varnish from a pressurized can from enough distance to prevent formation of bubbles, orient the parts on the slide while the lacquer is still wet, and label. A coverslip is not necessary. The mount is dry and usable within 30 minutes and can be stored flat in a dust-protected tray for future use.

Slides that show the various types of insect antennae can be purchased or may be prepared by this method.

2. **Demonstrations of adult and immature insects.** These may be selected to illustrate the types and growth stages—direct development and gradual and complete metamorphosis.

Exercise 13C: Insects—House Cricket

Materials

Crickets, alcohol-preserved
Crickets, living
Dissecting dish with wax base
Insect pins
Fine-tipped scissors
Fine-tipped forceps
Carbon dioxide cylinder and small jars for anesthetizing crickets
Ground dog food mixed with carmine and moistened (potato slices sprinkled with carmine make an acceptable substitute)
Plastic squeeze bottles
Methylene blue stain, 0.5% in insect or amphibian saline
Insect saline (7.5 g NaCl/liter) or amphibian saline (6.0 g NaCl/liter)
Dissecting microscopes and focused lighting

Notes

1. **Time required for this exercise.** About 2 hours is needed, with additional time for oral demonstrations. We combine this laboratory with Exercise 13A and most of 13B in one 4-hour laboratory period.

2. **The cricket is an ideal insect for this exercise** because of the ease with which it can be cultured and contained, its lack of distastefulness (compared with cockroaches), its relatively large size, and the ease of determining age and sex.

3. **Rearing crickets.** House crickets are omnivorous and are easily reared on chicken starter mash. Crickets can be kept in any container, such as a steel or fiberglass rat cage or an aquarium. No lid is necessary if stainless steel or glass containers are used because the crickets cannot climb the walls, unless the container is less than 20 cm high. If plywood is used, glue a 2- to 3-cm-wide strip of heavy-duty aluminum foil around the inner edge to form an escape-proof barrier. Do not cover the bottom of the container with sand, which tends to get wet and moldy. Egg cartons make good hiding places. Water is provided using plastic vials (55 to 75 ml capacity) with shallow slits cut in the top edge. These are filled with water and inverted over plastic Petri dish bottoms, which are then filled with small gravel to keep the crickets from drowning. Temperature should be 27° to 30°C for best growth. More information on rearing methods is found in the paper by Clifford et al. (1977).

4. **Dissecting dishes for this exercise** are 20 × 100 mm glass Petri dishes containing about 5 mm of dissecting pan wax.

5. **Reproductive behavior.** Two male crickets placed together in a jar usually display and sing competitively. Squeezing a male usually causes extrusion of a spermatophore, which, when placed in water, extrudes a stream of sperm that is visible with a microscope. If several pairs are separated in glass jars, a little patience will be rewarded with the female mounting the male and the transfer of a spermatophore.

All references to papers and publications mentioned in Appendix A are available on the website that accompanies *Animal Diversity,* sixth edition. Visit www.mhhe.com/hickmanad6e for more information.

Exercise 13D: Collection and Classification of Insects

Materials

For collecting:
    Insect nets (aerial, sweep, and water nets)
    Cheesecloth
    Collecting bottles
    Cellucotton or envelopes
For killing:
    Small screw-top bottles
    Ethyl acetate or carbon tetrachloride
    Cotton
Cardboard or blotting paper
For preserving:
    Mounting boxes
    Cotton
    Transparent cover (glass or acetate)
    Insect pins
    Cork pinning boards or insect spreading boards
    Labels
    70% alcohol
    KAAD (optional)

Notes

KAAD mixture for killing insect larvae can be prepared as follows:

Kerosene	10 ml
Glacial acetic acid	20 ml
95% ethyl alcohol	70 to 100 ml
Dioxane	10 ml

Larvae should be ready to transfer to alcohol for storage in 1/2 to 4 hours. For soft-bodied larvae, such as maggots, the amount of kerosene should be reduced.

Exercise 14: The Echinoderms

Exercise 14A: Class Asteroidea—Sea Stars

Materials

Living sea stars (*Asterias, Pisaster,* and others) in seawater
Preserved (or anesthetized) sea stars for dissection
Dishes for live material
Fresh or frozen seafood for feeding sea stars
Carmine suspension
Pieces of dried sea star tests
Fine artist's brushes
Dissecting pans and tools
Dissecting microscopes or hand lenses

Notes

1. **Sea star demonstrations.** Several different types of dried sea stars placed on demonstration will emphasize diversity and adaptive radiation within the group. The basic pentamerous body plan varies. Some species have numerous rays, basically multiples of 5, although above 10 rays, a sea star's arithmetic is not perfect and there are stars with 17 or 32 rays, for example. Point out that many sea stars are large and quite colorful, especially those of the West Coast, in contrast to most Atlantic Coast species, which tend toward more muted colors. If available, a

sea star with one or more stumpy rays illustrates the capacity to regenerate lost limbs.

2. **Microslides of asteroid larval forms.** Microslides of bipinnaria and brachiolaria larvae and of pedicellariae that may be used for microscope demonstrations are available from biological supply companies. See Appendix B for a list of suppliers of prepared microslides.

3. **Using living sea stars for external structure.** Observations of the aboral and oral surfaces can be accomplished to advantage on living sea stars. Place a sea star in a large culture dish or refrigerator dish, cover with seawater, and examine with a dissecting microscope. We use a wooden bridge to support the dish on the stage of a standard dissecting microscope.

4. **Demonstrating ampullar action with living sea stars.** If living sea stars are available, cut off the arm of a partly anesthetized animal (see note 5 below) and remove the dorsal surface. The disembodied arm will continue to move around in a dish of seawater for a long time, and students can see the action of the exposed ampulla. This provides a captivating demonstration that is well worth the mutilation of one animal. Further, if the sea star is large enough to have mature gonads, a smear of testis or ovary will reveal mature or developing sperm or eggs. Sea star spermatozoa have relatively large heads and long tails.

5. **Anesthetizing living sea stars for dissection.** Few schools are blessed with an easy supply of living sea stars, but for those that are, living material is obviously preferred to preserved. Sea stars *must* be anesthetized before dissection, using $MgCl_2$ or $MgSO_4$ dissolved in fresh water at a concentration of about 7%. We find it requires 40 to 60 minutes for full anesthetization. Sea stars will recover if returned to seawater. The exercise requires only slight modification for use with living material.

6. **Demonstrating action of podia.** Another effective demonstration of podia is provided by fastening a living sea star, oral side up, on a piece of plate glass by means of rubber bands. Lay a small piece of celluloid or thick polyethylene film on top of the tube feet and have students note the action of the tube feet and the direction the tube feet move.

7. **Demonstrating the action of coelomic cells.** Inject into the coelom of a live star 5 ml of carmine suspension in seawater. Set the animal aside for about 8 hours; then examine it under a dissecting microscope for the appearance of circulating particles in the skin gills. Pinch off some of the gills and examine them under a compound microscope. Some of the carmine particles will have been picked up by phagocytic cells of the coelom, and these may be seen migrating through the thin walls of the gills. Such cells appear to have an excretory function. Examine drops of coelomic fluid for the presence of other coelomic cells.

8. **Preparing echinoderm ossicles and pedicellariae for demonstration.** Put leftover skeletal parts (even whole specimens) into 5% to 10% potassium hydroxide solution or into full-strength commercial bleach. Warm over a Bunsen burner or boiling water to dissolve away the flesh. Let stand several hours; then decant carefully. Add fresh water, let stand again, and decant. The very tiny skeletal fragments settle slowly. Repeat until free of potassium hydroxide and debris. Cover with alcohol and again wash and decant, using two or three changes of alcohol. Now add 90% alcohol, shake, and put a drop on a clean slide. Quickly ignite the alcohol with a match. Add a drop or two of mounting medium and cover.

Exercise 14B: Class Ophiuroidea— Brittle Stars

Materials

Preserved brittle stars
Dissecting microscopes
Living brittle stars, if available
Fresh seafood for feeding, if living forms are used

Notes

1. **Observations on locomotion of living brittle stars.** This is best done in an aquarium or a tray large enough to allow free movement and containing enough seawater to enable the animal to right itself when turned over. Feeding is unpredictable. Some brittle stars will burrow if placed on a sandy substrate.

2. **Demonstrating brittle stars.** Dried brittle stars seldom survive student handling; plastic-embedded specimens are available. A dried basket star makes an interesting comparison with "conventional" brittle stars for demonstration. Basket stars (e.g., *Gorgonocephalus* sp. of the North Pacific), with their highly branched arms and tendril-like tips, belong to a separate order (Phrynophiurida) from that containing most other brittle stars (Ophiurida).

3. **How to visualize the vertebrae and vertebral articulations.** Place a brittle star arm in commercial bleach for several hours; then examine with a dissecting microscope.

Exercise 14C: Class Echinoidea— Sea Urchins

Materials

Living sea urchins in seawater
Preserved sea urchins
Dried or preserved sand dollars and/or sea biscuits
Dried tests of sea urchins
Glass plates
Large finger bowls
Carmine suspension
Dissecting pans and tools

Notes

1. **Demonstrations.** Provide dried or plastic-embedded echinoides, such as sand dollars (or key-hole urchins), sea biscuits, heart urchins, and various examples of sea urchins; preserved and/or dried Aristotle's lantern; microslides of sea urchin pedicellariae and pluteus larvae.

2. **Preparing the internal organs of a sea urchin for study.** Submerge only the aboral half of the animal in 2% nitric acid for 24 to 48 hours. Transfer to and submerge in water; cut a circle outside the periproct and extend the cuts through the ambulacral areas to the equator. The untreated portion of the test will hold the animal together for study.

3. **Preparing dried tests.** Submerge the animals in fresh commercial bleach (sodium hypochlorite). Complete removal of organic matter with a toothbrush.

4. **A study of living sand dollars.** If live sand dollars, such as the East Coast *Mellita,* Caribbean *Leodia,* or West Coast *Dendraster,* are available, they should be kept in sandy-bottomed containers. Place one in a bowl of seawater and examine with a hand lens or long-arm dissecting microscope. Have students examine spines

on the aboral surface. Are they movable? Note the petal-shaped ambulacra; they are called petaloids. The podia are adapted for gas exchange rather than locomotion. Look at the flattened oral surface. Are the spines movable? How do they differ from those of *Arbacia*? Note the central mouth and the five-toothed chewing apparatus. Return the animal to the sandy bottom and observe how it burrows under the sand, using its oral spines to move the sand. Sand dollars feed on minute organic particles from the sand. They are passed back by tiny, club-shaped aboral spines, caught in mucus, and carried by ciliary currents to food grooves on the oral side that lead to the mouth.

Exercise 14D: Class Holothuroidea— Sea Cucumbers

Materials

Living sea cucumbers in an aquarium or in bowls of seawater
Preserved or relaxed sea cucumbers for dissection
Prepared slides of holothurian ossicles

Notes

1. **How to relax sea cucumbers for dissection.** Inject living animals with 5 to 10 ml of 10% magnesium chloride an hour or so before use. Tentacles then may be gently forced out the anterior end.

2. **How to see sea cucumbers eviscerate themselves.** While sea cucumbers are famous for their radical evisceration, this is not readily accomplished merely by handling the animal roughly—at least not with *Thyone*. W. M. Reid (in Brown, op. cit.) suggests placing the animal for about a minute in 0.1% ammonium hydroxide made up in seawater. "Violent circular muscular contractions usually occur. If these movements do not begin spontaneously, hold the animal up by the posterior end for a few seconds. The entire anterior end of the animal with tentacles, cal-careous ring, and the digestive system is violently expelled." This works. These structures will regenerate in 2 to 3 weeks, but the regenerated animal is much smaller than before. According to Pierce and Maugel (1987), *Thyone* can be made to eviscerate by injecting potassium chloride into the coelomic fluid.

3. **Demonstrations.** Appropriate demonstrations to accompany this exercise include microslides of sea cucumber ossicles and holothurian larval forms (e.g., auricularia and doliolaria larvae).

4. **How to prepare holothurian ossicles for study.** Holothuroid taxonomy is based in large part on the structure of ossicles in the body wall. These consist of microscopic discs, rods, buttons, and "tables," many of them intricate in form. They make a fascinating demonstration for students when prepared as permanent slides. Ossicles are destroyed by formalin, so they must be prepared from alcohol-preserved sea cucumbers. To prepare ossicles for examination, cut a small piece from the body wall, place it in a small test tube, and add a few ml of Clorox. Let stand until all the flesh has dissolved (an hour or two), leaving a sediment of whitish particles on the bottom. Pour off the supernatant fluid carefully and add clean water. Allow to settle, decant the water carefully (or use a Pasteur pipette), and resuspend in a few ml of alcohol. Again allow to settle; then pipette some of the sediment onto a slide. Allow to dry. Add a drop of mounting medium to the dried sediment and add a coverslip. The same method can be used for the ossicles and pedicellariae of other echinoderms.

All references to papers and publications mentioned in Appendix A are available on the website that accompanies *Animal Diversity*, sixth edition. Visit www.mhhe.com/hickmanad6e for more information.

Exercise 15: Phylum Chordata

Exercise 15A: Subphylum Urochordata—*Ciona*, an Ascidian

Materials

Living (or preserved) *Ciona, Molgula, Corella,* or other small, translucent ascidian
Whole mounts of ascidian larvae
Carmine suspension in seawater
Finger bowls
Compound and dissecting microscopes

Notes

1. ***Ciona intestinalis*** has been the subject of several excellent morphological studies (see especially Miller, 1953, and Roule, 1884), so its anatomy is well understood. Goodbody (1974) should be consulted for ascidian physiology. The cosmopolitan *Ciona intestinalis* is probably the only species of the genus. The description in our exercise is largely "generic" in that it will describe genera other than *Ciona*.

2. **Whole mounts of ascidian tadpole larvae** are available from Carolina Biological, cat. 30-8260. The specimen used for the preparation of Figure 15.4 came from the series in Carolina's stock. They are far superior in quality to tadpole larvae that we sampled from other suppliers of microslides.

3. **Demonstrations.** Place on demonstration a variety of tunicates, both single and colonial (available from biological supply companies), and microslides of various stages of meta-morphosis of the ascidian tadpole. Good slides are difficult to find; you may have to order samples from several suppliers to get satisfaction.

4. **Living tunicates.** Both solitary and colonial tunicates are avail-able from marine suppliers. With transparent forms such as *Ciona intestinalis* or different species of *Clavelina*, the branchial sac can be viewed directly through the test.

5. **Behavioral observations.** Those suggested in the exercise are described in more detail by L. H. Kleinholz (in Brown, 1950, p. 554). The so-called crossed reflex is involved in coughing. If the inside of the siphon is tickled, the normal response is to close the *other* siphon and then contract the body. This would force out the source of an irritation (see Goodbody, 1974, p. 99). Sensory cells in *Ciona* have not been precisely identified but appear to be concentrated in the siphons.

Exercise 15B: Subphylum Cephalochordata—Amphioxus

Materials

Preserved mature amphioxus
Live animals can be obtained from Gulf Specimen Marine Labs (www.gulfspecimen.org)
Slides
　Stained and cleared whole mounts of immature amphioxus
　Stained cross sections of amphioxus
Watch glasses
Microscopes

Notes

Branchiostoma virginiae, commonly found along the south-eastern coast of the United States, is the species of amphioxus usually supplied as preserved and mounted material. *Branchiostoma californiense* occurs along the Pacific Coast from San Diego southward. Most of the detailed anatomical and physiological studies have been made on *Branchiostoma lanceolatum,* the European species. *Asymmetron,* the only other genus of cephalochordate, is so named because the gonads exist on the right side only.

Note that adult preserved specimens of amphioxus usually have blocks of gonads visible through the body wall; these are shown in Figure 15.5. However, the juveniles that are usually chosen for whole mounts because of their smaller size lack gonads.

Exercise 16: The Fishes

Exercise 16A: Class Petromyzontida—Lampreys (Ammocoete Larva and Adult)

Materials

Preserved material
    Ammocoetes (lamprey) larvae
    Adult lamprey specimens
    Longitudinal and transverse sections of lampreys
Watch glasses
Slides
    Stained whole mounts of ammocoetes
    Cross sections of ammocoetes
Compound and dissecting microscopes

Notes

1. **Life history set of a lamprey.** Ward's Natural Science Establishment offers a plastic-embedded life-history sequence of *Petromyzon* (eggs, several developmental stages, and section through head of the adult). Carolina offers embedded ammocoetes.

2. **Preparation of transverse sections of adult lampreys (or other large animals).** Freeze the preserved and injected animal; then cut it into 5-cm transverse sections with a sharp hacksaw. Clean the sections carefully and secure the viscera in place with insect pins. Place each section in a container that is slightly larger diameter than the section, and cover with a melted 2.5% solution of agar, being careful while the solution is still warm to place the organs properly and to expel trapped air bubbles. When they have cooled, cut away excess agar from the surface. These preparations can be stored in a formaldehyde solution for long periods. Longitudinal sections can be made by the same method.

Exercise 16B: Class Chondrichthyes—Cartilaginous Fishes

Materials

Preserved dogfish sharks
    or
Longitudinal and transverse sections of a shark

Exercise 16C: Class Actinopterygii—Bony Fishes

Materials

Preserved, injected perch (other species of teleosts may be substituted)
Living fishes, any kind, in aquarium
Stained slides of fish blood
Mounted fish skeletons and skeletons of other vertebrates
Longitudinal and cross sections of preserved perch

Notes

1. **Watching living fishes.** An aquarium containing living fish should be made available for observation of swimming and respiratory movements.

2. **Demonstrating diversity within bony fishes.** Diversity, especially within the teleosts, can be illustrated by putting on display a variety of preserved specimens.

3. **Demonstration of fish scales and chromatophores.** To demonstrate fish scales and the chromatophores overlying them, remove with fine forceps a few scales from a living fish (e.g., goldfish, minnow, or perch). The fish does not have to be anesthetized, although this can be done easily by immersing the fish in tricaine methanesulfonate (MS-222), 0.2 to 0.05 g/liter (there is a wide latitude of permissible dosage). Mount the scales in a drop of water on a slide under a coverslip and observe under low power. Note the different types of chromatophores and the amount of pigment dispersal.

    If desired, the effect of drugs or hormones on the chromatophores can be studied by adding a drop of the drug or hormone to the water on the slide. Any of the following can be used: epinephrine (1 mg/ml), acetylcholine (100 mg/ml), melatonin (0.5 mg/ml), or pituitary extract (1 g beef pituitary powder [whole gland] shaken up in 10 ml of frog Ringer solution and filtered after letting stand 30 minutes).

Experimenting in Zoology: Aggression in Paradise Fish, *Macropodus opercularis*

Materials

Male paradise fish
Glass aquaria (3–10 gal in size)
Aquaria lights
Stopwatch
Counters

Notes

When placed together, males of this species almost always interact in some fashion. Not all behaviors, however, will be obviously aggressive. Students should record more subtle interactions as well—for example, time spent close together, fin raises, flaring of the opercula. If the fish appear disinterested in interacting, they may be distracted by observers. Make sure that the room lights are off and that only the tank is illuminated. This minimizes the chances that the fish will be distracted by outside movement.

Experimenting in Zoology: Analysis of the Multiple Hemoglobin System in *Carassius auratus*, the Common Goldfish

Materials

Goldfish
0.9% saline
1.5-ml microcentrifuge tubes
1-ml syringes with 26-gauge needles
Powdered heparin
Microcentrifuge
Ice buckets
10 cm × 10 cm 8% native polyacrylamide gels (Novex)
2× polyacrylamide gel loading buffer
Tris-glycine running buffer
Low-voltage power supplies
20, 200, and 1000 μl micropipettes and disposable tips

Notes

1. **Animals: *Carassius auratus.*** The common goldfish may be purchased at any local aquarium supplier for modest cost. Animals in the 10 to 20 g size are recommended; however, very little blood is needed, so small animals will suffice. It is recommended that animals be treated in accordance with institutional and governmental guidelines for the humane care and treatment of laboratory animals for teaching and research.

   Anesthetized animals should be dispatched by medullary transection, and 1 ml of blood should be removed from the caudal vein or heart into a syringe that contains a small amount (about 1 mg) of heparin. Transfer the blood sample to a 1.5 ml microcentrifuge tube and centrifuge at 10,000 g for 5 minutes to pellet the red blood cells. Remove the serum and discard. Add 1.0 ml of 0.9% saline, resuspend the cells, and repeat the centrifugation. Remove the saline wash and discard. Add 200 μl (about 3 to 4 volumes) of distilled water to the red blood cell pellet and resuspend. The cells should lyse in 1 to 2 minutes. Centrifuge the sample at 12,000 g for 5 minutes to pellet the cell ghosts. Transfer the clear red supernatant to a clean tube and place on ice.

   Human hemoglobin may be collected from a finger prick using a sterile lancet after cleansing the fingertip with a 70% alcohol swab. Collect about 1 ml of blood into a syringe containing a small amount (about 1 mg) of heparin. (A needle is not necessary for this collection.) Prepare the hemoglobin lysate as described for the goldfish hemoglobin.

2. **Sample preparation.** For each sample, mix equal volumes of the hemoglobin lysate with polyacrylamide 2× gel loading buffer (100 mM Tris, pH 8.3, 10% glucerol, 0.0025% bromophenol blue) available from Novex, see note 3). Load 20-μl samples onto the gel using a micropipette.

3. **Vertical polyacrylamide gel electrophoresis of hemoglobin.** Necessary equipment includes a mini–vertical gel electrophoresis apparatus appropriate for running 10 cm × 10 cm polyacrylamide gels and a low-voltage power supply. Hemoglobin isoforms may be separated on an 8% native Tris glycine polyacrylamid gel. These gels may be purchased from

Invitrogen (www.catalog.invitrogen.com). The 1-mm thick, 10-well comb format is appropriate for this application (Novex cat. #EC6015). The gel is electrophoresed in 1× Tris-glycine running buffer (25 mM Tris, 192 mM glycine pH 8.3) for approximately 30 to 60 minutes. Running buffer stocks (10×) and sample buffer (2×) may also be purchased from Novex (cat. #LC2672 and LC2673).

Exercise 17: Class Amphibia

Exercise 17A: Behavior and Adaptations

Materials

Living frogs
Jars or bowls
Aquarium
Pond or dechlorinated water
Paper towel
Fruit flies (for feeding frogs)

Notes

1. **The animal rights question.** It may happen at this point in the zoology laboratory that one or more students will object to dissecting preserved frogs, perhaps asking, Do we have the right to use vertebrate animals in research? How can it be justified? The issue is addressed in a statement following the General Instructions for this manual (see p. x), but the objection may still arise. We can respond by asking whether a snake has the right to eat the frog or whether the frog has the right to eat an insect. Just as the snake eats the frog to survive, humans not only use animals as food for survival but also apply our abilities through biomedical research with animals to learn ways to minimize human pain, disease, and suffering—and to benefit other animals. It does not hurt to emphasize how we have all benefited from biomedical research on animals. Examples are listed in the statement on p. x.

   The matter is touchy because of several cases in which high school students have sued their schools over animal dissection requirements; there has also been at least one university case (see *Science,* 18 May 1990, p. 811). Yet, the direct dissection of vertebrate animals is the only way to understand vertebrate anatomy, and most biologists believe that such an understanding is essential for anyone oriented toward a career in animal biology or medicine. Direct examination of the organ systems of animals cannot be matched by any other approach. Computer simulation programs are *not* appropriate substitutes for direct dissection, despite claims to the contrary. To learn morphology from computer simulations would be rather like attempting to learn to play a musical instrument by watching videotapes of others playing the instrument. For more background on the animal rights issue, we recommend Laurence Pringle's book (1989) and the report of the Commission on Life Sciences, Committee on the Use of Laboratory Animals in Biomedical and Behavioral Research. As one of the committee members states, this report is the nearest thing we have to a national consensus and statement of policy on the animal use issue.

   How does an instructor deal with the occasional student who refuses to dissect? Try to find an alternative means for the student to learn the material, with as little fuss as possible. Perhaps the student will accept the option of observing while a partner does the dissection. If not, assign a library project that requires researching

All references to papers and publications mentioned in Appendix A are available on the website that accompanies *Animal Diversity,* sixth edition. Visit www.mhhe.com/hickmanad6e for more information.

some aspect of vertebrate adaptations. Stress that such a project is an alternative to dissection and not a penalty for refusing to dissect.

2. **This exercise may be supplemented** by providing demonstrations of living specimens of frogs, toads, and salamanders common to your area.

Exercise 17B: Skeleton

Materials

Frog and other vertebrate skeletons
Individual vertebrae
Frog skulls

Notes

1. **This exercise may be supplemented** by placing on display the skeletons of other vertebrates for comparison with the frog skeleton.

2. **Skeletons of embryos and *small* vertebrates may be stained *in situ*** by the following method. First, fix small specimens (skinned small frogs are excellent) in 95% alcohol for 2 to 4 days. Then place them in 2% potassium hydroxide solution until the bones are visible through the tissue. Check frequently to make sure the specimens are not macerated. Now transfer them to the following solution for 24 hours: 1 part of alizarin to 10,000 parts of 2% potassium hydroxide. Allow the stain to act until the desired intensity is obtained. It may take longer than 24 hours. Finally, clear the specimens in increasing concentrations of glycerin (10% to 50%). Excessively stained bones may be destained with 1% sulfuric acid made up in 95% alcohol.

Exercise 17C: Skeletal Muscles

Materials

Preserved frogs
Dissecting pans
Dissecting tools

Exercise 17D: Digestive, Respiratory, and Urogenital Systems

Materials

Preserved frogs
Prepared slides
    Frog kidney
    Frog testis
    Frog ovary
    Sperm smears

Notes

The following projects and demonstrations are appropriate supplements to this exercise.

1. **Transverse sections of a frog body.** Transverse sections should be made after the frog has been thoroughly frozen at a low temperature. Sections are easily cut with a hacksaw and completed according to directions given in the notes to Exercise 16A. Revealing relationships of organs may be seen from sections

made at the level of (a) a region a short distance anterior to the hindlegs and (b) a region just posterior to the forelegs.

2. **Study of transverse sections of intestine.** Compare slides of the transverse section of the human intestine with that of the frog.

3. **Peristalsis in a frog.** Pith the brain of a frog that has been fed an hour or two previously. Open the abdominal cavity and flood with warm 0.6% saline solution. Peristalsis should be observed. The students should understand what type of muscle is involved (smooth muscle) and in what directions the fibers run.

    Tie a thread tightly around the pyloric end of the stomach. Open the stomach near the cardiac end, and with a pipette introduce physiological salt solution into the lumen. Close the opening with a second ligature; then cut out the stomach. Suspend the stomach by the pyloric end in Ringer solution. Observe the wavelike peristaltic movement passing over the stomach. Record the rate per minute.

Exercise 17E: Circulatory System

Materials

Injected preserved frogs
Living frogs
Frog Ringer solution
Frog holders
Ice
Pins
Paper towel
Masking tape

Notes

1. **A frog's beating heart,** arranged to record on a chart recorder or smoked kymograph drum, can also be used to demonstrate the effects of temperature and of adrenaline and acetylcholine. The procedures are well understood by most instructors and will not be repeated here. An even simpler procedure is to demonstrate the durability of the isolated frog heart submerged in a dish of Ringer physiological saline. The saline must be well aerated. Adrenaline and acetylcholine may be added in concentrations of about 1/10,000. Flush with Ringer solution between each test.

2. **Capillary bed.** If a latex-injected frog is available, remove a piece of the skin for examination under a dissection microscope or the low power of a compound microscope. The capillary bed should show up well.

Exercise 18: The Nonavian Reptiles

Exercise 18: Painted Turtle

Materials

Turtle skeleton
Living painted turtles or other species
Preserved turtles

Notes

1. **Why turtles?** Turtles have long been chosen over lizards and snakes for the study of comparative anatomy for several good reasons: They are large, and their organs are easily observed (once the plastron is removed); the skeleton is stoutly built and resistant to the

abuses of student handling; they provide an interesting transition to birds, covered in Exercise 19 (turtles are "birds in shells" in that the two groups share, among other things, a highly flexible neck on a fused body frame, a toothless beak, and good vision); they are inexpensive; they represent a highly successful body plan, clumsy as they may appear; and students find them interesting and do not harbor any unreasonable fear of turtles (as many do of snakes).

2. **How much time?** This exercise will require 2 to 3 hours to complete in its entirety, but instructors may choose to limit the exercise to a 1-hour study of external structure and the skeleton.

3. **Singly injected turtles** are not required for this exercise, although having one or more available for the class will be helpful for visualizing the circulation.

4. **Exposing the viscera for dissection** is a cumbersome procedure using a bone saw and bone shears, and turtles should be prepared for the students before class. The bridge is partly sawed through with a bone saw, and the break completed with bone-cutting forceps (both instruments available from biological supply companies). Miniature saw blades that can be chucked in an electric drill are available (but difficult to find) and will vastly ease cutting through the bony bridge between the plastron and carapace. Singly injected turtles have had the plastron removed.

Exercise 19: The Birds

Exercise 19: Pigeon

Materials

Flight feathers
Other feather types as available for display
Pigeon skeleton
Preserved pigeons, plain
Preserved pigeons, air-sac-injected (optional, for demonstration)

Notes

1. **Allow about 2 1/2 hours** for this exercise if internal structure is reviewed as well as feathers and the skeleton. A study of the skeleton alone will reveal many of the adaptations for flight.

2. **Pigeon internal anatomy** can be covered by using prepared specimens or models available from biological supply houses, rather than having students dissect preserved specimens. Ward's offers a "Bio-Mount" of triple-injected and dissected pigeons (mounted on a plate and sealed in a museum jar). Models of chicken internal anatomy are also available from biological suppliers (although expensive).

Exercise 20: The Mammals

Exercise 20A: Skeleton

Materials

Skeletons, cat or dog and human
Sections of bones (if possible, a fresh joint)
Other vertebrate skeletons

All references to papers and publications mentioned in Appendix A are available on the website that accompanies *Animal Diversity*, sixth edition. Visit www.mhhe.com/hickmanad6e for more information.

Notes

Demonstrations appropriate to this exercise include (a) a fresh joint cut in longitudinal section with a bandsaw; (b) a long bone cut in longitudinal section; (c) a piece of skull or face bone available for examination of membrane bone; (d) skeletons of other vertebrates.

Exercise 20B: Muscular System

Materials

Fetal pig, embalmed

Although it is stated in the exercise, remind concerned students that fetal pigs are taken from the uteri of sows that have been slaughtered for market. Fetal pigs are commonly used in zoology laboratories because they are one of the by-products of the meat-packing industry. Consequently, they are inexpensive and serve as substitutes for other animals that would have to be killed to supply preserved material for dissection.

Notes

1. **Use the largest available fetal pigs** for this exercise; this facilitates muscle separation, never easy on a fetal pig. The muscle boundaries become more evident once the cutaneous muscle is removed. Sometimes it helps to use dry paper towel to rub off the cutaneous muscle—but caution the energetic student not to rub through the underlying musculature as well.

2. **Muscle names** make more sense to the students if you take a few minutes to explain what some of them mean. Refer also to Exercise 17C, where several of the Greek and Latin roots to common muscle names are explained.

3. **The text has been written to lead a student** through the dissection; distal musculature of the limbs has been omitted, and mention of the origin, insertion, and action has been placed in tabular form (Tables 20.1 and 20.2), so that the descriptive text is not burdened with information that many instructors will not require their students to learn.

Exercise 20C: Digestive System

Materials

Fetal pigs, embalmed
Prepared slides
    Human intestine, cross section
Frog intestine, cross section
    Mammalian liver
    Mammalian pancreas

Notes

1. **Demonstration of peristaltic movement.** About 30 to 45 minutes after a rat has finished feeding, anesthetize it in an ether jar, open the abdominal cavity, and submerge the contents in warm physiological saline solution. Note the peristaltic movement of the intestinal tract.

2. **Demonstration of the ruminant stomach.** The ruminants are cud-chewing animals (cattle, deer, camels, and their kin). All are members of order Artiodactyla, although not all Artiodactyla are ruminants (the pig, for example). If a ruminant stomach can be obtained, preferably fresh, it makes a fascinating demonstration. Food swallowed after brief mastication and generous addition of saliva, passes to the **rumen** for preliminary fermentation

by specialized microflora. Formed into small balls of cud, it is returned to the mouth for further mastication. Further fermentation follows in the rumen. When broken down to a pulp, it is passed to the **reticulum,** the second chamber, which has a honey-combed epithelium where fermentation continues. The pulp next passes to the **omasum,** where water, soluble food, and microbial products are absorbed. Finally, the smallest products pass to the **abomasum,** where proteolytic enzymes are added and normal digestion occurs in an acid environment.

3. **Demonstration slides.** Cross sections of mammalian esophagus, stomach, intestine, salivary glands, pancreas, and liver.

Exercise 20D: Urogenital System

Materials

Fetal pigs
Pregnant pig (or dog or cat) uteri for dissection or demonstration
Preserved sheep kidneys (optional)

Exercise 20E: Circulatory System

Materials

Fetal pigs
Pig or sheep hearts, fresh or preserved

Notes

1. **Microscope demonstrations.** These might include prepared slides of mammalian blood and of cross sections of artery and vein.

2. **The beef heart as a demonstration.** One reviewer of this exercise tells us that a fresh beef heart makes an excellent demonstration of heart valves. It is also possible to see the fossa ovalis and the ligamentum arteriosum (of the ductus arteriosus) in the beef heart.

3. **Demonstration of the action of heart valves.** If you are in the fortunate position of being able to procure a fresh sheep's heart with the roots of the great vessels uncut, the following method makes a fine demonstration of valve action. Tie glass tubes in the aorta, the pulmonary artery, and the veins of the atria. Suspend the heart by clamping the glass tube tied in the aorta to a stand. Water poured into the atria will rise through the pulmonary artery and aorta. By means of a long pipette or tubing inserted through the veins of the atria, remove some of the water from the ventricles. The water will still remain at the same height in the aorta and pulmonary arteries because of the action of the semilunar valves. Empty the heart of water. With a long pipette shoved down the aorta, place some water in the left ventricle. When the left ventricle is filled, the mitral valve prevents the water from entering the left atrium, and it will now rise up through the aorta.

APPENDIX B

Sources of Living Material and Prepared Microslides

Sources of Living Material

Carolina Biological Supply Co.
2700 York Road
Burlington, NC 27215
800-334-5551
www.carolina.com/

Connecticut Valley Biological
Supply Co., Inc.
P.O. Box 326
Southampton, MA 01073
800-628-7748
www.ctvalleybio.com/

Delta Biologicals
P.O. Box 2666
Tucson, AZ 85726
800-821-2502
www.deltabio.com/

Fisher Scientific Co.
485 S. Frontage Road
Burr Ridge, IL 60521
800-766-7000 (orders)
www.fishersci.com/

Gulf Specimen Co., Inc.
P.O. Box 237
Panacea, FL 32346
850-984-5297
www.gulfspecimen.org/

Leeches USA
300 Shames Drive
Westbury, NY 11590
800-645-3569
www.leechesusa.com/

NASCO (California)
4825 Stoddard Road
P.O. Box 3837
Modesto, CA 95352-3837
800-558-9595
www.enasco.com/

NASCO (Wisconsin)
901 Janesville Avenue
P.O. Box 901
Fort Atkinson, WI 53538-0901
800-558-9595
www.enasco.com/

Nebraska Scientific
3823 Leavenworth St.
Omaha, NE 68105-1180
800-228-7117
www.nebraskascientific.com/

Parco Scientific Co.
P.O. Box 189
Vienna, OH 44473
800-247-2726
www.parcoscientific.com/

Sargent-Welch
P.O. Box 5229
Buffalo Grove, IL 60089-5229
800-SARGENT (orders)
www.sargentwelch.com/

Science Kit and Boreal Laboratories
777 E. Park Dr.
Tonawanda, NY 14150
800-828-7777
www.sciencekit.com/

Ward's Natural Science Establishment,
Inc. (California)
P.O. Box 5010
San Luis Obispo, CA 93403
805-781-2700
www.wardsci.com/

Ward's Natural Science Establishment,
Inc. (New York)
P.O. Box 92912
Rochester, NY 14692-9019
800-962-2660 (orders)
www.wardsci.com/

Sources of Living Marine Material

AquaScience Research Group, Inc.
1100 Gentry Street
North Kansas City, MO 64116
816-842-5936
www.petsforum.com/aquascience/

Carolina Biological Supply Co.
See "Sources of Living Material"

Connecticut Valley Biological Supply Co.
See "Sources of Living Material"

Gulf Specimen Marine Labs
See "Sources of Living Material"

Marine Biological Laboratory
Supply Department
Woods Hole, MA 02543
508-548-3705
www.mbl.edu/

Sea Life Supply
740 Tioga Avenue
Sand City, CA 93955
831-394-0828 (orders)
Fax: 831-899-3399
www.sealifesupply.com/

Ward's Natural Science
See "Sources of Living Material"

Sources of Prepared Microslides

R. P. Cargille Laboratories, Inc.
Scientific Division
55 Commerce Rd.
Cedar Grove, NJ 07009
973-239-6633
Fax: 973-239-6096

Carolina Biological Supply Co.
2700 York Road
Burlington, NC 27215
800-334-5551
Fax: 800-222-7112
www.carolina.com/

Connecticut Valley Biological
Supply Co., Inc.
P.O. Box 326
Southampton, MA 01073
413-527-4030, 800-628-7748
www.ctvalleybio.com/

Dako Corp.
6392 Via Real
Carpinteria, CA 93013
805-566-6655, 800-235-5743
www.dakousa.com/

Fisher Scientific Co.
Science Education Division
485 S. Frontage Road
Burr Ridge, IL 60521
www.fishersci.com/

Frey Scientific Co.
P.O. Box 8101
Mansfield, OH 44901-8101
800-225-FREY
www.freyscientific.com/

Insect Lore Products
P.O. Box 1535
Shafter, CA 93263
800-548-3284 (orders)
Fax: 805-746-6047
www.insectlore.com/

NASCO
P.O. Box 222
Germantown, WI 53022
800-767-4288, 800-558-9595
www.enasco.com

Ward's Natural Science
Establishment, Inc.
See "Sources of Living Material"

CREDITS

Photo Credits

Part 1: Cleveland P. Hickman, Jr.

Part 2: Graham Edgar

Chapter 3

3.1: Courtesy of Leica, Inc.; **3.5:** © Carolina Biological Supply Co./Phototake.

Chapter 4

4.1: © Carolina Biological Supply Co./Phototake; **4.7:** © David M. Phillips/Visuals Unlimited, Inc.; **4.8:** © Eric V. Grave/Photo Researchers, Inc.; **4.14:** Cleveland P. Hickman, Jr.; **4.16:** © Karl Aufderheide/Visuals Unlimited, Inc.; **4.18:** © Phil A. Harrington/Peter Arnold, Inc./Photolibrary; **4.19–4.20:** © Walker England/Photo Researchers, Inc.

Chapter 5

5.3, 5.4b-c: © William C. Ober.**5.6a:** © Stanley Flegler/ Visuals Unlimited, Inc.; **5.6b, 5.7a:** © Donald Fawcett/Visuals Unlimited, Inc.; **5.7b:** Frances M. Hickman.

Chapter 6

6.2: © OSF/Animals Animals; **6.4a-b:** © Dr. D. P. Wilson/FLPA; **6.5:** © Robert Brons/Biological Photo Service; **6.8:** © William C. Ober.

Chapter 7

7.1: © Carolina Biological Supply Co./Phototake; **7.5:** © Cabisco/Visuals Unlimited, Inc.; **7.6a-b:** Cleveland P. Hickman, Jr.; **7.7:** © Stan Flegler/Visuals Unlimited, Inc.

Chapter 8

8.4: © Robert Calentine/Visuals Unlimited, Inc.; **8.6:** Courtesy Indiana University, School of Medicine.

Chapter 9

9.1: Frances M. Hickman; **9.6:** Cleveland P. Hickman, Jr.

Chapter 10

10.2a: © Daniel W. Gotshall/Visuals Unlimited, Inc.; **10.2b:** © William C. Ober; **10.4:** © G. L. Twiest/Visuals Unlimited, Inc.; **10.5:** © Carolina Biological Supply Co./Phototake; **10.6:** Cleveland P. Hickman, Jr.

Chapter 13

13.4: Larry S. Roberts.

Chapter 14

14.1: © Dr. D. P. Wilson/FLPA; **14.2:** Cleveland P. Hickman, Jr.; **14.5a-b, 14.7:** © William C. Ober; **14.9a-b:** © Rick Harbo/Marine Images.

Chapter 15

15.2, 15.4: Cleveland P. Hickman, Jr.

Chapter 16

16.5: © C. Stehr/Visuals Unlimited, Inc.